Optimal Control Theory with Applications in Economics

Optimal Control Theory with Applications in Economics

Thomas A. Weber

Foreword by A. V. Kryazhimskiy

The MIT Press
Cambridge, Massachusetts
London, England

© 2011 Massachusetts Institute of Technology

All rights reserved. No part of this book may be reproduced in any form by any electronic or mechanical means (including photocopying, recording, or information storage and retrieval) without permission in writing from the publisher.

For information about special quantity discounts, please email special_sales@mitpress.mit.edu.

This book was set in Palatino by Westchester Book Composition. Printed and bound in the United States of America.

Library of Congress Cataloging-in-Publication Data

Weber, Thomas A., 1969–
Optimal control theory with applications in economics / Thomas A. Weber; foreword by A. V. Kryazhimskiy.
   p. cm.
Includes bibliographical references and index.
ISBN 978-0-262-01573-8 (hardcover : alk. paper)
1. Economics—Mathematical models. 2. Control theory. 3. Mathematical optimization. 4. Game theory. I. Title.
HB135.W433   2011
330.01′515642—dc22   2010046482

10 9 8 7 6 5 4 3 2 1

For Wim

Contents

Foreword by A. V. Kryazhimskiy   ix
Acknowledgments   xi

1   Introduction   1
   1.1   Outline   3
   1.2   Prerequisites   5
   1.3   A Brief History of Optimal Control   5
   1.4   Notes   15

2   Ordinary Differential Equations   17
   2.1   Overview   17
   2.2   First-Order ODEs   20
   2.3   Higher-Order ODEs and Solution Techniques   67
   2.4   Notes   75
   2.5   Exercises   76

3   Optimal Control Theory   81
   3.1   Overview   81
   3.2   Control Systems   83
   3.3   Optimal Control—A Motivating Example   88
   3.4   Finite-Horizon Optimal Control   103
   3.5   Infinite-Horizon Optimal Control   113
   3.6   Supplement 1: A Proof of the Pontryagin Maximum Principle   119
   3.7   Supplement 2: The Filippov Existence Theorem   135
   3.8   Notes   140
   3.9   Exercises   141

4 Game Theory 149
- 4.1 Overview 149
- 4.2 Fundamental Concepts 155
- 4.3 Differential Games 188
- 4.4 Notes 202
- 4.5 Exercises 203

5 Mechanism Design 207
- 5.1 Motivation 207
- 5.2 A Model with Two Types 208
- 5.3 The Screening Problem 215
- 5.4 Nonlinear Pricing 220
- 5.5 Notes 226
- 5.6 Exercises 227

Appendix A: Mathematical Review 231
- A.1 Algebra 231
- A.2 Normed Vector Spaces 233
- A.3 Analysis 240
- A.4 Optimization 246
- A.5 Notes 251

Appendix B: Solutions to Exercises 253
- B.1 Numerical Methods 253
- B.2 Ordinary Differential Equations 258
- B.3 Optimal Control Theory 271
- B.4 Game Theory 302
- B.5 Mechanism Design 324

Appendix C: Intellectual Heritage 333

References 335
Index 349

Foreword

Since the discovery, by L. S. Pontryagin, of the necessary optimality conditions for the control of dynamic systems in the 1950s, mathematical control theory has found numerous applications in engineering and in the social sciences. T. A. Weber has dedicated his book to optimal control theory and its applications in economics. Readers can find here a succinct introduction to the basic control-theoretic methods, and also clear and meaningful examples illustrating the theory.

Remarkable features of this text are rigor, scope, and brevity, combined with a well-structured hierarchical approach. The author starts with a general view on dynamical systems from the perspective of the theory of ordinary differential equations; on this basis, he proceeds to the classical optimal control theory, and he concludes the book with more recent views of game theory and mechanism design, in which optimal control plays an instrumental role.

The treatment is largely self-contained and compact; it amounts to a lucid overview, featuring much of the author's own research. The character of the problems discussed in the book promises to make the theory accessible to a wide audience. The exercises placed at the chapter endings are largely original.

I am confident that readers will appreciate the author's style and students will find this book a helpful guide on their path of discovery.

A. V. Kryazhimskiy
Steklov Institute of Mathematics, Russian Academy of Sciences
International Institute for Applied Systems Analysis

Acknowledgments

This book is based on my graduate course on Applied Optimal Control Theory taught both at Moscow State University and Stanford University. The development of this course was made possible through funding from Moscow State University (MSU), the Steklov Mathematical Institute in Moscow, and Stanford University. In particular, a course development grant from James Plummer and Channing Robertson, dean and former vice dean of the School of Engineering at Stanford University, allowed Elena Rovenskaya to spend time at Stanford to work on problem sets and solutions. The numerous discussions with her were invaluable, and I am very grateful for her contributions. Nikolai Grigorenko, the deputy head of the Optimal Control Department at MSU, was instrumental in making this possible.

I am very grateful to Evgenii Moiseev, dean of the Faculty of Computational Mathematics and Cybernetics at Lomonosov Moscow State University, for his 2007 invitation to deliver a summer course on dynamic optimization with applications in economics. My deepest gratitude also goes to Arkady Kryazhimskiy and Sergey Aseev for their encouragement to write this book and their continuous support and friendship. They have fueled my interest in optimal control since 2001, when I was fortunate enough to meet them while participating in the Young Scientists Summer Program at the International Institute for Applied Systems Analysis in Laxenburg, Austria. They also invited me to the 2008 International Conference on Differential Equations and Topology in Moscow, dedicated to the centennial anniversary of Lev Pontryagin, where the discussions about the summer course continued.

I am indebted to the students in Moscow and Stanford for the many after-class discussions where they often taught me, perhaps unknowingly, just as much as I taught them. Kenneth Gillingham provided a set of impeccable handwritten notes from the course, which helped organize the thoughts for the book. Naveed Chehrazi was an excellent course assistant who contributed numerous insightful suggestions. Elena Rovenskaya and Denis Pivovarchuk organized the practice sessions in Moscow. Stefan Behringer, Andrei Dmitruk, and three MIT Press reviewers provided useful feedback on earlier versions of the book. Markus Edvall from TOMLAB was helpful in debugging several numerical algorithms. I would like to thank Alice Cheyer for her detailed copyediting, as well as Alexey Smirnov and my assistant, Marilynn Rose, for their help with editing earlier versions of the manuscript. Jane Macdonald at the MIT Press believed in this project from the moment I told her about it at the 2009 European Meeting of the Econometric Society in Barcelona. I am very grateful for her helpful advice, encouragement, and great support throughout the publishing process.

I should also like to acknowledge my indebtedness to the great teachers in control theory and economics, whom I encountered at MIT, the University of Pennsylvania, and Stanford University, in particular Kenneth Arrow, Dimitri Bertsekas, David Cass, Richard Kihlstrom, Alexandre Kirillov, Steve Matthews, and Ilya Segal. Richard Vinter introduced me to control theory in 1994 as a wonderful research advisor at Imperial College London. In my thinking and intellectual approach I also owe very much to my advisor at MIT, Alexandre Megretski, who did not believe in books and whose genius and critical mind I admire. Paul Kleindorfer, my dissertation advisor at Wharton, taught me so much, including the fact that broad interests can be a real asset. I am very grateful for his friendship and constant support.

My special thanks go to Sergey Aseev, Eric Clemons, David Luenberger, James Sweeney, and Andrew Whinston for their friendship, as well as Ann and Wim for their sacrifice, endless love, and understanding.

Stanford, California
May 2011

1 Introduction

Our nature consists in movement;
absolute rest is death.
—Blaise Pascal

Change is all around us. Dynamic strategies seek to both anticipate and effect such change in a given system so as to accomplish objectives of an individual, a group of agents, or a social planner. This book offers an introduction to continuous-time systems and methods for solving dynamic optimization problems at three different levels: single-person decision making, games, and mechanism design. The theory is illustrated with examples from economics. Figure 1.1 provides an overview of the book's hierarchical approach.

The first and lowest level, single-person decision making, concerns the choices made by an individual decision maker who takes the evolution of a system into account when trying to maximize an objective functional over feasible dynamic policies. An example would be an economic agent who is concerned with choosing a rate of spending for a given amount of capital, each unit of which can either accumulate interest over time or be used to buy consumption goods such as food, clothing, and luxury items.

The second level, games, addresses the question of finding predictions for the behavior and properties of dynamic systems that are influenced by a group of decision makers. In this context the decision makers (players) take each other's policies into account when choosing their own actions. The possible outcomes of the game among different players, say, in terms of the players' equilibrium payoffs and equilibrium actions, depend on which precise concept of equilibrium is applied. Nash (1950) proposed an equilibrium such that players' policies do not give any player an incentive to deviate from his own chosen policy, given

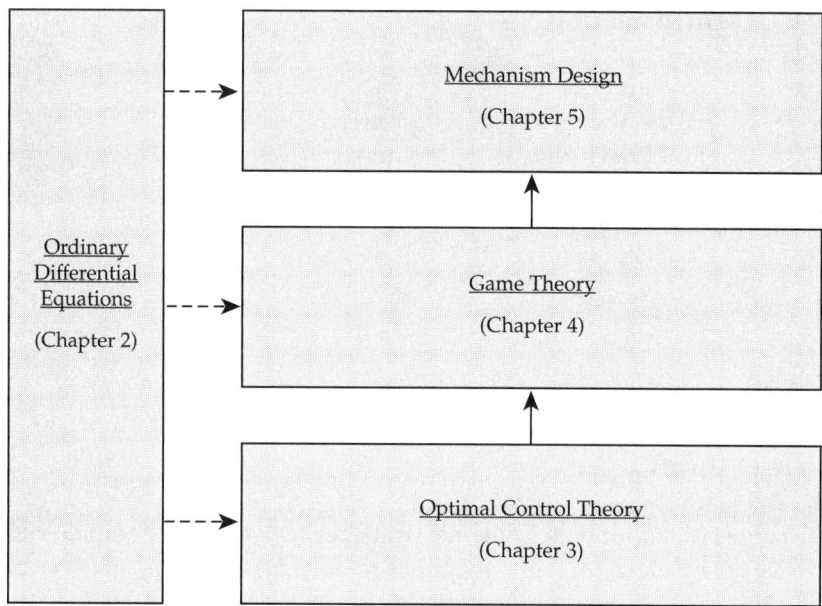

Figure 1.1
Topics covered in this book.

the other players' choices are fixed to the equilibrium policies. A classic example is an economy with a group of firms choosing production outputs so as to maximize their respective profits.

The third and highest level of analysis considered here is mechanism design, which is concerned with a designer's creation of an environment in which players (including the designer) can interact so as to maximize the designer's objective functional. Leading examples are the design of nonlinear pricing schemes in the presence of asymmetric information, and the design of markets. Arguably, this level of analysis is isomorphic to the first level, since the players' strategic interaction may be folded into the designer's optimization problem.

The dynamics of the system in which the optimization takes place are described in continuous time, using ordinary differential equations. The theory of ordinary differential equations can therefore be considered the backbone of the theory developed in this book.

Introduction

1.1 Outline

Ordinary Differential Equations (ODEs) Chapter 2 reviews basic concepts in the theory of ODEs. One-dimensional linear first-order ODEs can be solved explicitly using the Cauchy formula. The key insight from the construction of this formula (via variation of an integration constant) is that the solution to a linear initial value problem of the form

$$\dot{x} + g(t)x = h(t), \qquad x(t_0) = x_0,$$

for a given tuple of initial data (t_0, x_0) can be represented as the superposition of a homogeneous solution (obtained when $h = 0$) and a particular solution to the original ODE (but without concern for the initial condition). Systems of linear first-order ODEs,

$$\dot{x} = A(t)x + b(t), \tag{1.1}$$

with an independent variable of the form $x = (x_1, \ldots, x_n)$ and an initial condition $x(t_0) = x_0$ can be solved if a fundamental matrix $\Phi(t, t_0)$ as the solution of a homogeneous equation is available. Higher-order ODEs (containing higher-order derivatives) can generally be reduced to first-order ODEs. This allows limiting the discussion to (nonlinear) first-order ODEs of the form

$$\dot{x} = f(t, x), \tag{1.2}$$

for $t \geq t_0$. Equilibrium points, that is, points $\bar{x}$ at which a system does not move because $f(t, \bar{x}) = 0$, are of central importance in understanding a continuous-time dynamic model. The stability of such points is usually investigated using the method developed by Lyapunov, which is based on the principle that if system trajectories $x(t)$ in the neighborhood of an equilibrium point are such that a certain real-valued function $V(t, x(t))$ is nonincreasing (along the trajectories) and bounded from below by its value at the equilibrium point, then the system is stable. If this function is actually decreasing along system trajectories, then these trajectories must converge to an equilibrium point. The intuition for this finding is that the Lyapunov function V can be viewed as energy of the system that cannot increase over time. This notion of energy, or, in the context of economic problems, of value or welfare, recurs throughout the book.

Optimal Control Theory Given a description of a system in the form of ODEs, and an objective functional $J(u)$ as a function of a dynamic policy or control u, together with a set of constraints (such as initial conditions or control constraints), a decision maker may want to solve an *optimal control problem* of the form

$$J(u) = \int_{t_0}^{T} h(t, x(t), u(t))\, dt \longrightarrow \max_{u(\cdot)}, \qquad (1.3)$$

subject to $\dot{x}(t) = f(t, x(t), u(t))$, $x(t_0) = x_0$, and $u \in \mathcal{U}$, for all $t \in [t_0, T]$. Chapter 3 introduces the notion of a controllable system, which is a system that can be moved using available controls from one state to another. Then it takes up the construction of solutions (in the form of state-control trajectories $(x^*(t), u^*(t))$, $t \in [t_0, T]$) to such optimal control problems: necessary and sufficient optimality conditions are discussed, notably the Pontryagin maximum principle (PMP) and the Hamilton-Jacobi-Bellman (HJB) equation. Certain technical difficulties notwithstanding, it is possible to view the PMP and the HJB equation as two complementary approaches to obtain an understanding of the solution of optimal control problems. In fact, the HJB equation relies on the existence of a continuously differentiable value function $V(t, x)$, which describes the decision maker's optimal payoff, with the optimal control problem initialized at time t and the system in the state x. This function, somewhat similar to a Lyapunov function in the theory of ODEs, can be interpreted in terms of the value of the system for a decision maker. The necessary conditions in the PMP can be informally derived from the HJB equation, essentially by restricting attention to a neighborhood of the optimal trajectory.

Game Theory When more than one individual can make payoff-relevant decisions, game theory is used to determine predictions about the outcome of the strategic interactions. To abstract from the complexities of optimal control theory, chapter 4 introduces the fundamental concepts of game theory for simple discrete-time models, along the lines of the classical exposition of game theory in economics. Once all the elements, including the notion of a Nash equilibrium and its various refinements, for instance, via subgame perfection, are in place, attention turns to differential games. A critical question that arises in dynamic games is whether the players can trust each other's equilibrium strategies, in the sense that they are credible even after the game has started. A player may, after a while, find it best to deviate from a

Introduction

Nash equilibrium that relies on a "noncredible threat." The latter consists of an action which, as a contingency, discourages other players from deviating but is not actually beneficial should they decide to ignore the threat. More generally, in a Nash equilibrium that is not subgame-perfect, players lack the ability to commit to certain threatening actions (thus, noncredible threats), leading to "time inconsistencies."

Mechanism Design A simple economic mechanism, discussed in chapter 5, is a collection of a message space and an allocation function. The latter is a mapping from possible messages (elements of the message space) to available allocations. For example, a mechanism could consist of the (generally nonlinear) pricing schedule for bandwidth delivered by a network service provider. A mechanism designer, who is often referred to as the principal, initially announces the mechanism, after which the agent sends a message to the principal, who determines the outcome for both participants by evaluating the allocation function. More general mechanisms, such as an auction, can include several agents playing a game that is implied by the mechanism.

Optimal control theory becomes useful in the design of a static mechanism because of an information asymmetry between the principal and the various agents participating in the mechanism. Assuming for simplicity that there is only a single agent, and that this agent possesses private information that is encapsulated in a one-dimensional type variable θ in a type space $\Theta = [\underline{\theta}, \bar{\theta}]$, it is possible to write the principal's mechanism design problem as an optimal control problem.

1.2 Prerequisites

The material in this book is reasonably self-contained. It is recommended that the reader have acquired some basic knowledge of dynamic systems, for example, in a course on linear systems. In addition, the reader should possess a firm foundation in calculus, since the language of calculus is used throughout the book without necessarily specifying all the details or the arguments if they can be considered standard material in an introductory course on calculus (or analysis).

1.3 A Brief History of Optimal Control

Origins The human quest for finding extrema dates back to antiquity. Around 300 B.C., Euclid of Alexandria found that the minimal distance between two points A and B in a plane is described by the straight line $\overline{AB}$, showing in his *Elements* (Bk I, Prop. 20) that any two sides of a triangle together are greater than the third side (see, e.g., Byrne 1847, 20). This is notwithstanding the fact that nobody has actually ever seen a straight line. As Plato wrote in his Allegory of the Cave[1] (*Republic*, Bk VII, ca. 360 B.C.), perceived reality is limited by our senses (Jowett 1881). Plato's theory of forms held that ideas (or forms) can be experienced only as shadows, that is, imperfect images (W. D. Ross 1951). While Euclid's insight into the optimality of a straight line may be regarded merely as a variational inequality, he also addressed the problem of finding extrema subject to constraints by showing in his *Elements* (Bk VI, Prop. 27) that "of all the rectangles contained by the segments of a given straight line, the greatest is the square which is described on half the line" (Byrne 1847, 254). This is generally considered the earliest solved maximization problem in mathematics (Cantor 1907, 266) because

$$\frac{a}{2} \in \arg\max_{x \in \mathbb{R}}\{x(a-x)\},$$

for any $a > 0$. Another early maximization problem, closely related to the development of optimal control, is recounted by Virgil in his *Aeneid* (ca. 20 B.C.). It involves queen Dido, the founder of Carthage (located in modern-day Tunisia), who negotiated to buy as much land as she could enclose using a bull's hide. To solve her isoperimetric problem, that is, to find the largest area with a given perimeter, she cut the hide into a long strip and laid it out in a circle. Zenodorus, a Greek mathematician, studied Dido's problem in his book *On Isoperimetric Figures* and showed that a circle is greater than any regular polygon of equal contour (Thomas 1941, 2:387–395). Steiner (1842) provided five different proofs that any figure of maximal area with a given perimeter in the plane must be a circle. He omitted to show that there actually *exists* a solution to the isoperimetric problem. Such a proof was given later by Weierstrass (1879/1927).[2]

Remark 1.1 (Existence of Solutions) Demonstrating the existence of a solution to a variational problem is in many cases both important and nontrivial. Perron (1913) commented specifically on the gap left by Steiner in the solution of the isoperimetric problem regarding existence,

---

1. In the Allegory of the Cave, prisoners in a cave are restricted to a view of the real world (which exists behind them) solely via shadows on a wall in front of them.
2. Weierstrass's numerous contributions to the calculus of variations, notably on the existence of solutions and on sufficient optimality conditions, are summarized in his extensive lectures on *Variationsrechnung,* published posthumously based on students' notes.

and he provided several examples of variational problems without solutions (e.g., finding a polygon of given perimeter and maximal surface). A striking problem without a solution was posed by Kakeya (1917). He asked for the set of minimal measure that contains a unit line segment in all directions. One can think of such a Kakeya set (or Besicovitch set) as the minimal space that an infinitely slim car would need to turn around in a parking spot. Somewhat surprisingly, Besicovitch (1928) was able to prove that the measure of the Kakeya set cannot be bounded from below by a positive constant. □

The isoperimetric constraint appears naturally in economics as a budget constraint, which was recognized by Frisi in his written-in commentary on Verri's (1771) notion that a political economy shall be trying to maximize production subject to the available labor supply (Robertson 1949). Such budget-constrained problems are natural in economics.[3] For example, Sethi (1977) determined a firm's optimal intertemporal advertising policy based on a well-known model by Nerlove and Arrow (1962), subject to a constraint on overall expenditure over a finite time horizon.

Calculus of Variations The infinitesimal calculus (or later just *calculus*) was developed independently by Newton and Leibniz in the 1670s. Newton formulated the modern notion of a derivative (which he termed *fluxion*) in his *De Quadratura Curvarum*, published as an appendix to his treatise on *Opticks* in 1704 (Cajori 1919, 17–36). In 1684, Leibniz published his notions of derivative and integral in the *Acta Eruditorum*, a journal that he had co-founded several years earlier and that enjoyed a significant circulation in continental Europe. With the tools of calculus in place, the time was ripe for the calculus of variations, the birth of which can be traced to the June 1696 issue of the *Acta Eruditorum*. There, Johann Bernoulli challenged his contemporaries to determine the path from point A to point B in a vertical plane that minimizes the time for a mass point M to travel under the influence of gravity between A and B. This problem of finding a *brachistochrone* (figure 1.2) was posed

---

3. To be specific, let $C(t, x, u)$ be a nonnegative-valued cost function and $B > 0$ a given budget. Then along a trajectory $(x(t), u(t))$, $t \in [t_0, T]$, a typical *isoperimetric constraint* is of the form $\int_{t_0}^{T} C(t, x(t), u(t))\, dt \leq B$. It can be rewritten as $\dot{y}(t) = C(t, x(t), u(t))$, $y(t_0) = 0$, $y(T) \leq B$. The latter formulation falls squarely within the general optimal-control formalism developed in this book, so isoperimetric constraints do not need special consideration.

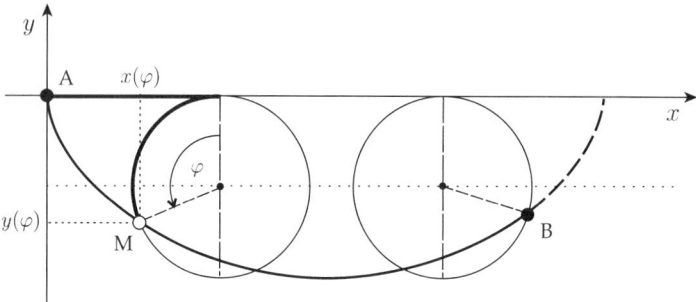

Figure 1.2
Brachistochrone connecting the points A and B in parametric form: $(x(\varphi), y(\varphi)) = (\alpha(\varphi - \sin(\varphi)), \alpha(\cos(\varphi) - 1))$, where $\varphi = \varphi(t) = \sqrt{g/\alpha}\, t$, and $g \approx 9.81$ meters per second squared is the gravitational constant. The parameter α and the optimal time $t = T^*$ are determined by the endpoint condition $(x(\varphi(T^*)), y(\varphi(T^*))) = B$.

earlier (but not solved) by Galilei (1638).[4] In addition to his own solution, Johann Bernoulli obtained four others, by his brother Jakob Bernoulli, Leibniz, de l'Hôpital, and Newton (an anonymous entry). The last was recognized immediately by Johann *ex ungue leonem* ("one knows the lion by his claw").

Euler (1744) investigated the more general problem of finding extrema of the functional

$$J = \int_0^T L(t, x(t), \dot{x}(t))\, dt, \tag{1.4}$$

subject to suitable boundary conditions on the function $x(\cdot)$. He derived what is now called the Euler equation (see equation (1.5)) as a necessary optimality condition used to this day to construct solutions to variational problems. In his 1744 treatise on variational methods, Euler did not create a name for his complex of methods and referred to variational calculus simply as the isoperimetric method. This changed with a 1755 letter from Lagrange to Euler informing the latter of his δ-calculus, with δ denoting variations (Goldstine 1980, 110–114). The name "calculus of variations" was officially born in 1756, when the minutes of meeting

---

4. Huygens (1673) discovered that a body which is bound to fall following a cycloid curve oscillates with a periodicity that is independent of the starting point on the curve, so he termed this curve *tautochrone*. The brachistochrone is also a cycloid and thus identical to the tautochrone, which led Johann Bernoulli to remark that "nature always acts in the simplest possible way" (Willems 1996).

Introduction

no. 441 of the Berlin Academy on September 16 note that Euler read "Elementa calculi variationum" (Hildebrandt 1989).

Remark 1.2 (Extremal Principles) Heron of Alexandria explained the equality of angles in the reflection of light by the principle that nature must take the shortest path, for "[i]f Nature did not wish to lead our sight in vain, she would incline it so as to make equal angles" (Thomas 1941, 2:497). Olympiodorus the younger, in a commentary (ca. 565) on Aristotle's *Meteora*, wrote, "[T]his would be agreed by all ... Nature does nothing in vain nor labours in vain" (Thomas 1941, 2:497).

In the same spirit, Fermat in 1662 used the *principle of least time* (now known as Fermat's principle) to derive the law of refraction for light (Goldstine 1980, 1–6). More generally, Maupertuis (1744) formulated the *principle of least action*, that in natural phenomena a quantity called action (denoting energy × time) is to be minimized (cf. also Euler 1744). The calculus of variations helped formulate more such extremal principles, for instance, *d'Alembert's principle*, which states that along any virtual displacement the sum of the differences between the forces and the time derivatives of the moments vanishes. It was this principle that Lagrange (1788/1811) chose over Maupertuis's principle in his *Mécanique Analytique* to firmly establish the use of differential equations to describe the evolution of dynamic systems. Hamilton (1834) subsequently established that the law of motion on a time interval $[t_0, T]$ can be derived as extremal of the functional in equation (1.4) (*principle of stationary action*), where L is the difference between kinetic energy and potential energy. Euler's equation in this variational problem is also known as the *Euler-Lagrange equation*,

$$\frac{d}{dt}\frac{\partial L(t, x(t), \dot{x}(t))}{\partial \dot{x}} - \frac{\partial L(t, x(t), \dot{x}(t))}{\partial x} = 0, \tag{1.5}$$

for all $t \in [t_0, T]$. With the Hamiltonian function $H(t, x, \dot{x}, \psi) = \langle \psi, \dot{x} \rangle - L(t, x, \dot{x})$, where $\psi = \partial L / \partial \dot{x}$ is an adjoint variable, one can show that (1.5) is in fact equivalent to the *Hamiltonian system*,[5]

---

5. To see this, note first that (1.6) holds by definition and that irrespective of the initial conditions,

$$0 = \frac{dH}{dt} - \frac{dH}{dt} = \frac{\partial H}{\partial t} + \left\langle \frac{\partial H}{\partial x}, \dot{x} \right\rangle + \left\langle \frac{\partial H}{\partial \psi}, \dot{\psi} \right\rangle - \left(\langle \dot{\psi}, \dot{x} \rangle + \langle \psi, \ddot{x} \rangle - \frac{\partial L}{\partial t} - \left\langle \frac{\partial L}{\partial x}, \dot{x} \right\rangle - \left\langle \frac{\partial L}{\partial \dot{x}}, \ddot{x} \right\rangle \right),$$

whence, using $\psi = \partial L/\partial \dot{x}$ and $\dot{x} = \partial H/\partial \psi$, we obtain

$$0 = \frac{\partial H}{\partial t} + \frac{\partial L}{\partial t} + \left\langle \frac{\partial H}{\partial x} + \frac{\partial L}{\partial x}, \dot{x} \right\rangle + \left\langle \frac{\partial H}{\partial \psi}, \dot{\psi} \right\rangle - \langle \dot{\psi}, \dot{x} \rangle = \left\langle \frac{\partial H}{\partial x} + \frac{\partial L}{\partial x}, \dot{x} \right\rangle.$$

Thus, $\partial H/\partial x = -\partial L/\partial x$, so the Euler-Lagrange equation (1.5) immediately yields (1.7).

$$\dot{x}(t) = \frac{\partial H(t, x(t), \dot{x}(t), \psi(t))}{\partial \psi}, \qquad (1.6)$$

$$\dot{\psi}(t) = -\frac{\partial H(t, x(t), \dot{x}(t), \psi(t))}{\partial x}, \qquad (1.7)$$

for all $t \in [t_0, T]$. To integrate the Hamiltonian system, given some initial data (t_0, x_0), Jacobi (1884, 143–157) proposed to introduce an action function,

$$V(t, x) = \int_{t_0}^{t} L(s, x(s), \dot{x}(s))\, ds,$$

on an extremal trajectory, which satisfies (1.6)–(1.7) on $[t_0, t]$ and connects the initial point (t_0, x_0) to the point (t, x). One can now show (see, e.g., Arnold 1989, 254–255) that

$$\frac{dV(t, x(t))}{dt} = \frac{\partial V(t, x(t))}{\partial t} + \frac{\partial V(t, x(t))}{\partial x} = \langle \psi(t), \dot{x}(t) \rangle - H(t, x(t), \dot{x}(t), \psi(t)),$$

so that $H = -\partial V / \partial t$ and $\psi = \partial V / \partial x$, and therefore the *Hamilton-Jacobi equation*,

$$-\frac{\partial V(t, x(t))}{\partial t} = H\left(t, x(t), \dot{x}(t), \frac{\partial V(t, x(t))}{\partial x}\right), \qquad (1.8)$$

holds along an extremal trajectory. This result is central for the construction of sufficient as well as necessary conditions for solutions to optimal control problems (see chapter 3). Extremal principles also play a role in economics. For example, in a Walrasian exchange economy, prices and demands will adjust so as to maximize a welfare functional. □

Remark 1.3 (Problems with Several Independent Variables) Lagrange (1760) raised the problem of finding a surface of minimal measure, given an intersection-free closed curve. The Euler-Lagrange equation for this problem expresses the fact that the mean curvature of the surface must vanish everywhere. This problem is generally referred to as *Plateau's problem*, even though Plateau was born almost half a century after Lagrange had formulated it originally. (Plateau conducted extended experiments with soap films leading him to discover several laws that were later proved rigorously by others.) Plateau's problem was solved independently by Douglas (1931) and Radó (1930). For historical details see, for instance, Fomenko (1990) and Struwe (1989). This book considers

Introduction

only problems where the independent variable is one-dimensional, so all systems can be described using ordinary (instead of partial) differential equations. □

In an article about beauty in problems of science the economist Paul Samuelson (1970) highlighted several problems in the calculus of variations, such as the brachistochrone problem, and connected those insights to important advances in economics. For example, Ramsey (1928) formulated an influential theory of saving in an economy that determines an optimal growth path using the calculus of variations. The Ramsey model, which forms the basis of the theory of economic growth, was further developed by Cass (1965) and Koopmans (1965).[6]

Feedback Control Before considering the notion of a control system, one can first define a *system* as a set of connected elements, where the connection is an arbitrary relation among them. The complement of this set is the *environment* of the system. If an element of the system is not connected to any other element of the system, then it may be viewed as part of the environment. When attempting to model a real-world system, one faces an age-old trade-off between veracity and usefulness. In the fourteenth century William of Occam formulated the *law of parsimony* (also known as Occam's razor), *entia non sunt multiplicanda sine necessitate*, to express the postulate that "entities are not to be multiplied without necessity" (Russell 1961, 453).[7] The trade-off between usefulness and veracity of a system model has been rediscovered many times, for instance, by Leonardo da Vinci ("simplicity it is the ultimate sophistication") and by Albert Einstein ("make everything as simple as possible, but not simpler").[8]

A *control system* is a system with an input (or *control*) $u(t)$ that can be influenced by human intervention. If the *state* $x(t)$ of the system can also be observed, then the state can be used by a *feedback law* $u(t) = \mu(t, x(t))$ to adjust the input, which leads to a *feedback control system* (figure 1.3).

There is a rich history of feedback control systems in technology, dating back at least to Ktesibios's float regulator in the third century B.C. for a water clock, similar to a modern flush toilet (Mayr 1970). Wedges

---
6. For more details on the modern theory of economic growth, see, e.g., Acemoglu (2009), Aghion and Howitt (2009), and Weitzman (2003).
7. For a formalization of Occam's razor, see Pearl (2000, 45–48).
8. Some "anti-razors" warn of oversimplification, e.g., Leibniz's *principle of plenitude* ("everything that can happen will happen") or Kant's insight that "[t]he variety of entities is not to be diminished rashly" (1781, 656).

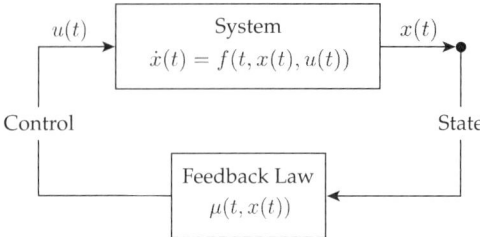

Figure 1.3
Feedback control system.

were inserted in the water flow to control the speed at which a floating device would rise to measure the time. In 1788, Watt patented the design of the centrifugal governor for regulating the speed of a rotary steam engine, which is one of the most famous early feedback control systems. Rotating flyballs, flung apart by centrifugal force, would throttle the engine and regulate its speed. A key difference between the Ktesibios's and Watt's machines is that the former does not use feedback to determine the control input (the number and position of the wedges), which is therefore referred to as *open-loop* control. Watt's flyball mechanism, on the other hand, uses the state of the system (engine rotations) to determine the throttle position that then influences the engine rotations, which is referred to as *closed-loop* (or feedback) control. Wiener (1950, 61) noted that "feedback is a method of controlling a system by reinserting into it the results of its past performance." He suggested the term *cybernetics* (from the Greek word $κυβερνητης$—governor) for the study of control and communication systems (Wiener 1948, 11–12).[9] Maxwell (1868) analyzed the stability of Watt's centrifugal governor by linearizing the system equation and showing that it is stable, provided its eigenvalues have strictly negative real parts. Routh (1877) worked out a numerical algorithm to determine when a characteristic equation (or equivalently, a system matrix) has stable roots. Hurwitz (1895) solved this problem independently, and to this day a stable system matrix A in equation (1.1) carries his name (see lemma 2.2). The stability of nonlinear systems of the form (1.2) was advanced by the seminal work of Lyapunov (1892), which showed that if an energy function $V(t, x)$ could be found such that it is bounded from below and decreasing along any

---

9. The term was suggested more than a hundred years earlier for the control of sociopolitical systems by Ampère (1843, 140–141).

Introduction

system trajectory $x(t)$, $t \geq t_0$, then the system is (asymptotically) stable, that is, the system is such that any trajectory that starts close to an equilibrium state converges to that equilibrium state. In variational problems the energy function $V(t,x)$ is typically referred to as a value function and plays an integral role for establishing optimality conditions, such as the Hamilton-Jacobi equation (1.8), or more generally, the Hamilton-Jacobi-Bellman equation (3.16).

In 1892, Poincaré published the first in a three-volume treatise on celestial mechanics containing many path-breaking advances in the theory of dynamic systems, such as integral invariants, Poincaré maps, the recurrence theorem, and the first description of chaotic motion. In passing, he laid the foundation for a geometric and qualitative analysis of dynamic systems, carried forward, among others, by Arnold (1988). An important alternative to system stability in the sense of asymptotic convergence to equilibrium points is the possibility of a limit cycle. Based on Poincaré's work between 1881 and 1885,[10] Bendixson (1901) established conditions under which a trajectory of a two-dimensional system constitutes a limit cycle (see proposition 2.13); as a by-product, this result implies that chaotic system behavior can arise only if the state-space dimension is at least 3. The theory of stability in feedback control systems has proved useful for the description of real-world phenomena. For example, Lotka (1920) and Volterra (1926) proposed a model for the dynamics of a biological predator-prey system that features limit cycles (see example 2.8).

In technological applications (e.g., when stabilizing an airplane) it is often sufficient to linearize the system equation and minimize a cost that is quadratic in the magnitude of the control and quadratic in the deviations of the system state from a reference state (or tracking trajectory)[11] in order to produce an effective controller. The popularity of this linear-quadratic approach is due to its simple closed-form solvability. Kalman and Bucy (1961) showed that the approach can also be very effective in dealing with (Gaussian) noise incorporating a state-estimation component, resulting in a continuous-time version of the Kalman filter, which was first developed by Rudolf Kalman for discrete-time systems. To deal with control constraints in a noisy

---

10. The relevant series of articles was published in the *Journal de Mathématiques*, reprinted in Poincaré (1928, 3–222); see also Barrow-Green (1997).
11. A linear-quadratic regulator is obtained by solving an optimal control problem of the form (1.3), with linear system function $f(t,x,u) = Ax + Bu$ and quadratic payoff function $h(t,x,u) = -x'Rx - u'Su$ (with R, S positive definite matrices); see example 3.3.

environment, the linear-quadratic approach has been used in receding-horizon control (or model predictive control), where a system is periodically reoptimized over the same fixed-length horizon.[12] More recently, this approach has been applied in financial engineering, for example, portfolio optimization (Primbs 2007).

Optimal Control In the 1950s the classical calculus of variations underwent a transformation driven by two major advances. Both advances were fueled by the desire to find optimal control interventions for given feedback control systems, in the sense that the optimal control trajectory $u^*(t)$, $t \in [t_0, T]$, would maximize an objective functional $J(u)$ by solving a problem of the form (1.3). The first advance, by Richard Bellman, was to incorporate a control function into the Hamilton-Jacobi variational equation, leading to the Hamilton-Jacobi-Bellman equation,[13]

$$-V_t(t,x) = \max_{u \in \mathcal{U}}\{h(t,x,u) + \langle V_x(t,x), f(t,x,u)\rangle\}, \tag{1.9}$$

which, when satisfied on the rectangle $[t_0, T] \times \mathcal{X}$ (where the state space $\mathcal{X}$ contains all the states), together with the endpoint condition $V(T,x) \equiv 0$, serves as a sufficient condition for optimality. The optimal feedback law $\mu(t,x)$ is obtained as the optimal value for u on the right-hand side of (1.9), so the optimal state trajectory $x^*(t)$, $t \in [t_0, T]$, solves the initial value problem (IVP)

$$\dot{x} = f(t, x, \mu(t,x)), \qquad x(t_0) = x_0,$$

which yields the optimal control

$$u^*(t) = \mu(t, x^*(t)),$$

for all $t \in [t_0, T]$. This approach to solving optimal control problems by trying to construct the value function is referred to as dynamic programming (Bellman 1957).[14] The second advance, by Lev Pontryagin and his students, is related to the lack of differentiability of the value function $V(t,x)$ in (1.9), even for the simplest problems (see, e.g., Pontryagin et al. 1962, 23–43, 69–73) together with the difficulties of actually solving the partial differential equation (1.9) when the value function is

---

12. Receding-horizon control has also been applied to the control of nonlinear systems, be they discrete-time (Keerthi and Gilbert 1988) or continuous-time (Mayne and Michalska 1990).
13. Subscripts denote partial derivatives.
14. The idea of dynamic programming precedes Bellman's work: for example, von Neumann and Morgenstern (1944, ch. 15) used backward induction to solve sequential decision problems in perfect-information games.

Introduction

differentiable. Pontryagin (1962), together with his students, provided a rigorous proof for a set of necessary optimality conditions for optimal control problems of the form (1.3). As shown in section 3.3, the conditions of the Pontryagin maximum principle (in its most basic version) can be obtained, at least heuristically, from the Hamilton-Jacobi-Bellman equation. A rigorous proof of the maximum principle usually takes a different approach, using needle variations introduced by Weierstrass (1879/1927). As Pontryagin et al. (1962) pointed out,

> The method of dynamic programming was developed for the needs of optimal control processes which are of a much more general character than those which are describable by systems of differential equations. Therefore, the method of dynamic programming carries a more universal character than the maximum principle. However, in contrast to the latter, this method does not have the rigorous logical basis in all those cases where it may be successfully made use of as a valuable heuristic tool. (69)

In line with these comments, the Hamilton-Jacobi-Bellman equation is often used in settings that are more complex than those considered in this book, for instance for the optimal control of stochastic systems. The problem with the differentiability of the value function was addressed by Francis Clarke by extending the notion of derivative, leading to the concept of nonsmooth analysis (Clarke 1983; Clarke et al. 1998).[15] From a practical point of view, that is, to solve actual real-world problems, nonsmooth analysis is still in need of exploration. In contrast to this, an abundance of optimal control problems have been solved using the maximum principle and its various extensions to problems with state-control constraints, pure state constraints, and infinite time horizons. For example, Arrow (1968) and Arrow and Kurz (1970a) provided an early overview of optimal control theory in models of economic growth.

1.4 Notes

An overview of the history and content of mathematics as a discipline can be found in Aleksandrov et al. (1969) and Campbell and Higgins (1984). Blåsjö (2005) illuminates the background of the isoperimetric problem. The historical development of the calculus of variations is summarized by Goldstine (1980) and Hildebrandt and Tromba (1985). For a history of technological feedback control systems, see Mayr (1970).

---

15. Vinter (2000) provided an account of optimal control theory in the setting of nonsmooth analysis.

2 Ordinary Differential Equations

Natura non facit saltus.
(Nature does not make jumps.)
—Gottfried Wilhelm Leibniz

2.1 Overview

An ordinary differential equation (ODE) describes the evolution of a variable $x(t)$ as a function of time t. The solution of such an equation depends on the initial *state* x_0 at a given time t_0. For example, $x(t)$ might denote the number of people using a certain product at time $t \geq t_0$ (e.g., a mobile phone). An ordinary differential equation describes how the (*dependent*) *variable* $x(t)$ changes as a function of time and its own current value. The change of state from $x(t)$ to $x(t+\delta)$ between the time instants t and $t+\delta$ as the increment δ tends to zero defines the *time derivative*

$$\dot{x}(t) = \lim_{\delta \to 0} \frac{x(t+\delta) - x(t)}{\delta}. \tag{2.1}$$

In an economic system such as a market, the change of a state can often be described as a function of time and the state at that time, in the form

$$\dot{x}(t) = f(t, x(t)), \tag{2.2}$$

where the *system function* f is usually given. The last relation is referred to as an ordinary differential equation. It is ordinary because the *independent variable* t is one-dimensional.[1] The descriptive question of how to find an appropriate representation f of the system is largely ignored

---

1. An equation that involves partial derivatives of functions with respect to components of a multidimensional independent variable is referred to as a partial differential equation (PDE). An example of such a PDE is the Hamilton-Jacobi-Bellman equation in chapter 3.

here because it typically involves observation and appropriate inference from data, requiring techniques that are different from optimal control, the main focus of this book.

Example 2.1 (Product Diffusion)   To see how to construct a system model in practice, let us consider the adoption of a new product, for example, a high-tech communication device. Let $x(t) \in [0,1]$ denote the installed base at time $t \geq t_0$ (for some given t_0), that is, the fraction of all potential adopters who at time t are in possession of the device. The fraction of new adopters between the instants t and $t + \delta$ as $\delta \to 0$ is referred to as the hazard rate,

$$h(t) \equiv \lim_{\delta \to 0} \frac{1}{\delta} \frac{x(t+\delta) - x(t)}{1 - x(t)} = \frac{\dot{x}(t)}{1 - x(t)},$$

which is defined using the concept of a derivative in equation (2.1). Based on empirical evidence on the adoption of television, Bass (1969) postulated an affine relation between the hazard rate and the installed base, such that

$$h(t) = \alpha x(t) + \beta,$$

referring to α as the coefficient of imitation and to β as the coefficient of innovation. A positive coefficient α can be attributed to a word-of-mouth effect, which increases the (conditional) likelihood of adoption proportional to the installed base. A positive coefficient β increases that likelihood irrespective of the installed base. The last relation implies a system equation of the form (2.2),

$$\dot{x}(t) = (1 - x(t))(\alpha x(t) + \beta),$$

for all $t \geq t_0$, where $f(t,x) = (1-x)(\alpha x + \beta)$ is in fact independent of t, or time-invariant. Despite its simplicity, the Bass diffusion model has often been shown to fit data of product-adoption processes astonishingly well, which may at least in part explain its widespread use (Bass et al. 1994). For more details on how to find the trajectories generated by the Bass model, see exercise 2.1c.   □

Section 2.2 discusses how to analyze a differential equation of the form (2.2). Indeed, when the system function f is sufficiently simple, it may be possible to obtain an explicit solution $x(t)$, which generally depends on a given initial state $x(t_0) = x_0$.[2] However, in many interesting

---
2. The Bass diffusion model introduced in example 2.1 is solved for $\beta = 0$ in example 2.3 and for the general case in exercise 2.1c.

applications, either the system function is too complex for obtaining a closed-form solution of the system, or the closed-form solution is too complicated to be useful for further analysis. In fact, the reader should not be shocked by this generic unsolvability of ordinary differential equations but should come to expect this as the modal case. Luckily, explicit solutions as a function of time are often not necessary to obtain important insights into the behavior of a system. For example, when the problem is well-posed (see section 2.2.3), then small variations of the system and the initial data will lead to small variations of the solution as well. This implies that the behavior of complicated systems that are sufficiently close to a simple system will be similar to the behavior of the simple system. This structural stability justifies the analysis of simple (e.g., linearized) systems and subsequent use of perturbation techniques to account for the effects of nonlinearities. An important system property that can be checked without explicitly solving the system equation (2.2) is the existence and stability of equilibria, which are points $\bar{x}$ at which the system function vanishes. The Bass diffusion model in example 2.1 for $\alpha > 0$ and $\beta = 0$ has two equilibria, at $x = 0$ and $x = 1$. When the initial state $x_0 = x(t_0)$ of the system at time t_0 coincides with one of these equilibrium states, the system will stay at rest there forever, that is, $x(t) = x_0$ for all $t \geq t_0$, because the rate of change $\dot{x}$ is zero. Yet, small perturbations of the initial state away from an equilibrium can have different consequences, giving rise to the notion of stability of an equilibrium. For example, choosing an initial state $x_0 = \varepsilon$ for a small $\varepsilon > 0$ will lead the state to move further away from zero, until the installed base $x(t)$ approaches saturation. Thus, $x_0 = 0$ is an unstable equilibrium. On the other hand, starting from an initial state $1 - \varepsilon$ for small positive ε, the system will tend to the state $x_0 = 1$, which is therefore referred to as stable equilibrium. Stability properties such as these, including the convergence of system trajectories to limit cycles, can often be analyzed by examining the monotonicity properties of a suitably defined energy or value function, usually referred to as a *Lyapunov function*. A generalized version of a Lyapunov function is used in chapter 3 to derive optimality conditions for optimal control problems.

In section 2.3 the framework for the analysis of first-order ordinary differential equations is extended to differential equations with higher-order time-derivatives by reducing the latter to systems of the former. That section also discusses a few more sophisticated solution techniques for systems of ordinary differential equations, such as the Laplace transform.

2.2 First-Order ODEs

2.2.1 Definitions

Let $n \geq 1$ be the dimension of the (real-valued) dependent variable

$$x(t) = (x_1(t), \ldots, x_n(t)),$$

which is also called state. A *(first-order) ordinary differential equation* is of the form

$$F(t, x(t), \dot{x}(t)) = 0,$$

where $\dot{x}(t)$ is the total derivative of $x(t)$ with respect to t, and $F : \mathbb{R}^{1+2n} \to \mathbb{R}^n$ is a continuously differentiable function. The differential equation is referred to as ordinary because the independent variable t is an element of the real line, $\mathbb{R}$. Instead of using the preceding *implicit* representation of an ODE, it is usually more convenient to use an *explicit* representation of the form

$$\dot{x}(t) = f(t, x(t)), \tag{2.3}$$

where $f : \mathcal{D} \to \mathbb{R}$ is a continuous function that directly captures how the derivative $\dot{x}$ depends on the state x and the independent variable t. The domain $\mathcal{D}$ is assumed to be a nonempty connected open subset of $\mathbb{R}^{1+n}$. Throughout this book it is almost always assumed that first-order ODEs are available in the explicit representation (2.3).[3]

Let $\mathcal{I} \subset \mathbb{R}$ be a nontrivial interval. A differentiable function $x : \mathcal{I} \to \mathbb{R}^n$ is called a *solution* to (or *integral curve* of) the ODE (2.3) (on $\mathcal{I}$) if

$$(t, x(t)) \in \mathcal{D} \quad \text{and} \quad \dot{x}(t) = f(t, x(t)), \quad \forall t \in \mathcal{I}. \tag{2.4}$$

Thus, $x(t)$ solves an *initial value problem* (IVP) relative to a given point $(t_0, x_0) \in \mathcal{D}$,[4] with $t_0 \in \mathcal{I}$, if in addition to (2.4), the *initial condition*

$$x(t_0) = x_0 \tag{2.5}$$

is satisfied (figure 2.1).

2.2.2 Some Explicit Solutions When $n = 1$

Let $n = 1$. For certain classes of functions f it is possible to obtain direct solutions to an IVP relative to a given point $(t_0, x_0) \in \mathcal{D} \subseteq \mathbb{R}^2$.

---

[3]. As long as $F_{\dot{x}} \neq 0$ at a point $(t, x(t), \dot{x}(t))$, by the implicit function theorem (proposition A.7 in appendix A) it is possible to solve for $\dot{x}(t)$, at least locally.
[4]. The IVP (2.4)–(2.5) is sometimes also referred to as the Cauchy problem (see footnote 9). Augustin-Louis Cauchy provided the first result on the existence and uniqueness of solutions in 1824 (see Cauchy 1824/1913, 399ff).

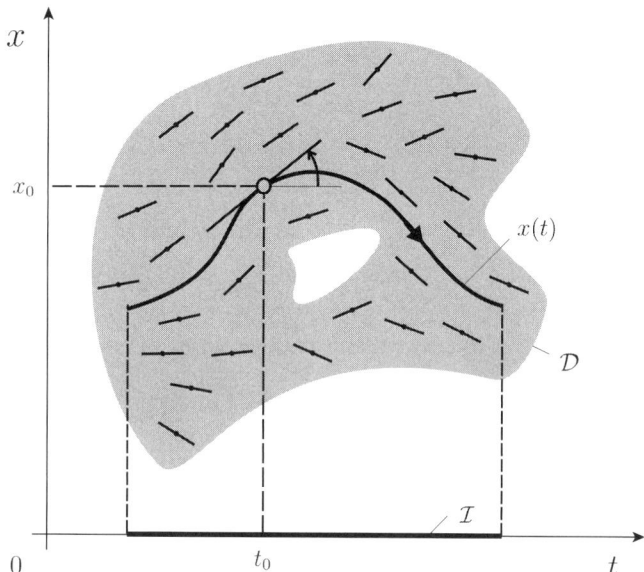

Figure 2.1
Solution to an ODE $\dot{x} = f(t,x)$ with initial condition $x(t_0) = x_0$ on $\mathcal{I}$.

Separability The function f is called *separable* if $f(t,x) = g(t)h(x)$ for all $(t,x) \in \mathcal{D}$. If $h(x)$ is nonzero for all $x \in \{\xi \in \mathbb{R} : \exists (s,\xi) \in \mathcal{D}\}$, then

$$H(x) \equiv \int_{x_0}^{x} \frac{d\xi}{h(\xi)} = \int_{t_0}^{t} g(s)\, ds \equiv G(t)$$

holds for all $(t,x) \in \mathcal{D}$, and $H(x)$ is invertible. Thus, $x(t) = H^{-1}(G(t))$ solves the IVP because in addition to (2.4), the initial condition $x(t_0) = H^{-1}(G(t_0)) = H^{-1}(0) = x_0$ is satisfied.

Example 2.2 (Exponential Growth) For a given parameter $\alpha > 0$, consider the ODE $\dot{x} = \alpha x$ with initial condition $x(t_0) = x_0$ for some $(t_0, x_0) \in \mathcal{D} = \mathbb{R}^2$ with $x_0 > 0$. Since the right-hand side of the ODE is separable,

$$\ln(x) - \ln(x_0) = \int_{x_0}^{x} \frac{d\xi}{\xi} = \alpha \int_{t_0}^{t} ds = \alpha(t - t_0),$$

so $x(t) = x_0 e^{\alpha(t - t_0)}$ is the unique solution to the IVP for all $t \in \mathbb{R}$. □

Example 2.3 (Logistic Growth) The initial size of a population is $x_0 > 0$. Let $x(t)$ denote the size of this population at time $t \geq t_0$, which evolves

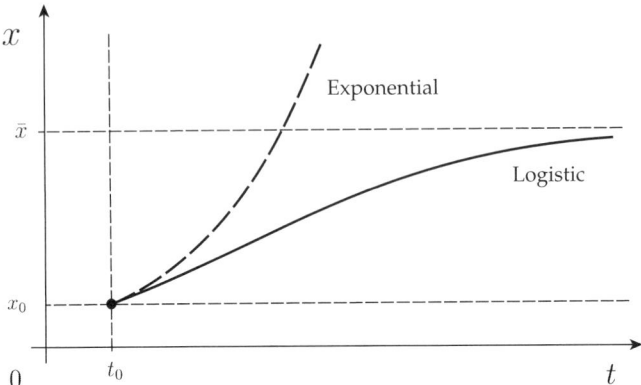

Figure 2.2
Exponential and logistic growth (see examples 2.2 and 2.3).

according to

$$\dot{x} = \alpha \left(1 - \frac{x}{\bar{x}}\right) x, \qquad x(t_0) = x_0,$$

where $\alpha > 0$ is the (maximum) relative growth rate, and $\bar{x} > x_0$ is a carrying capacity, which is a tight upper bound for $x(t)$. Analogous to example 2.2,

$$\frac{1}{\bar{x}}\left[\ln\left(\frac{x}{\bar{x}-x}\right) - \ln\left(\frac{x_0}{\bar{x}-x_0}\right)\right] = \int_{x_0}^{x} \frac{d\xi}{(\bar{x}-\xi)\xi} = \frac{\alpha}{\bar{x}}\int_{t_0}^{t} ds = \frac{\alpha}{\bar{x}}(t-t_0),$$

so $x(t)/(\bar{x} - x(t)) = x_0/(\bar{x} - x_0)e^{\alpha(t-t_0)}$ for all $t \geq t_0$. Hence,

$$x(t) = \frac{x_0 \bar{x}}{x_0 + (\bar{x} - x_0)e^{-\alpha(t-t_0)}}, \qquad t \geq t_0,$$

solves the IVP with logistic growth (figure 2.2). □

Homogeneity The function f is called *homogeneous (of degree zero)* if $f(t,x) = \rho(x/t)$.[5] Set $\varphi(t) = x(t)/t$ (for $t \geq t_0 > 0$); then $\dot{\varphi} = \dot{x}/t - x/t^2 = (\dot{x} - \varphi)/t$. Thus, the ODE (2.3) becomes $\dot{x} = t\dot{\varphi} + \varphi = \rho(\varphi)$, so

$$\dot{\varphi} = (1/t)\left(\rho(\varphi) - \varphi\right) \equiv g(t)h(\varphi)$$

---

5. The function $f : \mathcal{D} \to \mathbb{R}^n$ is *homogeneous of degree* $k \geq 0$ if for any $\alpha > 0$ and any $(t,x) \in \mathcal{D}$ the relation $f(\alpha t, \alpha x) = \alpha^k f(t,x)$ holds. Thus, for $k = 0$ and $\alpha = 1/t > 0$, we obtain that $f(t,x) = \rho(x/t)$ as long as $\rho(x/t) = f(1, x/t)$.

is a first-order ODE with separable right-hand side. A solution $\varphi(t)$ to the corresponding IVP with initial condition $\rho(t_0) = x_0/t_0$ implies a solution $x(t) = t\varphi(t)$ to the original IVP with initial condition (2.5).

Generalized Homogeneity If $f(t,x) = \rho(\frac{at+bx+c}{\alpha t+\beta x+\gamma})$, where $a, b, c, \alpha, \beta, \gamma$ are constants, then it is possible to find a solution using the previous methods after a suitable *affine coordinate transformation*. To see this, let $A = \begin{bmatrix} a & b \\ \alpha & \beta \end{bmatrix}$ and assume that $(a,b) \neq 0$ and that $\gamma \neq 0$, without loss of generality.[6] Depending on whether the matrix A is singular or not, two cases can be distinguished:

- Case 1. $\det A = 0$, that is, there is a λ such that $(\alpha, \beta) = \lambda(a,b)$. Since in that case

$$\frac{at+bx+c}{\alpha t+\beta x+\gamma} = \frac{at+bx+c}{\lambda(at+bx)+\gamma} = \frac{1+c/(at+bx)}{\lambda+\gamma/(at+bx)},$$

consider only $f(t,x) = \rho(at+bx)$. Set $\varphi = at+bx$; then

$$\dot\varphi = a+b\rho(\varphi) \equiv h(\varphi)$$

is an ODE with separable right-hand side.

- Case 2. $\det A \neq 0$, which implies that it is possible to find a reference point $(\hat{t}, \hat{x})$ such that[7]

$$\frac{at+bx+c}{\alpha t+\beta x+\gamma} = \frac{a(t-\hat{t})+b(x-\hat{x})}{\alpha(t-\hat{t})+\beta(x-\hat{x})} = \frac{a+b\left(\frac{x-\hat{x}}{t-\hat{t}}\right)}{\alpha+\beta\left(\frac{x-\hat{x}}{t-\hat{t}}\right)} = \frac{a+b\left(\frac{\xi}{\tau}\right)}{\alpha+\beta\left(\frac{\xi}{\tau}\right)},$$

where $\tau = t-\hat{t}$ and $\xi(\tau) = x(t)-\hat{x}$. Thus, $\dot\xi(\tau) = \dot x(t)$, which implies that the original ODE, using $\hat\rho(\xi/\tau) = \rho\left(\frac{a+b(\frac{\xi}{\tau})}{\alpha+\beta(\frac{\xi}{\tau})}\right)$, can be replaced by

$$\dot\xi(\tau) = \hat\rho(\xi/\tau),$$

an ODE with homogeneous right-hand side. Given a solution $\xi(\tau)$ to that ODE, a solution to the original ODE is then $x(t) = \xi(t-\hat{t}) + \hat{x}$.

---

6. Otherwise, if f is not already separable or homogeneous, simply switch the labels of (a,b,c) with (α, β, γ) and use a suitable definition of ρ.

7. Indeed, $\begin{bmatrix} \hat{t} \\ \hat{x} \end{bmatrix} = A^{-1} \begin{bmatrix} c \\ \gamma \end{bmatrix}$.

Linear First-Order ODE An important special case is when $f(t, x) = -g(t)x + h(t)$. The resulting (first-order) *linear* ODE is usually written in the form

$$\dot{x} + g(t)x = h(t). \tag{2.6}$$

The linear ODE is called *homogeneous* if $h(t) \equiv 0$. A so-called homogeneous solution $x_h(t; C)$ for that case is obtained immediately by realizing that $f(t, x) = -g(t)x$ is separable:

$$x_h(t; C) = C \exp\left[-\int_{t_0}^{t} g(s)\, ds\right], \tag{2.7}$$

where C is a suitable (nonzero) constant, which is determined by an initial condition. The solution $x(t)$ to the linear ODE (2.6) subject to the initial condition (2.5) can be provided as the sum of the homogeneous solution $x_h(t; C)$ in (2.7) and any particular solution $x_p(t)$ of (2.6). To construct the particular solution, one can use the so-called *variation-of-constants method*, dating back to Lagrange (1811),[8] where one takes the homogeneous solution $x_h(t; C)$ in (2.7) but allows the constant to vary with t. That is, one sets $x_p(t) = x_h(t; C(t))$. Substituting this in (2.6), one obtains the ODE

$$\dot{C}(t) = h(t) \exp\left[\int_{t_0}^{t} g(s)\, ds\right],$$

which implies (by separability of the right-hand side) that

$$C(t) = C_0 + \int_{t_0}^{t} h(s) \exp\left[\int_{t_0}^{s} g(\theta)\, d\theta\right] ds,$$

where $C(t_0) = C_0$. Without any loss of generality one can set $C_0 = 0$, which entails that $x_p(t_0) = 0$. Moreover,

$$x(t) = x_h(t; x_0) + x_p(t).$$

This is often referred to as the Cauchy formula.

Proposition 2.1 (Cauchy Formula) The unique solution to the linear IVP

---

8. Joseph-Louis Lagrange communicated the method in 1808 to the French Academy of Sciences; in concrete problems it was applied earlier by Leonhard Euler and Daniel Bernoulli.

Ordinary Differential Equations

$$\dot{x} + g(t)x = h(t), \qquad x(t_0) = x_0, \tag{2.8}$$

is given by[9]

$$x(t) = \left(x_0 + \int_{t_0}^{t} h(s) \exp\left[\int_{t_0}^{s} g(\theta) \, d\theta\right] ds\right) \exp\left[-\int_{t_0}^{t} g(s) \, ds\right]. \tag{2.9}$$

The formula (2.9) is very helpful in practice, and it is used frequently in this book.

Remark 2.1 (Duhamel Principle) The Cauchy formula (2.9) can be written in the form

$$x(t) = x_h(t; x_0) + (k * h)(t), \tag{2.10}$$

where the second term on the right-hand side is referred to as a convolution product with kernel $k(t, s) = x_h(t; 1) \left(x_h(s; 1)\right)^{-1}$,

$$(k * h)(t) = \int_{t_0}^{t} k(t, s) h(s) \, ds = \int_{t_0}^{t} \exp\left[-\int_{s}^{t} g(\theta) \, d\theta\right] h(s) \, ds.$$

Equation (2.10) may be easier to remember than the Cauchy formula (2.9). It makes plain that the solution to the linear IVP (2.8) is obtained as a *superposition* of the homogeneous solution (which depends only on the function $g(t)$) and a solution that is directly generated by the disturbance function $h(t)$. □

Figure 2.3 summarizes the methods used to solve several well-known classes of IVPs for $n = 1$.

Example 2.4 (Effect of Advertising on Sales) (Vidale and Wolfe 1957) Let $x(t)$ represent the sales of a certain product at time $t \geq 0$, and let initial sales $x_0 \in [0, \bar{x}]$ at time $t = 0$ be given, where $\bar{x} > 0$ is an estimated saturation level. A well-known model for the response of sales to a continuous rate of advertising expenditure $u(t)$, $t \geq 0$, can be written in the form of a linear IVP,

$$\dot{x} = r\left(1 - \frac{x}{\bar{x}}\right) u(t) - \lambda x, \qquad x(t_0) = x_0,$$

where $r \in (0, 1]$ is the response coefficient and $\lambda > 0$ a sales decay constant. The former describes how effective advertising expenditure is in

---

9. The term *Cauchy formula* is adopted for convenience. Instead, one can also refer to equation (2.9), which, after all, was obtained by Lagrange's variation-of-constants method, as the "solution formula to the (linear) Cauchy problem" (see footnote 4).

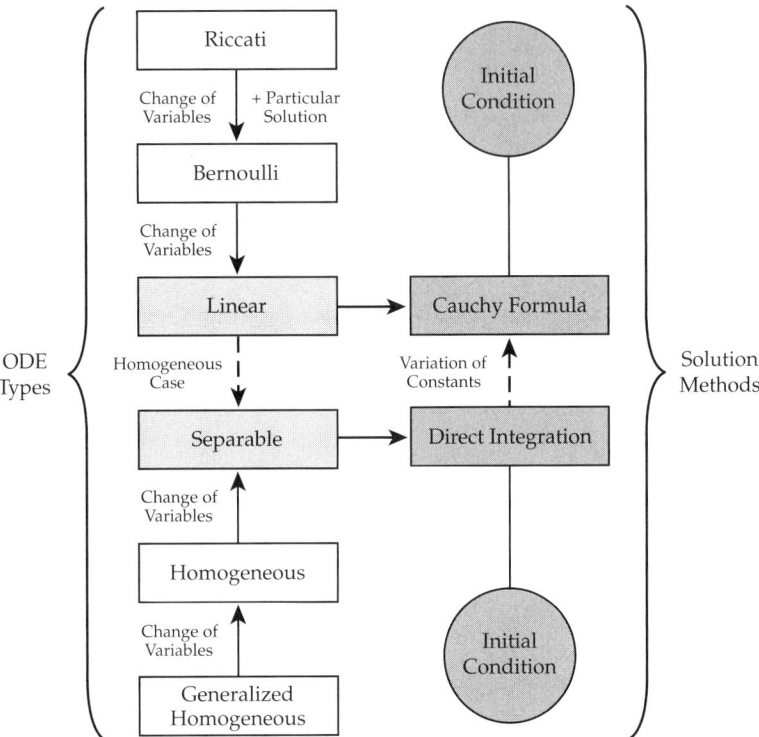

Figure 2.3
Solution of well-known types of IVPs when $n = 1$.

generating sales, and the latter defines the sales response when advertising is stopped altogether. Setting $g(t) = \lambda + ru(t)/\bar{x}$ and $h(t) = ru(t)$, the Cauchy formula (2.9) yields that

$$x(t) = \left(x_0 \exp\left[-\frac{r}{\bar{x}} \int_0^t u(\theta)\, d\theta \right] \right.$$
$$\left. + r \int_0^t u(s) \exp\left[\lambda s - \frac{r}{\bar{x}} \int_s^t u(\theta)\, d\theta \right] ds \right) e^{-\lambda t},$$

for all $t \geq 0$. For example, if the rate of advertising expenditure is equal to the constant $u_0 > 0$ on the time interval $[0, T]$ for some campaign horizon T, and zero thereafter, then the previous expression specializes to

$$x(t) = \begin{cases} x_0 e^{-(\lambda + ru_0/\bar{x})t} + \bar{x}\, \frac{ru_0}{\lambda \bar{x} + ru_0} \left(1 - e^{-(\lambda + ru_0/\bar{x})t} \right) & \text{if } t \in [0, T], \\ x(T) e^{-\lambda(t-T)} & \text{if } t > T. \end{cases}$$

Ordinary Differential Equations

Thus, an infinite-horizon advertising campaign (as $T \to \infty$) with expenditure rate u_0 would cause sales $x(t)$ to approach the level $x_\infty \equiv (1 + \frac{\lambda \bar{x}}{r u_0})^{-1} \bar{x} < \bar{x}$ for $t \to \infty$. □

Bernoulli Equation The nonlinear ODE

$$\dot{x} + g(t)x + h(t)x^\alpha = 0, \qquad \alpha \neq 1, \tag{2.11}$$

can be transformed into a linear ODE. Indeed, multiplying (2.11) by $x^{-\alpha}$ yields

$$x^{-\alpha}\dot{x} + g(t)x^{1-\alpha} + h(t) = 0.$$

With the substitution of $\varphi = x^{1-\alpha}/(1-\alpha)$ the last ODE becomes linear,

$$\dot{\varphi} + (1-\alpha)g(t)\varphi + h(t) = 0, \tag{2.12}$$

and can be solved using the Cauchy formula (see exercise 2.1a).

Riccati Equation The nonlinear ODE

$$\dot{x} + \rho(t)x + h(t)x^2 = \sigma(t) \tag{2.13}$$

cannot generally be solved explicitly. Yet, if a particular solution x_p is known, then it is possible to compute all other solutions. Let x be another solution to (2.13). Then the difference, $\Delta = x - x_p$, satisfies the ODE

$$\dot{\Delta} + \rho(t)\Delta + h(t)\underbrace{(x^2 - x_p^2)}_{(x+x_p)(x-x_p)} = \dot{\Delta} + \rho(t)\Delta + h(t)(\Delta + 2x_p)\Delta = 0.$$

In other words, the difference Δ satisfies the Bernoulli equation

$$\dot{\Delta} + g(t)\Delta + h(t)\Delta^2 = 0,$$

where $g(t) = \rho(t) + 2x_p(t)h(t)$. Thus, using the substitution $\varphi = -(1/\Delta)$, one obtains the linear ODE (2.12) with $\alpha = 2$, which can be solved using the Cauchy formula. Any particular solution x_p to the Riccati equation (2.13) therefore implies all other solutions in the form

$$x = x_p - \frac{1}{\varphi},$$

where φ is any solution to (2.12). Example 2.14 shows how to solve a matrix Riccati equation with some special structure by reducing it to a system of linear ODEs. The Riccati equation plays an important role for the optimal control of a linear system with quadratic objective functional, which is often referred to as a linear-quadratic regulator (see example 3.3).

2.2.3 Well-Posed Problems

In accord with a notion by Hadamard (1902) for mathematical models of physical phenomena, an IVP is said to be *well-posed* if the following three requirements are satisfied:

- There exists a solution.
- Any solution is unique.
- The solution depends continuously on the available data (parameters and initial condition).

Problems that do not satisfy at least one of these requirements are called *ill-posed*.[10] A deterministic economic system is usually described here in terms of a well-posed IVP. This ensures that system responses can be anticipated as unique consequences of outside intervention and that these responses remain essentially unaffected by small changes in the system, leading to a certain robustness of the analysis with respect to modeling and identification errors. In what follows, easy-to-verify conditions are established under which an IVP is well-posed.

Existence and Uniqueness The existence of a solution to an IVP is guaranteed when the right-hand side of the ODE (2.3) is continuous, as assumed from the outset.

Proposition 2.2 (Existence) (Peano 1890) Let $f \in C^0(\mathcal{D})$. For any $(t_0, x_0) \in \mathcal{D}$ the IVP

$$\dot{x} = f(t, x), \qquad x(t_0) = x_0,$$

has a solution on a nontrivial interval $\mathcal{I} \subset \mathbb{R}$ that contains t_0. Any such solution can be extended (in the direction of both positive and negative times t) such that it comes arbitrarily close to the boundary of $\mathcal{D}$.

Proof See, for example, Walter (1998, 73–78).[11] ∎

---

10. A well-known class of ill-posed problems is that of inverse problems, where a model is to be determined from data. To reduce the sensitivity of solutions to data errors, one can use regularization methods, e.g., the one by Tikhonov (1963) for linear models (see, e.g., Kress 1998, 86–90).

11. The proof of this result is nonconstructive and is therefore omitted.

Ordinary Differential Equations

The phrase "arbitrarily close to the boundary of $\mathcal{D}$" in proposition 2.2 means that any solution $x(t)$ can be extended for all times $t \in \mathbb{R}$, unless a boundary of $\mathcal{D}$ can be reached in finite time, in which case it is possible to approach that escape time arbitrarily closely. A solution to the IVP that has been fully extended to the closure $\bar{\Omega} \subset \mathcal{D}$ of a nonempty connected open set Ω such that $\bar{\Omega}$ contains the initial point (t_0, x_0) is said to be *maximal* on Ω.

Example 2.5 (Nonuniqueness) Let $\mathcal{D} = \mathbb{R}^2$, and consider the IVP

$$\dot{x} = \sqrt{|x|}, \quad x(0) = 0.$$

Note first that the function $f(t, x) = \sqrt{|x|}$ on the right-hand side is continuous. Proposition 2.2 guarantees the existence of a solution to this IVP, but one cannot expect uniqueness, the second of Hadamard's requirements for well-posedness, to hold. Indeed, since the right-hand side is symmetric, with any solution $x(t)$ of the IVP the function $-x(-t)$ is also a solution. Note also that the function $x(t) = 0$ is a solution for all $t \in \mathbb{R}$. Since the system function f is separable, one can compute another solution by direct integration,

$$2\sqrt{x} = \int_0^x \frac{d\xi}{\sqrt{\xi}} = \int_0^t ds = t,$$

for $(t, x) \geq 0$. Using the aforementioned symmetry, this results in a second solution to the IVP, of the form $\hat{x}(t) = t|t|/4$, which is defined for all $t \in \mathbb{R}$. Besides $x(t) \equiv 0$ and $\hat{x}(t)$, there are (infinitely) many more solutions,

$$x(t; C_1, C_2) = \begin{cases} -\frac{(t+C_1)^2}{4} & \text{if } t \leq -C_1, \\ 0 & \text{if } t \in [-C_1, C_2], \\ \frac{(t-C_2)^2}{4} & \text{if } t \geq C_2, \end{cases}$$

indexed by the nonnegative constants C_1, C_2, with the two earlier solutions at diametrically opposed extremes, such that

$$x(t; 0, 0) = \hat{x}(t), \quad \text{and} \quad \lim_{C_1, C_2 \to \infty} x(t; C_1, C_2) = 0 \text{ (pointwise)},$$

for all $t \in \mathbb{R}$ (figure 2.4). □

The function $f(t, x)$ is said to be *Lipschitz* (with respect to x, on $\mathcal{D}$) if there exists a nonnegative constant L such that

$$\|f(t, \hat{x}) - f(t, x)\| \leq L\|\hat{x} - x\|, \quad \forall (t, \hat{x}), (t, x) \in \mathcal{D}, \tag{2.14}$$

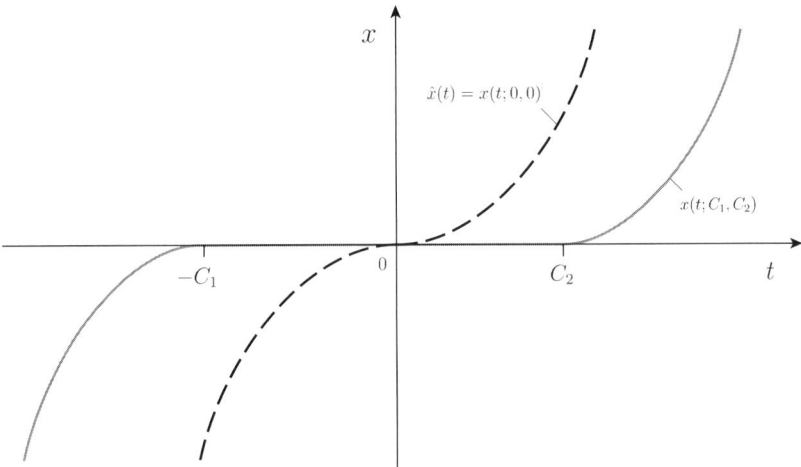

Figure 2.4
Nonuniqueness of solutions to the IVP $\dot{x} = \sqrt{|x|}$, $x(0) = 0$ (see example 2.5).

where $\|\cdot\|$ is any suitable norm on $\mathbb{R}^n$. If, instead, for any point $(t, x) \in \mathcal{D}$ there exist $\varepsilon_x > 0$ and $L_x \geq 0$ such that

$$\|f(t, \hat{x}) - f(t, x)\| \leq L_x \|\hat{x} - x\|,$$
$$\forall (t, \hat{x}) \in \{(t, \xi) \in \mathcal{D} : \|(t, x) - (t, \xi)\| < \varepsilon_x\},$$
(2.15)

then the function $f(t, x)$ is called *locally Lipschitz* (with respect to x, on $\mathcal{D}$). Clearly, if a function is Lipschitz, then it is also locally Lipschitz, but not vice versa.[12] On the other hand, if $f(t, x)$ is locally Lipschitz, then it is Lipschitz on any compact subset of $\mathcal{D}$. Both properties are, broadly speaking, implied if the function $f(t, x)$ is continuously differentiable in x on $\mathcal{D}$.

Lemma 2.1 Assume that the function $f(t, x)$ is continuously differentiable in x on $\mathcal{D}$. (1) The function $f(t, x)$ is locally Lipschitz with respect to x on $\mathcal{D}$. (2) If $\mathcal{D}$ is convex and the Jacobian matrix $f_x = [\partial f_i / \partial x_j]$ is bounded on $\mathcal{D}$, then the function $f(t, x)$ is Lipschitz with respect to x on $\mathcal{D}$.

Proof (1) Fix any point $(t, x) \in \mathcal{D}$, and select $\varepsilon_x > 0$ such that the ball

$$\mathcal{B}_{\varepsilon_x}(t, x) = \{(\tau, \xi) \in \mathbb{R}^{1+n} : \|(t, x) - (\tau, \xi)\| \leq \varepsilon_x\}$$

is contained in $\mathcal{D}$. This is possible, since $\mathcal{D}$ is an open set. Consider $(t, \hat{x}) \in \mathcal{B}_{\varepsilon_x}(t, x)$, and apply the mean-value theorem (proposition A.14) to the ith

---

12. Consider $f(t, x) = tx^2$ on $\mathcal{D} = \mathbb{R}^2$, which is locally Lipschitz but does not satisfy (2.14).

coordinate function f_i, $i \in \{1, \ldots, n\}$, of $f = (f_1, \ldots, f_n)'$. Then

$$f_i(t, \hat{x}) - f_i(t, x) = \langle f_{i,x}(t, \lambda_i \hat{x} + (1 - \lambda_i)x), \hat{x} - x \rangle,$$

for some $\lambda_i \in [0, 1]$. Since the derivative $f_{i,x}$ is continuous on the compact set $\mathcal{B}_{\varepsilon_x}(t, x)$, it is also bounded there. Hence, there exists a constant $L_{i,x} \geq 0$ such that

$$\|f_i(t, \hat{x}) - f_i(t, x)\| \leq L_{i,x} \|\hat{x} - x\|.$$

Combining this argument for the n coordinate functions (e.g., using the maximum norm together with the norm equivalence in remark A.1) yields that there exists a constant $L_x \geq 0$ such that

$$\|f(t, \hat{x}) - f(t, x)\| = \|f_x(t, \lambda \hat{x} + (1 - \lambda)x)(\hat{x} - x)\| \leq L_x \|\hat{x} - x\|,$$

for all $(t, \hat{x}) \in \mathcal{B}_{\varepsilon_x}(t, x)$, which in turn implies that $f(t, x)$ is locally Lipschitz with respect to x on $\mathcal{D}$.

(2) Since $\mathcal{D}$ is convex, for any two points $(t, \hat{x})$ and (t, x) in $\mathcal{D}$, $(t, \lambda \hat{x} + (1 - \lambda)x) \in \mathcal{D}$ for all $\lambda \in [0, 1]$. As in part (1) (with $\mathcal{B}_{\varepsilon_x}(t, x)$ replaced by $\mathcal{D}$), applying the mean-value theorem yields that $f(t, x)$ is Lipschitz with respect to x on $\mathcal{D}$. ■

Proposition 2.3 (Local Existence and Uniqueness) If f is locally Lipschitz with respect to x on $\mathcal{D}$, then for any given $(t_0, x_0) \in \mathcal{D}$ the IVP

$$\dot{x} = f(t, x), \quad x(t_0) = x_0, \tag{2.16}$$

has a unique solution on a nontrivial interval $\mathcal{I} \subset \mathbb{R}$ which contains t_0.

Proof Let $\mathcal{I} = [t_0, t_0 + \delta]$ be a nontrivial interval for a suitably small positive δ. If $x(t)$ is a solution to the IVP (2.16) on $\mathcal{I}$, then by integration one obtains the integral equation

$$x(t) = x_0 + \int_{t_0}^{t} f(s, x(s)) \, ds, \tag{2.17}$$

for all $t \in \mathcal{I}$. Since (2.17) in turn implies (2.16), this integral equation is in fact an equivalent *integral representation* of the IVP. The statements for proving (2.17) follow.

(1) *Existence.* Consider the Banach space[13] $\mathcal{X} = C^0(\mathcal{I})$ of all continuous functions $x : \mathcal{I} \to \mathbb{R}^n$, equipped with the maximum norm $\|\cdot\|_\infty$ that is defined by

---

13. A *Banach space* is a linear space that is equipped with a norm and that is *complete* in the sense that any Cauchy sequence converges. See appendix A.2 for more details.

$$\|x\|_\infty = \max_{t\in[t,t_0+\delta]} \|x(t)\|$$

for all $x \in \mathcal{X}$. The right-hand side of the integral equation (2.17) maps any function $x \in \mathcal{X}$ to another element Px in $\mathcal{X}$, where $P : \mathcal{X} \to \mathcal{X}$ is a continuous *functional*. For any given time $t \in \mathcal{I}$ the right-hand side of (2.17) evaluates to $(Px)(t)$. With this notation, the integral equation (2.17) on $\mathcal{I}$ can be equivalently rewritten as a fixed-point problem of the form

$$x = Px. \tag{2.18}$$

Let $\mathcal{S}_r = \{x \in \mathcal{X} : \|x - x_0\|_\infty \leq r\}$ be a closed ball of radius $r > 0$ in the Banach space $\mathcal{X}$ centered on the constant function x_0. Now note that for a small enough r, $P : \mathcal{S}_r \to \mathcal{S}_r$ (i.e., P maps the ball $\mathcal{S}_r$ onto itself) *and* moreover P is a *contraction mapping* on $\mathcal{S}_r$, that is, it satisfies the Lipschitz condition

$$\|P\hat{x} - Px\|_\infty \leq K\|\hat{x} - x\|_\infty, \tag{2.19}$$

for all $\hat{x}, x \in \mathcal{S}_r$, with a Lipschitz constant $K < 1$. This would allow the application of the Banach fixed-point theorem (proposition A.3), which guarantees the existence of a unique solution to the fixed-point problem (2.18). In addition, the fixed point can be obtained by successive iteration from an arbitrary starting point in $\mathcal{S}_r$.

For this, let $M_f = \|f(\cdot, x_0)\|_\infty$ be the maximum of $\|f(t, x_0)\|$ over all $t \in \mathcal{I}$. Take an arbitrary function $x \in \mathcal{S}_r$. Since

$$(Px)(t) - x_0 = \int_{t_0}^{t} f(s, x(s))\, ds$$

$$= \int_{t_0}^{t} \left(f(s, x(s)) - f(s, x_0) + f(s, x_0) \right) ds,$$

this implies (using the assumption that f is Lipschitz with respect to x with constant L) that

$$\|(Px)(t) - x_0\| \leq \int_{t_0}^{t} \|f(s, x(s))\|\, ds$$

$$= \int_{t_0}^{t} \left(\|f(s, x(s)) - f(s, x_0)\| + \|f(s, x_0)\| \right) ds$$

$$\leq \int_{t_0}^{t} \left(L\|x(s) - x_0\| + M_f \right) ds$$

$$\leq (t-t_0)(Lr+M_f)$$
$$\leq \delta(Lr+M_f),$$

for all $t \in \mathcal{I}$. Thus, as long as $\delta \leq r/(Lr+M_f)$,

$$\|Px-x_0\|_\infty = \max_{t\in[t_0,t_0+\delta]} \|(Px)(t)-x_0\| \leq \delta(Lr+M_f) \leq r,$$

which implies that $Px \in \mathcal{S}_r$. Hence, it has been shown that when δ is small enough P maps $\mathcal{S}_r$ into itself. Now select any $\hat{x}, x \in \mathcal{S}_r$. Then

$$\|(P\hat{x})(t) - (Px)(t)\| = \left\|\int_{t_0}^t (f(s,\hat{x}(s)) - f(s,x(s)))\,ds\right\|$$
$$\leq \int_{t_0}^t \|f(s,\hat{x}(s)) - f(s,x(s))\|\,ds$$
$$\leq \int_{t_0}^t L\|\hat{x}(s) - x(s)\|\,ds \leq L\delta\|\hat{x}-x\|_\infty,$$

for all $t \in [t_0, t_0+\delta]$, which implies that

$$\|P\hat{x} - Px\|_\infty \leq L\delta\|\hat{x}-x\|_\infty,$$

that is, the Lipschitz condition (2.19) for $K = L\delta < 1$, provided that $\delta < 1/L$. Hence, as long as

$$\delta < \min\left\{\frac{r}{Lr+M_f}, \frac{1}{L}\right\}, \tag{2.20}$$

the continuous functional $P: \mathcal{S}_r \to \mathcal{S}_r$ is a contraction mapping on the convex set $\mathcal{S}_r$. The hypotheses of the Banach fixed-point theorem are satisfied, so (2.18) has a unique solution x, which is obtained as the pointwise limit of the sequence $\{x^k\}_{k=0}^\infty$, where $x^0 = x_0$ and $x^{k+1} = Px^k$ for all $k \geq 0$, that is,

$$x(t) = \lim_{k\to\infty} x^k(t) = \lim_{k\to\infty} (P^k x_0)(t),$$

for all $t \in \mathcal{I}$, where P^k denotes the k-fold successive application of P.

(2) *Uniqueness.* To show that the solution $x(t)$ established earlier is unique, it is enough to demonstrate that $x(t)$ must stay inside the ball $\mathcal{B}_r(x_0) = \{\xi \in \mathbb{R}^n : \|\xi - x_0\| \leq r\}$. Let $t_0 + \Delta$ be the first intersection time, so that $x(t_0 + \Delta) - x_0 = r$. If $\Delta > \delta$, then there is nothing to prove. Thus, suppose that $\Delta \leq \delta$. Then

$$\|x(t_0 + \Delta) - x_0\| = r,$$

and for all $t \in [t_0, t_0 + \Delta]$, as before,

$$\|x(t) - x_0\| \leq \Delta(Lr + M_f).$$

In particular, $r = \|x(t_0 + \Delta) - x_0\| \leq \Delta(Lr + M_f)$, so by virtue of (2.20) it is

$$\delta \leq \frac{r}{Lr + M_f} \leq \Delta,$$

which implies that the solution $x(t)$ cannot leave the ball $\mathcal{B}_r(x_0)$ for all $t \in \mathcal{I}$. Therefore any continuous solution x must be an element of $\mathcal{S}_r$, which by the Banach fixed-point theorem implies that the solution is unique.

This completes the proof of proposition 2.3. ∎

Remark 2.2 (Picard-Lindelöf Error Estimate) Consider the interval $\mathcal{I} = [t_0, t_0 + \delta]$, where the constant δ satisfies (2.20). By the Banach fixed-point theorem (proposition A.3) the iteration scheme featured in the proof of proposition 2.3 (often referred to as *successive approximation*) converges to a unique solution. If, starting from any initial function $x^0 \in \mathcal{S}_r$,

$$\|x^1(t) - x^0(t)\| \leq M|t - t_0| + \mu$$

with appropriate positive constants μ, M for all $t \in \mathcal{I}$ (e.g., $M = M_f$ and $\mu = r$), then

$$\|x^k(t) - x^{k-1}(t)\| = \|(Px^{k-1})(t) - (Px^{k-2})(t)\|$$

$$\leq L \left| \int_{t_0}^{t} \|x^{k-1}(s) - x^{k-2}(s)\| ds \right|$$

$$\leq \cdots \leq L^{k-1} \int_{t_0}^{t} (M|t - t_0| + \mu) \, ds$$

$$= L^{k-1} M \frac{|t - t_0|^2}{2} + L^{k-1} \mu |t - t_0|,$$

for all $k \geq 2$. Carrying the recursion for the integration toward the left in this chain of inequalities yields

$$\|x^k(t) - x^{k-1}(t)\| \leq \frac{M}{L} \frac{(L|t - t_0|)^k}{k!} + \mu \frac{(L|t - t_0|)^{k-1}}{(k-1)!}.$$

Ordinary Differential Equations

In addition,

$$x = x^n + \sum_{k=n+1}^{\infty} (x^k - x^{k-1}),$$

for all $n \geq 0$, which, using the previous inequality, implies (by summing over k) the Picard-Lindelöf error estimate

$$\|x^n(t) - x(t)\| \leq \frac{M}{L} \sum_{k=n+1}^{\infty} \frac{(L(t-t_0))^k}{k!} + \mu \sum_{k=n}^{\infty} \frac{(L|t-t_0|)^{k-1}}{(k-1)!}$$

$$\leq \left(\frac{M}{L} \frac{(L|t-t_0|)^{n+1}}{(n+1)!} + \mu \frac{(L|t-t_0|)^n}{n!} \right) e^{L|t-t_0|}$$

$$\leq \left(\frac{M\delta}{n+1} + \mu \right) \frac{(L\delta)^n e^{L\delta}}{n!},$$

for all $t \in \mathcal{I}$. □

Proposition 2.4 (Global Existence and Uniqueness) If f is uniformly bounded and Lipschitz with respect to x on $\mathcal{D}$, then for any given (t_0, x_0) the IVP (2.16) has a unique solution.[14]

Proof By proposition 2.3 there exists a local solution on $[t_0, t_0 + \delta]$ for some small (but finite) $\delta > 0$. Since f is Lipschitz on the domain $\mathcal{D}$ and uniformly bounded, the constants r, M_f, L, K in the proof of the local result become independent of the point x_0. Hence, it is possible to extend the solution forward starting from the initial data $(t_0 + \delta, x(t_0 + \delta))$ on the interval $\mathcal{I}_2 = [t_0 + \delta, t_0 + 2\delta]$, and so forth, on $\mathcal{I}_k = [t_0 + (k-1)\delta, t_0 + k\delta]$ for all $k \geq 2$ until the solution approaches the boundary of $\mathcal{D}$. The same can be done in the direction of negative times. ■

A system with *finite escape time* is such that it leaves *any* compact subset Ω of an unbounded state space $\mathcal{X}$ in finite time (assuming that $\mathcal{D} = [t_0, \infty) \times \mathcal{X}$). Under the assumptions of proposition 2.4 a system cannot have a finite escape time.

Example 2.6 (Finite Escape Time) Consider the IVP $\dot{x} = x^2$, $x(t_0) = x_0$, for some $(t_0, x_0) \in \mathcal{D} = \mathbb{R}^2_{++}$. The function x^2 is locally (but not globally) Lipschitz with respect to x on $\mathcal{D}$. Thus, by proposition 2.3 there is a unique solution,

---

14. More specifically, the maximal solution on any $\Omega \subset \mathcal{D}$, as in the definition of a maximal solution following proposition 2.2, exists and is unique.

$$x(t) = \frac{x_0}{1 - x_0(t - t_0)},$$

which can be computed by direct integration (using separability of the right-hand side of the system equation). Since

$$\lim_{t \to t_e^-} x(t) = \infty,$$

for $t_e = t_0 + (1/x_0) < \infty$, the system has finite escape time (equal to t_e). □

Continuous Dependence Continuous dependence, the third condition for well-posedness of an IVP, requires that small changes in initial data as well as in the system equation have only a small impact on the solution (figure 2.5). Assume that $\alpha \in \mathbb{R}^p$ is an element of a p-dimensional Euclidean parameter space, where $p \geq 1$ is a given integer. Now consider perturbations of a *parameterized* IVP,

$$\dot{x} = f(t, x, \alpha), \qquad x(t_0) = x_0, \qquad (2.21)$$

where the function $f : \mathcal{D} \times \mathbb{R}^p$ is assumed to be continuous and $(t_0, x_0) \in \mathcal{D}$. For a given $\alpha = \alpha_0 \in \mathbb{R}^p$ the parameterized IVP is called a *nominal* IVP, relative to which small model perturbations can be examined.

Proposition 2.5 (Continuous Dependence) If the function $f(t, x, \alpha)$ is locally Lipschitz with respect to x on $\mathcal{D}$, then for any $\varepsilon > 0$ there exists $\delta_\varepsilon > 0$ such that

$$|\hat{t}_0 - t_0| + \|\hat{x}_0 - x_0\| + \|\hat{\alpha} - \alpha_0\| \leq \delta_\varepsilon \Rightarrow \|\hat{x}(t, \hat{\alpha}) - x(t, \alpha_0)\| \leq \varepsilon,$$

for all t in an open interval $\mathcal{I}$, where $x(t, \alpha_0)$ is the solution to the IVP (2.21) for $\alpha = \alpha_0$ and $\hat{x}(t, \hat{\alpha})$ is the solution to (2.21), for $\alpha = \hat{\alpha}$ and $(t_0, x_0) = (\hat{t}_0, \hat{x}_0)$.

Proof Since $f(t, x, \alpha_0)$ is locally Lipschitz, by applying proposition 2.3 in both directions of time, one can find $\Delta > 0$ such that there is a unique solution $x(t, \alpha_0)$ of the (nominal) IVP (2.21) for $\alpha = \alpha_0$ on the interval $[t_0 - \Delta, t_0 + \Delta]$. Fix $\varepsilon_1 \in (0, \min\{\Delta, \varepsilon, \bar{\varepsilon}_1\})$ such that the ε_1-neighborhood of the corresponding trajectory is contained in $\mathcal{D}$, that is, such that

$$\mathcal{N} = \{(\tau, \xi) \in \mathcal{D} \cap [t_0 - \Delta, t_0 + \Delta] \times \mathbb{R}^n : \|x(\tau, \alpha_0) - \xi\| \leq \varepsilon_1\} \subset \mathcal{D}.$$

The constant $\bar{\varepsilon}_1$ is determined in equation (2.23). Note that the set $\mathcal{N}$ (which can be interpreted geometrically as a tube containing the nominal trajectory) is compact. Fix an arbitrary $a > 0$. Without loss of generality,

Ordinary Differential Equations

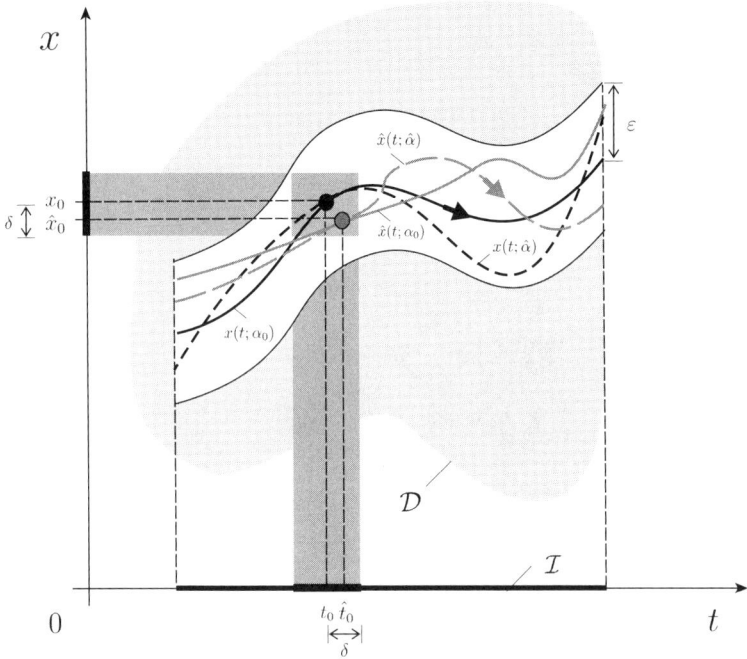

Figure 2.5
Continuous dependence on initial conditions and parameter α of solutions to the IVP $\dot{x} = f(t, x)$ with initial conditions $x(t_0; \alpha) = x_0$ and $x(\hat{t}_0; \alpha) = \hat{x}_0$.

one can restrict attention to parameters α that lie in a closed ball $\mathcal{B}_a(\alpha_0)$ of radius $a > 0$, with the nominal parameter $\alpha_0 \in \mathbb{R}^p$ at its center, so that

$$\|\alpha - \alpha_0\| \leq a.$$

Since $f(t, x, \alpha)$ is by assumption continuous on $\mathcal{D} \times \mathbb{R}^p$, it is uniformly continuous on the compact subset $\mathcal{N} \times \mathcal{B}_a(\alpha_0)$. Thus, there exists a $\rho \in (0, a]$ such that

$$\|\hat{\alpha} - \alpha\| \leq \rho \Rightarrow \|f(t, x, \hat{\alpha}) - f(t, x, \alpha)\| \leq \varepsilon_1. \tag{2.22}$$

Using an integral representation of the solution to the parameterized IVP (2.21), analogous to (2.17) in the proof of proposition 2.3, it is

$$\hat{x}(t, \hat{\alpha}) - x(t, \alpha_0) = \hat{x}_0 + \int_{\hat{t}_0}^{t} f(s, \hat{x}(s, \hat{\alpha}), \hat{\alpha}) \, ds$$

$$- \left(x_0 + \int_{t_0}^{t} f(s, x(s, \alpha_0), \alpha_0) \, ds \right)$$

$$= \hat{x}_0 - x_0 + \int_{\hat{t}_0}^{t_0} f(s, \hat{x}(s, \hat{\alpha}), \hat{\alpha}) \, ds$$

$$+ \int_{t_0}^{t} \left(f(s, \hat{x}(s, \hat{\alpha}), \hat{\alpha}) - f(s, x(s, \alpha_0), \alpha_0) \right) ds,$$

for all $t \in [t_0 - \Delta, t_0 + \Delta]$. Hence, provided that

$$|\hat{t}_0 - t_0| + \|\hat{x}_0 - x_0\| + \|\hat{\alpha} - \alpha_0\| \leq \delta$$

for some $\delta \in (0, \min\{\varepsilon_1, \rho\}]$, and taking into account (2.22),[15]

$$\|\hat{x}(t, \hat{\alpha}) - x(t, \alpha_0)\| \leq \delta + \delta M_f$$

$$+ \left| \int_{t_0}^{t} \|f(s, \hat{x}(s, \hat{\alpha}), \hat{\alpha}) - f(s, x(s, \alpha_0), \alpha_0)\| ds \right|$$

$$\leq (1 + M_f)\delta + \varepsilon_1 |t - t_0|$$

$$+ \left| \int_{t_0}^{t} \|f(s, \hat{x}(s, \hat{\alpha}), \alpha_0) - f(s, x(s, \alpha_0), \alpha_0)\| ds \right|$$

$$\leq (1 + M_f)\delta + \varepsilon_1 \Delta + L \left| \int_{t_0}^{t} \|\hat{x}(s, \hat{\alpha}) - x(s, \alpha_0)\| ds \right|,$$

for all $t \in [t_0 - \Delta, t_0 + \Delta]$, where $M_f = \max_{(t,x,\alpha) \in \mathcal{N} \times \mathcal{B}_a(\alpha_0)} \|f(t, x, \alpha)\|$ and $L \geq 0$ is the (local) Lipschitz constant of f with respect to x. Applying the Gronwall-Bellman inequality (proposition A.9) to the last inequality yields that

$$\|\hat{x}(t, \hat{\alpha}) - x(t, \alpha_0)\| \leq \left((1 + M_f)\delta + \varepsilon_1 \Delta \right) e^{L|t - t_0|} \leq \left((1 + M_f)\delta + \varepsilon_1 \Delta \right) e^{L\Delta},$$

for all $t \in [t_0 - \Delta, t_0 + \Delta]$. Since by construction $\delta \leq \varepsilon_1$, it is

$$\|\hat{x}(t, \hat{\alpha}) - x(t, \alpha_0)\| \leq \varepsilon,$$

for all $t \in [t_0 - \Delta, t_0 + \Delta]$, as long as

$$\varepsilon_1 \leq \bar{\varepsilon}_1 \equiv \min \left\{ \frac{\varepsilon e^{-L\Delta}}{1 + M_f + \Delta}, \Delta \right\}, \tag{2.23}$$

and $\delta = \delta_\varepsilon \leq \min\{\varepsilon_1, \rho\}$, which completes the proof. ∎

---

15. Note also that

$$\|f(s, \hat{x}, \hat{\alpha}) - f(s, x, \alpha_0)\| \leq \|f(s, \hat{x}, \hat{\alpha}) - f(s, \hat{x}, \alpha_0)\| + \|f(s, \hat{x}, \alpha_0) - f(s, x, \alpha_0)\|$$

$$\leq \varepsilon_1 + \|f(s, \hat{x}, \alpha_0) - f(s, x, \alpha_0)\|.$$

Remark 2.3 (Sensitivity Analysis) If, in addition to the assumptions in proposition 2.5, the function $f(t, x, \alpha)$ is continuously differentiable in (x, α), then it is possible to differentiate the solution

$$x(t, \alpha) = x_0 + \int_{t_0}^{t} f(s, x(s, \alpha), \alpha) \, ds$$

to the parameterized IVP (2.21) in a neighborhood of (t_0, α_0) with respect to α,

$$x_\alpha(t, \alpha) = \int_{t_0}^{t} \left(f_x(s, x(s, \alpha), \alpha) x_\alpha(s, \alpha) + f_\alpha(s, x(s, \alpha), \alpha) \right) ds,$$

and subsequently with respect to t. Thus, the *sensitivity matrix* $S(t) = x_\alpha(t, \alpha_0)$ (with values in $\mathbb{R}^{n \times p}$) satisfies the linear IVP

$$\dot{S}(t) = A(t)S(t) + B(t), \quad S(t_0) = 0,$$

where $A(t) = f_x(t, x(t, \alpha_0), \alpha_0) \in \mathbb{R}^{n \times n}$ and $B(t) = f_\alpha(t, x(t, \alpha_0), \alpha_0) \in \mathbb{R}^{n \times p}$. This results in the approximation[16]

$$x(t, \alpha) - x(t, \alpha_0) = x_\alpha(t, \alpha_0)(\alpha - \alpha_0) + O\left((\alpha - \alpha_0)^2\right)$$

$$= S(t)(\alpha - \alpha_0) + O\left((\alpha - \alpha_0)^2\right), \tag{2.24}$$

for (t, α) in a neighborhood of (t_0, α_0). □

2.2.4 State-Space Analysis

Instead of viewing the trajectory of the solution to an IVP on $\mathcal{I}$ as a graph

$$\{(t, x(t)) : t \in \mathcal{I}\} \subset \mathbb{R}^{1+n}$$

that lies in the $(1 + n)$-dimensional domain $\mathcal{D}$ of $f(t, x)$, it is often convenient to consider the projections $\mathcal{D}(t) = \{x \in \mathbb{R} : (t, x) \in \mathcal{D}\}$ and $\mathcal{D}_0 = \{t \in \mathbb{R} : \exists\, (t, x) \in \mathcal{D}\}$, and restrict attention to the graph

$$\{x(t) \in \mathcal{D}(t) : t \in \mathcal{D}_0\} \subset \mathbb{R}^n$$

that lies in the *state space* $\mathbb{R}^n$. This is especially true when $n \in \{2, 3\}$ because then it is possible to graphically represent the state trajectories in the state space. When viewed in the state space, the right-hand side of the ODE $\dot{x} = f(t, x)$ defines a *vector field*, and the unique solution to any

---

16. The *Landau notation* $O(\cdot)$ describes the limiting behavior of the function in its argument. For example, $\Delta(\xi) = O(\|\xi\|_2^2)$ if and only if there exists $M > 0$, such that $\|\Delta(\xi)\| \leq M\|\xi\|_2^2$ in a neighborhood of the origin, i.e., for all ξ such that $\|\xi\| < \varepsilon$ for some $\varepsilon > 0$.

well-posed IVP is described by the *flow* $\phi : \mathbb{R} \times \mathcal{D} \to \mathbb{R}^n$ of this vector field, which for a time increment τ maps the initial data $(t_0, x_0) \in \mathcal{D}$ to the state $x(t_0 + \tau)$. Furthermore,

$$x(t_0 + \tau) = \phi(\tau, t_0, x_0) = x_0 + \int_{t_0}^{t_0+\tau} f(s, x(s))\, ds$$

solves the IVP (2.16) on some time interval. The flow is often a convenient description of a system trajectory subject to an initial condition. It emphasizes the role of an ODE in transporting initial conditions to endpoints of trajectories (as solutions to the associated well-posed IVPs).

Remark 2.4 (Group Laws) The flow $\phi(\tau, t, x)$ of the system $\dot{x} = f(t, x)$ satisfies the group laws [17]

$$\phi(0, t, x) = x \quad \text{and} \quad \phi(\tau + \sigma, t, x) = \phi(\sigma, t + \tau, \phi(\tau, t, x)) \qquad (2.25)$$

for all $(t, x) \in \mathcal{D}$ and increments τ, σ. □

Remark 2.5 (Flow of Autonomous System) When considering an ODE of the form $\dot{x} = f(x)$, where the function $f : \mathbb{R}^n \to \mathbb{R}^n$ does not depend on t and is Lipschitz and bounded, one can set the initial time t_0 to zero without any loss in generality. The flow, with simplified notation $\phi(t, x)$, describes the time-t value of a solution to an IVP starting at the point x. The group laws (2.25) for the flow of this autonomous system can be written in the more compact form

$$\phi(0, x) = x \quad \text{and} \quad \phi(t + s, x) = \phi(s, \phi(t, x)), \qquad (2.26)$$

for all $x \in \mathbb{R}^n$ and all $s, t \in \mathbb{R}$ (figure 2.6). □

2.2.5 Exact ODEs and Potential Function

A vector-valued function $v = (v_0, v_1, \ldots, v_n) : \mathcal{D} \to \mathbb{R}^{1+n}$ has a *potential* (*on* $\mathcal{D}$) if there exists a real-valued potential function $V : \mathcal{D} \to \mathbb{R}$ such that

$$(V_t(t, x), V_x(t, x)) = v(t, x), \qquad \forall (t, x) \in \mathcal{D}.$$

---

[17]. In algebra, a *group* $(\mathcal{G}, \circ)$ consists of a set $\mathcal{G}$ together with an operation $\circ$ that combines any two elements a, b of $\mathcal{G}$ to form a third element $a \circ b$, such that for all $a, b, c \in \mathcal{G}$ the following four conditions (group axioms) are satisfied: (1) (closure) $a \circ b \in \mathcal{G}$; (2) (associativity) $(a \circ b) \circ c = a \circ (b \circ c)$; (3) (identity) $\exists e \in \mathcal{G} : e \circ a = a \circ e = e$; (4) (inverse) $\exists a^{-1} \in \mathcal{G}$ such that $a^{-1} \circ a = a \circ a^{-1} = e$, where e is the identity. For the group $(\mathcal{D}, +)$ with $\mathcal{D} = \mathbb{R}^{1+n}$ the function $\hat{\phi} : \mathbb{R} \times \mathcal{D} \to \mathcal{D}$ with $(\tau, (t, x)) \mapsto \hat{\phi}(\tau, (t, x)) = (t + \tau, \phi(\tau, t, x))$ forms a *one-parameter group action*, with identity $\hat{\phi}(0, \cdot)$ and such that $\hat{\phi}(\tau + \sigma, \cdot) = \hat{\phi}(\sigma, \hat{\phi}(\tau, \cdot))$.

Ordinary Differential Equations

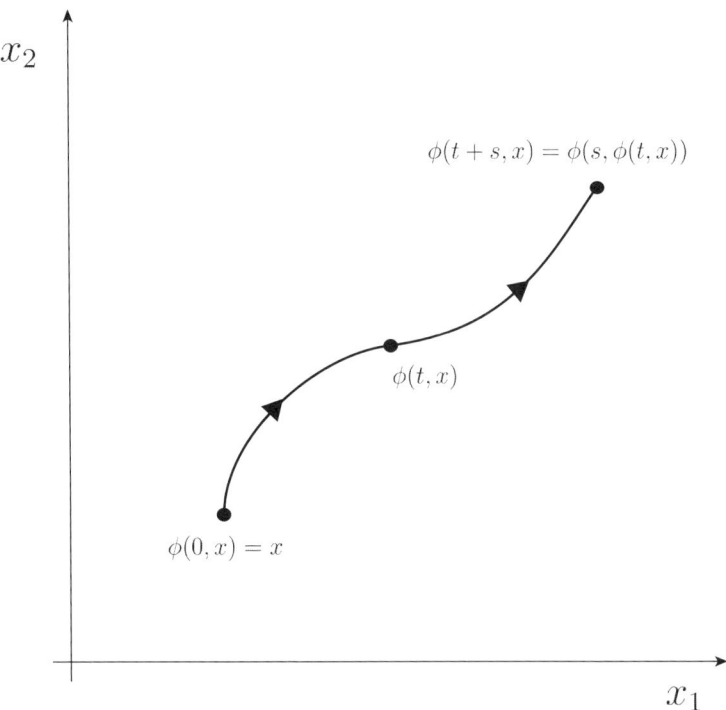

Figure 2.6
Group laws for the flow of a two-dimensional autonomous system.

Consider now the ODE (in implicit representation)

$$g(t,x) + \langle h(t,x), \dot{x}\rangle = 0, \tag{2.27}$$

where $g : \mathcal{D} \to \mathbb{R}$ and $h : \mathcal{D} \to \mathbb{R}^n$ are given continuously differentiable functions. This ODE is called *exact* if the function $v(t,x) = (g(t,x), h(t,x))$ has a potential on $\mathcal{D}$.

Proposition 2.6 (Poincaré Lemma) Let $v = (v_0, v_1, \ldots, v_n) : \mathcal{D} \to \mathbb{R}^{1+n}$ be a continuously differentiable function, defined on a contractible domain $\mathcal{D}$.[18] The function v has a potential (on $\mathcal{D}$) if and only if

$$\frac{\partial v_0}{\partial x_i} = \frac{\partial v_i}{\partial t} \quad \text{and} \quad \frac{\partial v_i}{\partial x_j} = \frac{\partial v_j}{\partial x_i}, \quad \forall i,j \in \{1,\ldots,n\}, \tag{2.28}$$

on $\mathcal{D}$.

---

18. A domain is *contractible* if it can be deformed to a point using a suitable continuous mapping (which is referred to as a homotopy).

Proof
⇒: Let V be a potential of the function v on $\mathcal{D}$ so that the gradient $V_{(t,x)} = v$. Since v is by assumption continuously differentiable, the potential function V is twice continuously differentiable. This implies that the Hessian matrix of second derivatives of V is symmetric on $\mathcal{D}$ (see, e.g., Zorich 2004, 1:459–460), that is, condition (2.28) holds.
⇐: See, for example, Zorich (2004, 2:353–354). ∎

Remark 2.6 (Computation of Potential Function) If a function

$$v = (v_0, v_1, \ldots, v_n)$$

has a potential V on the simply connected domain $\mathcal{D}$, then

$$V(t,x) - V(t_0, x_0) = \int_{(t_0,x_0)}^{(t,x)} \langle v(\tau, \xi), d(\tau, \xi) \rangle = \int_0^1 \langle v(\gamma(s)), d\gamma(s) \rangle, \quad (2.29)$$

where the integration is carried out along any differentiable path $\gamma : [0,1] \to \mathcal{D}$ which is such that $\gamma(0) = (t_0, x_0)$ and $\gamma(1) = (t, x)$. □

An exact ODE can be written equivalently in the form

$$\frac{dV(t,x)}{dt} = V_t(t,x) + \langle V_x(t,x), \dot{x} \rangle = 0.$$

As a result, $V(t, x(t)) \equiv V(t_0, x_0)$ for any solution $x(t)$ of an exact ODE that satisfies the initial condition $x(t_0) = x_0$. The potential function for an exact ODE is often referred to as a *first integral* (of (2.27)). Note that a first integral confines trajectories to an $(n-1)$-dimensional subset (manifold) of the state space. For a given ODE of the form (2.3) it may be possible to find $n-1$ different exact ODEs of the form (2.27) with first integrals, in which case the vector field f in (2.3) is called *completely integrable*. The complete integrability of vector fields is related to the controllability of nonlinear systems.[19]

Remark 2.7 (Integrating Factor/Euler Multiplier) Given a function v which does not have a potential, it is sometimes possible to find a (continuously differentiable) integrating factor (or *Euler multiplier*) $\mu : \mathcal{D} \to \mathbb{R}$ such that μv has a potential. By the Poincaré lemma this is the case if and only if

$$\frac{\partial(\mu v_0)}{\partial x_i} = \frac{\partial(\mu v_i)}{\partial t} \quad \text{and} \quad \frac{\partial(\mu v_i)}{\partial x_j} = \frac{\partial(\mu v_j)}{\partial x_i}, \quad \forall i, j \in \{1, \ldots, n\}. \quad (2.30)$$

---

19. For details see the references in section 3.2.2.

Ordinary Differential Equations

In order to find an integrating factor, it is often useful to assume that μ is separable in the different variables, for instance, $\mu(t,x) = \mu_0(t)\mu_1(x_1)\cdots\mu_n(x_n)$. □

Example 2.7 (Potential for Linear First-Order ODE) Consider the linear first-order ODE (2.6) on the simply connected domain $\mathcal{D} = \mathbb{R}^2$, which can equivalently be written in the form

$$g(t)x - h(t) + \dot{x} = v_0(t,x) + v_1(t,x)\dot{x} = 0,$$

where $v_0(t,x) = g(t)x - h(t)$ and $v_1(t,x) = 1$. With the use of an Euler multiplier $\mu(t)$, condition (2.30) becomes

$$\mu(t)g(t) = \dot{\mu}(t).$$

The latter is satisfied as long as the (nonzero) integrating factor is of the form

$$\mu(t) = \mu_0 \exp\left[\int_{t_0}^t g(s)\,ds\right],$$

where $\mu_0 \neq 0$. Thus, integrating the exact ODE $\mu v_0 + \mu v_1 \dot{x} = 0$ along any path from the initial point $(t_0, x_0) \in \mathcal{D}$ to the point $(t,x) \in \mathcal{D}$ as in (2.29) yields the potential function

$$V(t,x) = \int_{t_0}^t v_0(s,x_0)\mu(s)\,ds + \int_{x_0}^x v_1(t,\xi)\mu(t)\,d\xi$$

$$= \mu(t)x - \mu_0 x_0 - \int_{t_0}^t \mu(s)h(s)\,ds.$$

On any solution $x(t)$ to the linear IVP (2.8), the potential function $V(t,x(t))$ stays constant, so that $V(t,x(t)) \equiv V(t_0,x_0) = 0$, which implies the Cauchy formula in proposition 2.1,

$$x(t) = \frac{\mu_0 x_0}{\mu(t)} + \int_{t_0}^t \frac{\mu(s)}{\mu(t)} h(s)\,ds$$

$$= x_0 \exp\left[-\int_{t_0}^t g(s)\,ds\right] + \int_{t_0}^t \exp\left[-\int_s^t g(\theta)\,d\theta\right] h(s)\,ds,$$

for all $t \geq t_0$. □

2.2.6 Autonomous Systems

An *autonomous system* is such that the right-hand side of the ODE (2.3) does not depend on t. It is represented by the ODE

$$\dot{x} = f(x),$$

where $f : \mathcal{D} \subset \mathbb{R}^n \to \mathbb{R}^n$. Note that the domain $\mathcal{D}$ of f does not contain time and is therefore a nonempty connected open subset of $\mathbb{R}^n$. In the special case where $n = 2$,

$$\frac{dx_2}{dx_1} = \frac{f_2(x)}{f_1(x)}$$

describes the *phase diagram*.

Example 2.8 (Predator-Prey Dynamics) Consider the evolution of two interacting populations, prey and predator. The relative growth of the predator population depends on the availability of prey. At the same time, the relative growth of the prey population depends on the presence of predators. Lotka (1920) and Volterra (1926) proposed a simple linear model of such predator-prey dynamics. Let $\xi_1(\tau)$ and $\xi_2(\tau)$ be sizes of the prey and predator populations at time $\tau \geq 0$, respectively, with given positive initial sizes of ξ_{10} and ξ_{20}. The Lotka-Volterra predator-prey IVP is of the form

$$\dot{\xi}_1 = \xi_1(a - b\xi_2), \qquad \xi_1(0) = \xi_{10},$$
$$\dot{\xi}_2 = \xi_2(c\xi_1 - d), \qquad \xi_2(0) = \xi_{20},$$

where a, b, c, d are given positive constants. To reduce the number of constants to what is necessary for an analysis of the system dynamics, it is useful to first *de-dimensionalize* the variables using a simple linear transformation.[20] For this, one can set

$$t = a\tau, \qquad x_1(t) = (c/d)\xi_1(\tau), \qquad x_2(t) = (b/a)\xi_2(t),$$

which yields the following equivalent but much simplified IVP:

---

20. A fundamental result in dimensional analysis is the Theorem Π by Buckingham (1914), which (roughly speaking) states that if in a mathematical expression n variables are measured in k ($\leq n$) independent units, then $n - k$ variables in that expression may be rendered dimensionless (generally, in more than one way). For the significance of this result in the theory of ODEs, see Bluman and Kumei (1989). Exercise 2.2a gives another example of how to de-dimensionalize a system.

Ordinary Differential Equations

$$\dot{x}_1 = x_1(1 - x_2), \qquad x_1(0) = x_{10}, \tag{2.31}$$

$$\dot{x}_2 = \alpha x_2(x_1 - 1), \qquad x_2(0) = x_{20}, \tag{2.32}$$

where $\alpha = d/a > 0$, and $x_0 = (x_{10}, x_{20}) = (c\xi_{10}/d, b\xi_{20}/b)$.

Solving the IVP (2.31)–(2.32) is difficult, but one can use a simple state-space analysis to reduce the dimensionality of the system to 1. Then, by finding a first integral, it is possible to determine the state-space trajectories of the predator-prey system. Provided that $x_2 \neq 1$, the ODE

$$\frac{dx_2}{dx_1} = \alpha \left(\frac{x_1 - 1}{x_1} \right) \left(\frac{x_2}{1 - x_2} \right)$$

has a separable right-hand side, so (via straightforward integration)

$$V(x_1, x_2) \equiv \alpha x_1 + x_2 - \ln x_1^\alpha x_2 = C, \tag{2.33}$$

where $C \geq 1 + \alpha$ is a constant, describes a phase trajectory (figure 2.7).[21] Note that $V(x_1, x_2)$ in (2.33) is a potential function for the exact ODE derived from (2.31)–(2.32),

$$\frac{\mu(x_1, x_2)}{x_1(1 - x_2)} \dot{x}_1 - \frac{\mu(x_1, x_2)}{\alpha x_2(x_1 - 1)} \dot{x}_2 = 0,$$

where $\mu(x_1, x_2) = \alpha(x_1 - 1)(1 - x_2)$ is an integrating factor. □

2.2.7 Stability Analysis

An *equilibrium* (also referred to as *steady state* or *stationary point*) of the ODE $\dot{x} = f(t, x)$ (with domain $\mathcal{D} = \mathbb{R}_+ \times \mathcal{X}$) is a point $\bar{x} \in \{\xi : (\tau, \xi) \in \mathcal{D}$ for some $\tau \in \mathbb{R}_+\} = \mathcal{D}_0$ such that

$$f(t, \bar{x}) = 0, \qquad \forall\, t \in \{\tau : (\tau, \bar{x}) \in \mathcal{D}\} = \mathcal{D}_0. \tag{2.34}$$

If instead of (2.34) there exists a nontrivial interval $\mathcal{I}$ such that $\mathcal{I} \times \{\bar{x}\} \subset \mathcal{D}$ and $f(t, \bar{x}) = 0$ for all $t \in \mathcal{I}$, then the point $\bar{x}$ is a *(temporary) equilibrium on the time interval* $\mathcal{I}$. For autonomous systems, where the right-hand side of the ODE does not depend on time, any temporary equilibrium is also an equilibrium. To obtain a good qualitative understanding of the behavior of a given dynamic system, it is important to examine the trajectories in the neighborhood of its equilibria, which is often referred to as *(local) stability analysis*. For practical examples of stability analyses, see exercises 2.2–2.4.

---

21. The lowest possible value for C can be determined by minimizing the left-hand side of (2.33). It is approached when $(x_1, x_2) \to (1, 1)$.

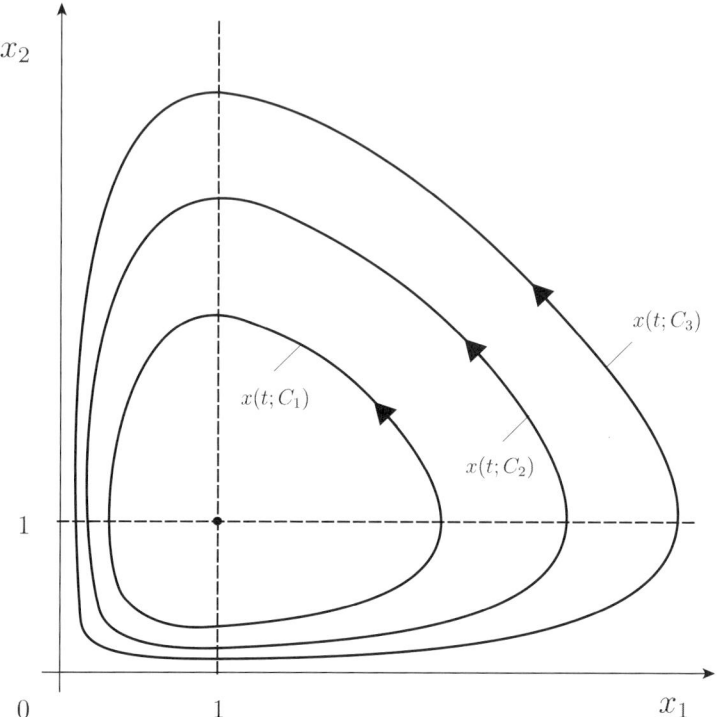

Figure 2.7
Periodic solutions $x(t; C_1)$, $x(t; C_2)$, $x(t; C_3)$ of the Lotka-Volterra predator-prey IVP (2.31)–(2.32), characterized by a constant potential $V(x) = C \in \{C_1, C_2, C_3\}$ in (2.33) with $1 + \alpha < C_1 < C_2 < C_3 < \infty$.

Intuitively, at an equilibrium the system described by an ODE can come to a rest, so that (at least for some time interval) a solution of the system stays at the same state.[22] An equilibrium can be found by setting all components of the system function $f(x) = (f_1(x), \ldots, f_n(x))$ to zero. Each such equation $f_i(x) = 0$, $i \in \{1, \ldots, n\}$, defines a *nullcline*. For example, when $n = 2$ each nullcline usually corresponds to a line in the plane, with system equilibria appearing at intersection points. In exercises 2.2–2.4, which all deal with two-dimensional systems, nullclines play an important role.

---

22. As seen in example 2.5, the solution to an ODE may not be unique. According to the definition of a (temporary) equilibrium $\bar{x}$, the IVP for some initial data $(t_0, \bar{x})$ has the solution $x(t) = \bar{x}$ (on a nontrivial time interval that contains t_0), but it may have other solutions as well. Requiring IVPs to be well-posed (see section 2.2.3) leads to uniqueness of trajectories, increasing the significance of an equilibrium for the description of the system dynamics.

Ordinary Differential Equations

We assume that the system function $f(t,x)$ is locally Lipschitz in x and that $\bar{x}$ with $[c, \infty) \times \{\bar{x}\} \subset \mathcal{D}$, for some $c \in \mathbb{R}$, is an equilibrium. The equilibrium $\bar{x}$ is said to be *stable* if for any $t_0 \geq c$ and any solution $x(t)$ of the ODE (2.3) for $t \geq t_0$,

$$\forall \varepsilon > 0, \exists \delta > 0 : \quad \|x(t_0) - \bar{x}\| < \delta \implies \|x(t) - \bar{x}\| < \varepsilon, \qquad \forall t \geq t_0.$$

Otherwise the equilibrium $\bar{x}$ is called *unstable*. Finally, if there exists a $\delta > 0$ (independent of t_0) such that

$$\|x(t_0) - \bar{x}\| < \delta \implies \lim_{t \to \infty} x(t) = \bar{x},$$

then the equilibrium $\bar{x}$ is called *asymptotically stable*;[23] an asymptotically stable equilibrium $\bar{x}$ is said to be *exponentially stable* if there exist $\delta, C, \lambda > 0$ (independent of t_0) such that

$$\|x(t_0) - \bar{x}\| < \delta \implies \|x(t_0) - \bar{x}\| \leq C e^{-\lambda(t-t_0)}, \qquad \forall t \geq t_0.$$

Example 2.9 In the Vidale-Wolfe advertising model (see example 2.4) it is easy to verify that the sales limit x_∞ of a stationary advertising policy is an exponentially stable equilibrium. It is in fact *globally* exponentially stable, since $\lim_{t \to \infty} \phi(t, x_0) = x_\infty$ for any initial sales level $x_0 \geq 0$. □

Figure 2.8 provides an overview of possible system behaviors close to an equilibrium. To establish the stability properties of a given equilibrium, it is useful to find continuously differentiable real-valued functions $V : \mathcal{D} \to \mathbb{R}$ that are monotonic along system trajectories, at least in a neighborhood of the equilibrium point. It was noted in section 2.2.5 that if such a function is constant along system trajectories (and otherwise nonconstant), it is a first integral of the ODE (2.3) and can be interpreted as a potential function of an associated exact ODE. On the other hand, if V is nonincreasing along system trajectories in the neighborhood of an equilibrium $\bar{x}$ where V also has a local minimum, then it is referred to as a *Lyapunov function* (figure 2.9). Lyapunov (1892) realized that the existence of such functions provides valuable information about the stability of an equilibrium.

Stability in Autonomous Systems Now consider the stability of a given equilibrium for an *autonomous system*, described by the ODE

$$\dot{x} = f(x), \tag{2.35}$$

---

23. For nonautonomous systems it is common to add the word *uniformly* to the terms *stable* and *asymptotically stable* to emphasize the fact that the definitions are valid independent of the starting time t_0.

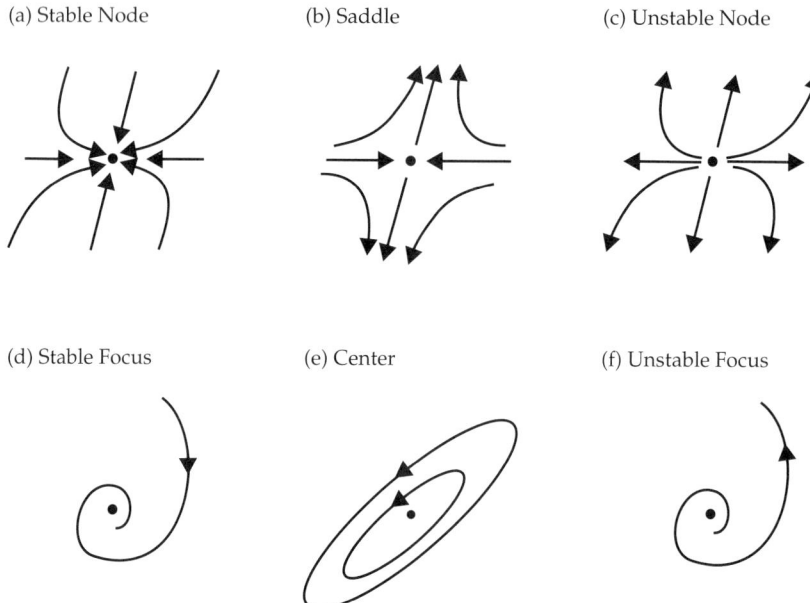

Figure 2.8
Classification of the local system behavior in the neighborhood of an equilibrium $\bar{x}$ according to the eigenvalues λ_i, $i \in \{1,2\}$, of the system matrix $A = f_x(\bar{x})$: (a) stable node (λ_i real and negative); (b) saddle (λ_i real, with different signs); (c) unstable node (λ_i real and positive); (d) stable focus (λ_i conjugate complex, with negative real parts); (e) center (λ_i, with zero real parts); (f) unstable focus (λ_i conjugate complex, with positive real parts).

where $f : \mathcal{D} \subset \mathbb{R}^n \to \mathbb{R}^n$ is the system function. The key to establishing stability results for autonomous systems is to find a function $V(x)$, defined on $\mathcal{D}$ or a suitable neighborhood of the equilibrium, which is such that its values decrease along a system trajectory $x(t)$, that is,

$$\dot{V}(x(t)) \equiv \langle V_x(x(t)), \dot{x}(t) \rangle = \langle V_x(x(t)), f(x(t)) \rangle$$

is nonpositive, or even negative, in a neighborhood of the equilibrium.

Proposition 2.7 (Local Stability) (Lyapunov 1892) Let $\bar{x}$ be an equilibrium point of an autonomous system, and let $V : \mathcal{D} \to \mathbb{R}$ be a continuously differentiable function.[24] (1) If there exists $\varepsilon > 0$ such that

---

24. Note that $\dot{V}(x) = \langle V_x(x), f(x) \rangle$ is interpreted as a function of $x \in \mathcal{D}$ (i.e., it does not depend on t). Thus, while $V(x)$ can have a local minimum at the point $\bar{x}$, it is at the same time possible that $\dot{V}(x) \leq 0$ in a neighborhood of $\bar{x}$. This implies that $V(x(t))$ is nonincreasing along system trajectories in the neighborhood of $\bar{x}$.

Ordinary Differential Equations

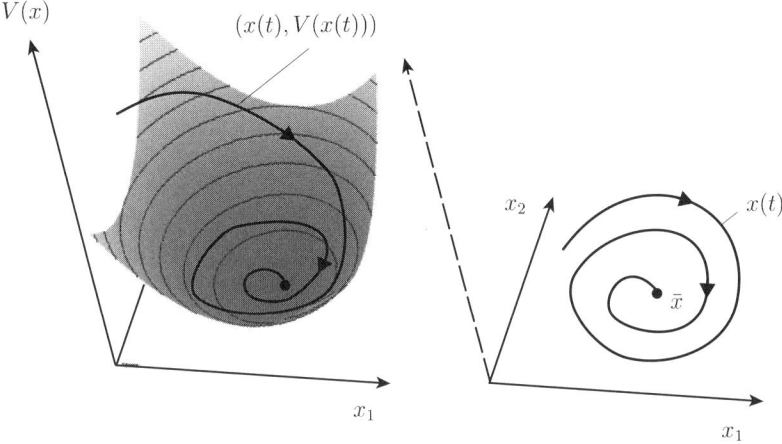

Figure 2.9
A Lyapunov function $V(x)$ decreases along a trajectory $x(t)$ of an (autonomous) system $\dot{x} = f(x)$ in the neighborhood of an asymptotically stable equilibrium point $\bar{x}$, so $\dot{V}(x(t)) = \langle V_x(x(t)), f(x(t)) \rangle < 0$.

$$0 < \|x - \bar{x}\| < \varepsilon \Rightarrow V(\bar{x}) < V(x) \quad \text{and} \quad \dot{V}(x) \leq 0,$$

then $\bar{x}$ is locally stable. (2) If there exists $\varepsilon > 0$ such that

$$0 < \|x - \bar{x}\| < \varepsilon \Rightarrow V(\bar{x}) < V(x) \quad \text{and} \quad \dot{V}(x) < 0,$$

then $\bar{x}$ is asymptotically stable. (3) If for all $\varepsilon > 0$ the set $\mathcal{S}_\varepsilon = \{x \in \mathbb{R}^n : \|x - \bar{x}\| < \varepsilon, V(x) > V(\bar{x})\}$ is nonempty and $\dot{V} > 0$ on $\bar{\mathcal{S}}_\varepsilon \setminus \{\bar{x}\}$,[25] then $\bar{x}$ is unstable (Četaev 1934).

Proof (1) Without loss of generality, let $\varepsilon > 0$ be such that the closed ball $\mathcal{B}_\varepsilon(\bar{x}) = \{\xi \in \mathbb{R}^n : \|\xi - \bar{x}\| \leq \varepsilon\}$ is contained in $\mathcal{D}$; otherwise just select an appropriate $\varepsilon_1 \in (0, \varepsilon]$. Furthermore, let m_V be equal to the minimum of $V(x)$ on the boundary of $\mathcal{B}_\varepsilon(\bar{x})$, so that

$$m_V = \min_{x \in \partial \mathcal{B}_\varepsilon(\bar{x})} V(x) > V(\bar{x}),$$

since $V(x) > V(\bar{x})$ for $x \neq \bar{x}$ by assumption. Now let $\mu = (m_V - V(\bar{x}))/2$, which implies that there exists $\delta \in (0, \varepsilon)$ such that

$$\mathcal{B}_\delta(\bar{x}) \subseteq \{\xi \in \mathcal{B}_\varepsilon(\bar{x}) : V(\xi) \leq \mu\} \subsetneq \mathcal{B}_\varepsilon(\bar{x}),$$

the last (strict) inclusion being implied by the fact that $\mu < m_V$. Since $\dot{V}(x) \leq 0$, it is not possible that any trajectory starting at an arbitrary

---

25. The set $\bar{\mathcal{S}}_\varepsilon$ denotes the closure of $\mathcal{S}_\varepsilon$.

point $x \in \mathcal{B}_\delta(\bar{x})$ leaves the set $\mathcal{B}_\varepsilon(\bar{x})$, which implies that the equilibrium $\bar{x}$ is stable.

(2) Along any trajectory $x(t) = \phi(t,x)$ starting at a point $x \in \mathcal{B}_\varepsilon(\bar{x})$ (with ε small enough as in part (1)) the function $V(x(t))$ decreases. Since V is bounded from below by $V(\bar{x})$, it follows that $\lim_{t\to\infty} V(x(t)) = V_\infty$ exists and that $V_\infty \geq V(\bar{x})$. If $V_\infty > V(\bar{x})$, then let

$$\nu = \max\{\dot{V}(\xi) : V_\infty \leq V(\xi), \xi \in \mathcal{B}_\varepsilon(\bar{x})\}$$

be the largest gradient $\dot{V}$ evaluated at points at which V is not smaller than V_∞ (which excludes a neighborhood of $\bar{x}$). Thus, $\nu < 0$ and, using the fundamental theorem of calculus (proposition A.6),

$$V_\infty - V(x) \leq V(\phi(t,x)) - V(x) = \int_0^t \dot{V}(\phi(s,x))\,ds \leq \nu t < 0,$$

which leads to a contradiction for times $t > (V(x) - V_\infty)/(-\nu) \geq 0$. Hence, $V_\infty = V(\bar{x})$, which by strictness of the minimum of V at $\bar{x}$ implies that $\lim_{t\to\infty} \phi(t,x) = \bar{x}$ for all $x \in \mathcal{B}_\varepsilon(\bar{x})$; thus $\bar{x}$ is asymptotically stable.

(3) Let $\varepsilon > 0$ be small enough, so that $\mathcal{S}_\varepsilon \subset \mathcal{D}$. The set $\mathcal{S}_\varepsilon$ is open, since with any point $x \in \mathcal{S}_\varepsilon$ points close enough to $\bar{x}$ also satisfy the strict inequalities in the definition of that set. Thus, fixing any point $x_0 \in \mathcal{S}_\varepsilon$, the flow $\phi(t,x_0)$ stays in $\mathcal{S}_\varepsilon$ for some time $\tau > 0$, and it is $V(\phi(\tau,x_0)) > V(\bar{x})$, since $\dot{V} > 0$ on $\mathcal{S}_\varepsilon$. Set $\nu = \inf\{\dot{V}(x) : V(x) \geq V(x_0), x \in \mathcal{S}_\varepsilon\}$; then $\nu > 0$, and

$$V(\phi(t,x_0)) - V(x_0) = \int_0^t \dot{V}(\phi(s,x_0))\,ds \geq \nu t > 0,$$

for all $t > 0$, so that, because V is bounded on $\mathcal{S}_\varepsilon$, the trajectory must eventually leave the set $\mathcal{S}_\varepsilon$. Since $V(\phi(t,x_0)) > V(x_0) > V(\bar{x})$, this cannot be through the boundary where $V(x) = V(\bar{x})$ but must occur through the boundary of $\mathcal{S}_\varepsilon$ where $\|x - \bar{x}\| = \varepsilon$. This is true for all (small enough) $\varepsilon > 0$, so the equilibrium $\bar{x}$ must be unstable. ∎

Note that proposition 2.7 can be applied without the need for explicit solutions to an ODE.

Example 2.10 Let $n = 1$, and consider an autonomous system, described by the ODE $\dot{x} = f(x)$, where $xf(x) < 0$ for all $x \neq 0$ (which implies that $f(0) = 0$). Then

Ordinary Differential Equations

$$V(x) = -\int_0^x f(\xi)\,d\xi$$

is a Lyapunov function for the equilibrium point $\bar{x} = 0$, since $V(x) > V(\bar{x}) = 0$ for all $x \neq \bar{x}$. Furthermore,

$$\dot{V}(x) = V_x(x)f(x) = -f^2(x) < 0$$

for all $x \neq \bar{x}$, which by virtue of proposition 2.7 implies asymptotic stability of $\bar{x}$. □

Example 2.11 (Stability of Linear System) Consider the linear ODE (with constant coefficients)

$$\dot{x} = Ax, \tag{2.36}$$

where $A \in \mathbb{R}^{n \times n}$ is a given nonsingular system matrix. To examine the stability properties of the only equilibrium point $\bar{x} = 0$,[26] consider a quadratic Lyapunov function $V(x) = x'Qx$, where Q is a symmetric positive definite matrix. This guarantees that $V(x) > V(\bar{x}) = 0$ for all $x \neq \bar{x}$. Then

$$\dot{V}(x) = x'Q\dot{x} + \dot{x}'Qx = x'QAx + x'A'Qx = x'\left(QA + A'Q\right)x.$$

Thus, if Q is chosen such that the (symmetric) matrix $R = -(QA + A'Q)$ is positive definite, then $\dot{V}(x) < 0$ for all $x \neq \bar{x}$, and the origin is an asymptotically stable equilibrium. As shown in section 2.3.2 (see equation 2.57), the flow of a linear system is

$$\phi(t,x) = e^{At}x = [Pe^{Jt}P^{-1}]x = P\left[\sum_{i=1}^r e^{J_i t}\right]P^{-1}x, \quad \forall t \geq 0, \tag{2.37}$$

where $P \in \mathbb{C}^{n \times n}$ is a (possibly complex) nonsingular matrix that transforms the system into its *Jordan canonical form*,[27]

$$J = P^{-1}AP = \text{block diag}(J_1, \ldots, J_r).$$

---

26. If the system matrix A is singular (i.e., $\det(A) = 0$), then A has a nontrivial null space, every point of which is an equilibrium. In particular, the set of equilibria is itself a linear subspace and in every neighborhood of an equilibrium there exists another equilibrium (i.e., no equilibrium can be isolated).

27. For any matrix A and any nonsingular matrix P, the matrix $P^{-1}AP$ has the same eigenvalues as A, which is why this procedure is referred to as *similarity transform*. Note also that the Jordan canonical form is a purely conceptual tool and is not used in actual computations, as the associated numerical problem tends to be ill-conditioned.

For an eigenvalue λ_i (of multiplicity m_i) in A's set $\{\lambda_1, \ldots, \lambda_r\}$ of $r \in \{1, \ldots, n\}$ distinct complex eigenvalues (such that $m_1 + \cdots + m_r = n$), the corresponding Jordan block J_i is

$$J_i = \begin{bmatrix} \lambda_i & 1 & 0 & \cdots & \cdots & 0 \\ 0 & \lambda_i & 1 & 0 & \cdots & 0 \\ \vdots & & \ddots & & & \vdots \\ \vdots & & & \ddots & & 0 \\ \vdots & & & & \ddots & 1 \\ 0 & \cdots & \cdots & \cdots & 0 & \lambda_i \end{bmatrix} \in \mathbb{C}^{m_i \times m_i},$$

and therefore

$$e^{J_i t} = e^{\lambda_i t} \begin{bmatrix} 1 & t & \frac{t^2}{2!} & \cdots & \frac{t^{m_i-1}}{(m_i-1)!} \\ 0 & 1 & t & \cdots & \frac{t^{m_i-2}}{(m_i-2)!} \\ \vdots & & \ddots & & \vdots \\ 0 & \cdots & 0 & 1 & t \\ 0 & 0 & \cdots & 0 & 1 \end{bmatrix}. \tag{2.38}$$

For convenience, the linear system (2.36) or the system matrix A is called stable (asymptotically stable, unstable) if the equilibrium $\bar{x} = 0$ is stable (asymptotically stable, unstable). The representation (2.37) directly implies the following characterization for the stability of A in terms of its eigenvalues being located in the left half of the complex plane or not.

Lemma 2.2 (Stability of Linear System) Let $\{\lambda_1, \ldots, \lambda_r\}$ be the set of r distinct eigenvalues of the nonsingular system matrix $A \in \mathbb{R}^{n \times n}$. (1) A is stable if and only if $\text{Re}(\lambda_i) \leq 0$ for all $i \in \{1, \ldots, r\}$, and $\text{Re}(\lambda_i) = 0 \Rightarrow J_i = \lambda_i$. (2) A is asymptotically stable (or A is Hurwitz) if and only if $\text{Re}(\lambda_i) < 0$ for all $i \in \{1, \ldots, r\}$.

Proof (1) If A has an eigenvalue λ_i with positive real part, then it is clear from (2.37) that $e^{J_i t}$ is unbounded for $t \to \infty$. Hence, a necessary condition for stability is that $\text{Re}(\lambda_i) \leq 0$ for all $i \in \{1, \ldots, r\}$. Part (2) of this proof implies that when all these inequalities are strict, one obtains stability. Consider now the case where $\text{Re}(\lambda_i) = 0$ for some eigenvalue λ_i with multiplicity $m_i > 1$. Then, as can be seen from (2.38), the term $e^{J_i t}$ becomes unbounded for $t \to \infty$, so the implication $\text{Re}(\lambda_i) = 0 \Rightarrow J_i = \lambda_i$ (i.e., $m_i = 1$) is necessary for the stability of A as well.

(2) Asymptotic stability of A obtains if and only if $\phi(t, x) = e^{At}x \to 0$ as $t \to \infty$ for any starting point $x \in \mathbb{R}^n$. From the representation (2.37) of

Ordinary Differential Equations

the flow in Jordan canonical form it is immediately clear that the origin is an asymptotically stable equilibrium, provided that all eigenvalues of A have negative real parts. If there exists an eigenvalue λ_i with $\text{Re}(\lambda_i) \geq 0$, then not all trajectories are converging to the origin. ∎

For a linear system asymptotic stability is equivalent to global asymptotic stability, in the sense that $\lim_{t \to \infty} \phi(t, x) = 0$ independent of the initial state x. The following result links the stability of the system matrix A to the Lyapunov functions in proposition 2.7.

Lemma 2.3 (Characterization of Hurwitz Property) Let R be an arbitrary symmetric positive definite matrix (e.g., $R = I$). The system matrix A of the linear system (2.36) is Hurwitz if and only if there is a symmetric positive definite matrix Q that solves the Lyapunov equation,

$$QA + A'Q + R = 0. \tag{2.39}$$

Furthermore, if A is Hurwitz, then (2.39) has a unique solution.

Proof Let the symmetric positive definite matrix R be given.

$\Leftarrow$: If Q solves the Lyapunov equation (2.39), then, as shown earlier, $V(x) = x'Qx$ is a Lyapunov function and by proposition 2.7(2) the origin $\bar{x} = 0$ is asymptotically stable, so that by lemma 2.2 the system matrix A is Hurwitz.

$\Rightarrow$: Let

$$Q = \int_0^\infty e^{A't} R e^{At} dt.$$

Since A is Hurwitz, one can see from the Jordan canonical form in (2.37) that the matrix Q is well-defined (the integral converges); it is also positive semidefinite and symmetric. Furthermore,

$$QA + A'Q = \int_0^\infty \left(e^{A't} R e^{At} A + A' e^{A't} Q e^{At} \right) dt$$

$$= \int_0^\infty \frac{d}{dt} e^{A't} R e^{At} dt = \left[e^{A't} R e^{At} \right]_0^\infty = -R,$$

so that the Lyapunov equation (2.39) is satisfied. Last, the matrix Q is positive definite because otherwise there is a nonzero vector x such that $x'Qx = 0$, which implies that $e^{At}x \equiv 0$ on $\mathbb{R}_+$. But this is possible only if $x = 0$, a contradiction, so Q is positive definite.

Uniqueness: Given two symmetric positive definite solutions $\hat{Q}, Q$ to the Lyapunov equation (2.39), it is

$$(\hat{Q} - Q)A + A'(\hat{Q} - Q) = 0,$$

so

$$0 = e^{A't}[(\hat{Q} - Q)A + A'(\hat{Q} - Q)]e^{At} = \frac{d}{dt}e^{A't}(\hat{Q} - Q)e^{At},$$

for all $t \geq 0$. Thus, $e^{A't}(\hat{Q} - Q)e^{At}$ is constant for all $t \geq 0$, so

$$e^{A't}(\hat{Q} - Q)e^{At} = e^{A'0}(\hat{Q} - Q)e^{A0} = \hat{Q} - Q = \lim_{t \to \infty} e^{A't}(\hat{Q} - Q)e^{At} = 0,$$

which implies that $\hat{Q} = Q$. ∎

When $f(x)$ is differentiable, it is possible to linearize the system in a neighborhood of an equilibrium $\bar{x}$ by setting

$$A = f_x(\bar{x}),$$

so that $f(x) = A(x - \bar{x}) + O\left((x - \bar{x})^2\right)$, and then to consider the linearized system

$$\dot{x} = A(x - \bar{x})$$

instead of the nonlinear system, in a neighborhood of the equilibrium $\bar{x}$. Figure 2.8 classifies an equilibrium $\bar{x}$ in the Euclidean plane according to the eigenvalues λ_1, λ_2 of the system matrix A, for $n = 2$.

The following linearization criterion, which allows statements about the local stability of a nonlinear system based on the eigenvalues of its linearized system matrix at an equilibrium, is of enormous practical relevance. Indeed, it proves very useful for the stability analyses in exercises 2.2–2.4.

Proposition 2.8 (Linearization Criterion) Assume that $f(\bar{x}) = 0$. (1) If $f_x(\bar{x})$ is Hurwitz, then $\bar{x}$ is an asymptotically stable equilibrium. (2) If $f_x(\bar{x})$ has an eigenvalue with positive real part, then $\bar{x}$ is an unstable equilibrium.

Proof (1) If the matrix $A = f_x(\bar{x})$ is Hurwitz, then by lemma 2.3 for $R = I$ there exists a unique symmetric positive definite solution Q to the Lyapunov equation (2.39), so that $V(\xi) = \xi'Q\xi$, with $\xi = x - \bar{x}$, is a natural Lyapunov-function candidate for the nonlinear system in a neighborhood of the equilibrium $\bar{x}$. Indeed, if one sets $\Delta(\xi) = f(\xi) - A\xi = O(\|\xi\|_2^2)$, its total derivative with respect to time is

Ordinary Differential Equations

$$\dot{V}(\xi) = \xi' Q f(\xi) + f'(\xi) Q \xi$$
$$= \xi' Q \left(A\xi + \Delta(\xi) \right) + \left(\xi' A' + \Delta'(\xi) \right) Q \xi$$
$$= \xi' (QA + A'Q) \xi + 2\xi' Q \Delta(\xi)$$
$$= -\xi' R \xi + 2\xi' Q \Delta(\xi)$$
$$= -\|\xi\|_2^2 + O(\|\xi\|_2^3),$$

that is, negative for $\xi \neq 0$ in a neighborhood of the origin. Therefore, by proposition 2.7 the equilibrium $\xi = 0$ is asymptotically stable, that is, $\bar{x}$ in the original coordinates is asymptotically stable.

(2) Assuming initially no eigenvalues on the imaginary axis, the main idea is to decompose the system into a stable and an unstable part by a similarity transform (to a real-valued Jordan canonical form) such that

$$SAS^{-1} = \begin{bmatrix} -\hat{A}_+ & 0 \\ 0 & \hat{A}_- \end{bmatrix},$$

with $S \in \mathbb{R}^{n \times n}$ nonsingular and the square matrices A_+, A_- both Hurwitz (corresponding to the eigenvalues with positive and negative real parts, respectively). Using the new dependent variable

$$\eta = \begin{bmatrix} \eta_+ \\ \eta_- \end{bmatrix} = S\xi,$$

where η_+, η_- are compatible with the dimensions of $\hat{A}_+, \hat{A}_-$, respectively, one obtains

$$\dot{\eta} = SAS^{-1}\eta + S\Delta(S^{-1}\eta) = \begin{bmatrix} -\hat{A}_+ & 0 \\ 0 & \hat{A}_- \end{bmatrix} \begin{bmatrix} \eta_+ \\ \eta_- \end{bmatrix} + \begin{bmatrix} \hat{\Delta}_+(\eta) \\ \hat{\Delta}_-(\eta) \end{bmatrix},$$

so

$$\dot{\eta}_+ = -A_+ \eta_+ + \hat{\Delta}_+(\eta) = -A_+ \eta_+ + O(\|\eta\|_2^2),$$
$$\dot{\eta}_- = A_- \eta_- + \hat{\Delta}_-(\eta) = A_- \eta_- + O(\|\eta\|_2^2).$$

Since A_+ and A_- are Hurwitz, by lemma 2.3 there exist unique symmetric positive definite matrices $\hat{Q}_+, \hat{Q}_-$ such that $\hat{Q}_+ \hat{A}_+ + \hat{A}'_+ \hat{Q}_+ = \hat{Q}_- \hat{A}_- + \hat{A}'_- \hat{Q}_- = -I$. With these matrices, introduce the function

$$V(\eta) = \eta' \begin{bmatrix} \hat{Q}_+ & 0 \\ 0 & \hat{Q}_- \end{bmatrix} \eta = \eta'_+ \hat{Q}_+ \eta_+ - \eta'_- \hat{Q}_- \eta_-,$$

which (for $\eta \neq 0$) is positive on the subspace $\{\eta \in \mathbb{R}^n : \eta_- = 0\}$ and negative on the subspace $\{\eta \in \mathbb{R}^n : \eta_+ = 0\}$. Hence, for all $\varepsilon > 0$ the set $\mathcal{S}_\varepsilon = \{\eta \in \mathbb{R}^n : \|\eta\|_2 < \varepsilon, V(\eta) > V(0) = 0\}$ is nonempty, and (with computations as in part (1) of the proof),

$$\dot{V}(\eta) = -\eta'_+(\hat{Q}_+\hat{A}_+ + \hat{A}'_+\hat{Q}_+)\eta_+ - \eta'_-(\hat{Q}_-\hat{A}_- + \hat{A}'_-\hat{Q}_-)\eta_+$$
$$+ 2(\eta'_+\hat{Q}_+\hat{\Delta}_+(\eta) - \eta'_-\hat{Q}_-\hat{\Delta}_-(\eta))$$
$$= \|\eta_+\|_2^2 + \|\eta_-\|_2^2 + O(\|\eta\|_2^3) = \|\eta\|_2^2 + O(\|\eta\|_2^3) > 0$$

on $\bar{\mathcal{S}}_\varepsilon \setminus \{0\}$ as long as $\varepsilon > 0$ is small enough. Thus, by proposition 2.7(3), $\eta = 0$, and thus $\xi = 0$, or equivalently, $x = \bar{x}$ in the original coordinates is unstable. The case where A has eigenvalues on the imaginary axis (in addition to the ones with positive real parts) can be treated in the same way as before by considering $A_\delta = A - \delta I$ instead for some small $\delta \neq 0$, so A_δ does not have any eigenvalues on the imaginary axis. Then all the previous arguments remain valid and all (strict) inequalities continue to hold for $\delta \to 0$, which concludes the proof. ∎

Example 2.12 (Generic Failure of Linearization Criterion) Consider the nonlinear system $\dot{x} = \alpha x^3$. The point $\bar{x} = 0$ is the only equilibrium. Linearization at that point yields $A = f_x(\bar{x}) = 3\alpha x^2|_{x=\bar{x}} = 0$, so nothing can be concluded from proposition 2.8. Using the Lyapunov function $V(x) = x^4$ gives $\dot{V} = 4\alpha x^6$. Thus, by proposition 2.7 the equilibrium $\bar{x}$ is stable if $\alpha \leq 0$, unstable if $\alpha > 0$, and asymptotically stable if $\alpha < 0$. □

A (nonempty) set $\mathcal{S} \subset \mathcal{X}$ is called *invariant* (with respect to the autonomous system (2.35)) if any trajectory starting in $\mathcal{S}$ remains there for all future times, that is, if

$$x(t_0) = x_0 \in \mathcal{S} \Rightarrow \phi(t, x_0) \in \mathcal{S}, \qquad \forall t \geq t_0.$$

For example, given any equilibrium point $\bar{x}$ of the system, the singleton $\mathcal{S} = \{\bar{x}\}$ must be invariant. The following remark shows that, more generally, the set of points from which trajectories converge to $\bar{x}$ (which includes those trajectories) is invariant.

Remark 2.8 (Region of Attraction) The *region of attraction* $\mathcal{A}(\bar{x})$ of an equilibrium $\bar{x}$ is defined as the set of points from which a trajectory would asymptotically approach $\bar{x}$. That is,

$$\mathcal{A}(\bar{x}) = \{\xi \in \mathbb{R}^n : \lim_{t \to \infty} \phi(t, \xi) = \bar{x}\}.$$

Ordinary Differential Equations

Clearly, if $\bar{x}$ is not asymptotically stable, then its region of attraction is a singleton containing only $\bar{x}$. On the other hand, *if $\bar{x}$ is asymptotically stable, then $\mathcal{A}(\bar{x})$ is an invariant open set and its boundary is formed by system trajectories.* The fact that $\mathcal{A}(\bar{x})$ is invariant is trivial. To show that the region of attraction is open, pick an arbitrary point $x_0 \in \mathcal{A}(\bar{x})$. Since $\bar{x}$ is asymptotically stable, there exists $\varepsilon > 0$ such that

$$\|x - \bar{x}\| < \varepsilon \Rightarrow x \in \mathcal{A}(\bar{x}). \tag{2.40}$$

Since $\lim_{t \to \infty} \phi(t, x_0) = \bar{x}$, there exists a time $T > 0$ such that $\|\phi(T, x_0) - \bar{x}\| < \varepsilon/2$ (figure 2.10). By continuous dependence on initial conditions (proposition 2.5) there exists $\delta > 0$ such that

$$\|\hat{x}_0 - x_0\| < \delta \Rightarrow \|\phi(T, \hat{x}_0) - \phi(T, x_0)\| \leq \varepsilon/2,$$

so

$$\|\hat{x}_0 - x_0\| < \delta \Rightarrow \|\phi(T, \hat{x}_0) - \bar{x}\| \leq \|\phi(T, \hat{x}_0) - \phi(T, x_0)\|$$
$$+ \|\phi(T, x_0) - \bar{x}\| < \varepsilon. \tag{2.41}$$

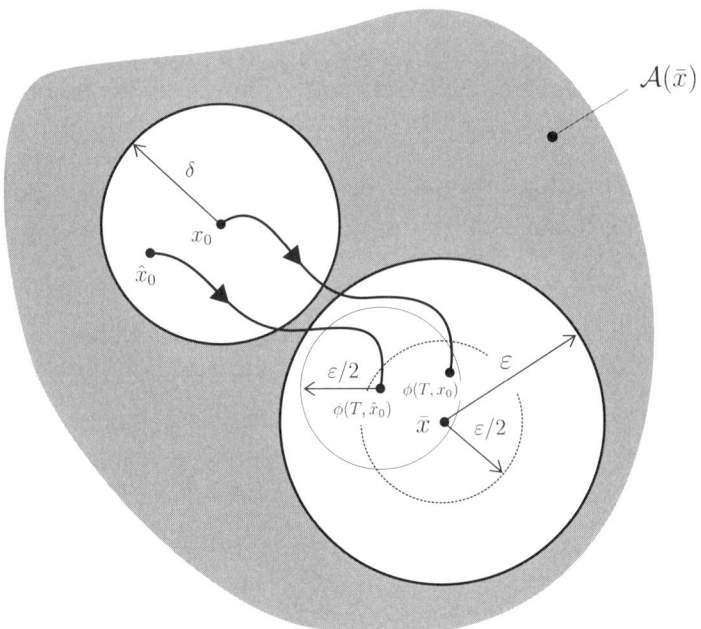

Figure 2.10
The region of attraction $\mathcal{A}(\bar{x})$ of an asymptotically stable equilibrium $\bar{x}$ is an open set.

Hence, from (2.40) and (2.41) it follows that with any $x_0 \in \mathcal{A}(\bar{x})$ an open δ-neighborhood of x_0 is also contained in $\mathcal{A}(\bar{x})$, so the region of attraction is indeed an open set. The fact that the boundary of $\mathcal{A}(\bar{x})$ is formed by trajectories can be seen as follows. Consider any boundary point $\hat{x} \in \partial \mathcal{A}(\bar{x})$. Since $\hat{x} \notin \mathcal{A}(\bar{x})$, by the definition of the region of attraction $\mathcal{A}(\bar{x})$, $\phi(t,\hat{x}) \notin \mathcal{A}(\bar{x})$ for all $t \geq 0$. On the other hand, taking any sequence of points, $\{x^k\}_{k=0}^{\infty} \subset \mathcal{A}(\bar{x})$ such that $x^k \to \hat{x}$ as $k \to \infty$, by the continuity of the flow (which is guaranteed by proposition 2.5) it is

$$\lim_{k\to\infty} \phi(t, x^k) = \phi(t, \hat{x}), \quad \forall t \geq 0,$$

which implies that $\phi(t, \hat{x}) \in \partial \mathcal{A}(\bar{x})$ for all $t \geq 0$. □

Proposition 2.9 (Global Asymptotic Stability) (Barbashin and Krasovskii 1952) Assume that $\bar{x}$ is an equilibrium of an autonomous system, and let $V : \mathbb{R}^n \to \mathbb{R}$ be a continuously differentiable function, which is *coercive* in the sense that $V(x) \to \infty$ as $\|x\| \to \infty$. If $V(x) > V(\bar{x})$ and $\dot{V}(x) < 0$ for all $x \neq \bar{x}$, then $\mathcal{A}(\bar{x}) = \mathbb{R}^n$, that is, $\bar{x}$ is globally asymptotically stable.

Proof For any initial point $x \in \mathbb{R}^n$ the set $\Omega = \{\xi \in \mathbb{R}^n : V(\xi) \leq V(x)\}$ is compact, because V is coercive. It is also invariant; no trajectory starting inside the set can increase the value of V. The rest of the proof is analogous to part (2) of the proof of proposition 2.7. ∎

Stability in Time-Variant Systems Let $\mathcal{D} = \mathbb{R}_+ \times \mathcal{X}$. Now consider the stability properties of equilibria of the time-variant system

$$\dot{x} = f(t, x), \tag{2.42}$$

where $f : \mathcal{D} \to \mathbb{R}^n$ is continuous and locally Lipschitz with respect to x. The discussion is limited to asymptotic stability of an equilibrium.

Proposition 2.10 (Asymptotic Stability) Let $\bar{x}$ be an equilibrium of the time-variant system (2.42), and let $V : \mathcal{D} \to \mathbb{R}$ be a continuously differentiable function. If there exist $\delta > 0$ and continuous,[28] increasing functions $\rho_i : [0, \delta) \to \mathbb{R}_+$ satisfying $\rho_i(0) = 0$, $i \in \{1, 2, 3\}$, such that

$$\rho_1(\|x - \bar{x}\|) \leq V(t, x) \leq \rho_2(\|x - \bar{x}\|)$$

and

---

28. Instead of being continuous on their domain, it is enough if all ρ_i, $i \in \{1, 2, 3\}$, are right-continuous at the origin.

Ordinary Differential Equations

$$\dot{V}(t,x) = V_t(t,x) + \langle V_x(t,x), f(t,x) \rangle \leq -\rho_3(\|x - \bar{x}\|)$$

for all $(t,x) \in \mathcal{D}$ with $t \geq t_0$ and $\|x - \bar{x}\| < \delta$, then $\bar{x}$ is asymptotically stable.

Proof Fix $\varepsilon \in [0, \delta)$, and for any time $t \geq 0$, let $\Omega_{t,\varepsilon} = \{x \in \mathcal{X} : V(t,x) \leq \rho_1(\varepsilon)\}$ be the set of states $x \in \mathcal{X}$ at which the function $V(t,x)$ does not exceed $\rho_1(\varepsilon)$. By assumption $\rho_1 \leq \rho_2$ on $[0, \varepsilon]$, so

$$\rho_2(\|x - \bar{x}\|) \leq \rho_1(\varepsilon) \Rightarrow x \in \Omega_{t,\varepsilon}$$

$$\Rightarrow \rho_1(\|x - \bar{x}\|) \leq V(t,x) \leq \rho_1(\varepsilon)$$

$$\Rightarrow \|x - \bar{x}\| \leq \varepsilon,$$

for all $(t,x) \in \mathcal{D}$. Since $\dot{V}(t,x) < 0$ for all $x \in \mathcal{X}$ with $0 < \|x - \bar{x}\| \leq \varepsilon$, for any initial data $(t_0, x_0) \in \mathcal{D}$ the set $\Omega_{t_0,\varepsilon}$ is *invariant*, that is,

$$x_0 \in \Omega_{t_0,\varepsilon} \Rightarrow \phi(t - t_0, t_0, x_0) \in \Omega_{t_0,\varepsilon}, \qquad \forall t \geq t_0,$$

where $\phi(t - t_0, t_0, x_0)$ denotes the flow of the vector field $f(t,x)$ starting at (t_0, x_0). Without loss of generality assume that $\rho_2(\|x_0 - \bar{x}\|) \leq \rho_1(\varepsilon)$ and $x(t_0) = x_0$ for some $t_0 \geq 0$. Thus, using the invariance of $\Omega_{t_0,\varepsilon}$ and the fact that

$$\rho_2^{-1}(V(t,x(t))) \leq \|x(t) - \bar{x}\|$$

gives

$$\dot{V}(t, x(t)) \leq -\rho_3(\|x(t) - \bar{x}\|) \leq -\rho_3(\rho_2^{-1}(V(t,x(t)))), \qquad \forall t \geq t_0. \quad (2.43)$$

Let $\rho : [0, \rho_1(\varepsilon)] \to \mathbb{R}_+$ be an increasing, locally Lipschitz function that satisfies $\rho(0) = 0$ and $\rho \leq \rho_3 \circ \rho_2^{-1}$. From (2.43) it follows that if $v(t)$ is a solution to the autonomous IVP

$$\dot{v} = -\rho(v), \qquad v(t_0) = v_0, \qquad (2.44)$$

for $t \geq t_0$, and $v_0 = V(t_0, x_0) \in [0, \rho_1(\varepsilon)]$, then it is $V(t, x(t)) \leq v(t)$ for all $t \geq t_0$. The ODE in (2.44) has a separable right-hand side, so (see section 2.2.2)

$$v(t) = \begin{cases} H^{-1}(t - t_0; v_0) & \text{if } v_0 > 0, \\ 0 & \text{if } v_0 = 0, \end{cases}$$

where $H(v; v_0) = \int_v^{v_0} \frac{d\xi}{\rho(\xi)} > 0$ is a decreasing function on $(0, v_0)$ for $0 < v_0 \leq \rho_1(\varepsilon)$, so its inverse, $H^{-1}(\cdot; v_0)$, exists. Note that since $\lim_{v \to 0^+}$

$H(v; v_0) = \infty,$[29] any trajectory $v(t)$ starting at a point $v_0 > 0$ converges to the origin as $t \to \infty$, that is,

$$\lim_{t \to \infty} v(t) = \lim_{t \to \infty} H^{-1}(t - t_0; v_0) = 0.$$

Hence, it has been shown that

$$\|\phi(t - t_0, t_0, x_0) - \bar{x}\| = \|x(t) - \bar{x}\| \leq \rho_1^{-1}(V(t, x(t))) \leq \rho_1^{-1}(v(t))$$

$$\to \rho_1^{-1}(0) = 0$$

as $t \to \infty$, that is, $\lim_{t \to \infty} \phi(t - t_0, t_0, x_0) = \bar{x}$, which implies that $\bar{x}$ is asymptotically stable. ∎

Remark 2.9 (Global Asymptotic Stability) If the assumptions of proposition 2.10 are satisfied for all $\delta > 0$, and ρ_1, ρ_2 are coercive in the sense that $\lim_{\xi \to \infty} \rho_i(\xi) = \infty$ for $i \in \{1, 2\}$, then $\lim_{\xi \to \infty} \rho_2^{-1}(\rho_1(\xi)) = \infty$, so any $x \in \mathcal{X}$ is contained in $\Omega_{t,\varepsilon}$ for large enough $\varepsilon > 0$, which implies that the equilibrium $\bar{x}$ is *globally asymptotically stable*. Moreover, if $\rho_i(\xi) = \alpha_i \xi^c$ for some positive c and α_i, $i \in \{1, 2, 3\}$, then $\bar{x}$ is *(globally) exponentially stable*, since (see the proof of proposition 2.10) $\rho(\xi) = \rho_3(\rho_2^{-1}(\xi)) = (\alpha_3/\alpha_2)\xi$ is locally Lipschitz and

$$\|\phi(t - t_0, t_0, x_0) - \bar{x}\| \leq \left(\frac{\alpha_3}{\alpha_1}\right)^{1/c} \|x_0\| \exp\left[-\frac{\alpha_3}{c\alpha_2}(t - t_0)\right],$$

for all $t \geq t_0$. □

Example 2.13 (Linear System) Consider the linear ODE (with variable coefficients)

$$\dot{x} = A(t)x, \tag{2.45}$$

for $t \geq t_0 \geq 0$, where the matrix function $A(t)$ is continuous with values in $\mathbb{R}^{n \times n}$ for all $t \geq 0$. The point $\bar{x} = 0$ is an equilibrium of (2.45), and in analogy to the autonomous case (see example 2.11), one tries the Lyapunov function candidate $V(t, x) = x'Q(t)x$, for all $(t, x) \in \mathcal{D}$, where $Q(t)$ is continuous, with symmetric positive definite values. Then

$$\dot{V}(t, x) = x'\left(\dot{Q}(t) + Q(t)A(t) + A'(t)Q(t)\right)x$$

---

29. The notation $g(\xi_0^+) = \lim_{\xi \to \xi_0^+} g(\xi) = \lim_{\varepsilon \to 0} g(\xi_0 + \varepsilon^2)$ denotes a *right-sided limit* as $\xi \in \mathbb{R}$ approaches the point ξ_0 on the real line from the right. A *left-sided limit* $g(\xi_0^-)$ is defined analogously.

Ordinary Differential Equations

is negative if $Q(t)$ solves the *Lyapunov differential equation*[30]

$$\dot{Q} + QA(t) + A'(t)Q + R(t) = 0 \tag{2.46}$$

for some continuous symmetric positive definite matrix $R(t)$ (e.g., $R(t) \equiv I$). If in addition there exist positive constants $\alpha_1, \alpha_2, \alpha_3$ such that $\alpha_1 I \leq Q(t) \leq \alpha_2 I$ and $R(t) \geq \alpha_3 I$ for all $t \geq t_0$, then $\alpha_1 \|x\|_2^2 \leq V(t,x) \leq \alpha_2 \|x\|_2^2$ and $\dot{V}(t,x) \leq -\alpha_3 \|x\|_2^2$ for all $(t,x) \in \mathcal{D}$, so by proposition 2.10 and remark 2.9 the equilibrium $\bar{x} = 0$ is globally asymptotically stable. □

Proposition 2.11 (Generalized Linearization Criterion) Let $\bar{x}$ be an equilibrium of the time-variant system (2.42). Assume that f is continuously differentiable in a neighborhood of $\bar{x}$ and that the Jacobian matrix f_x is bounded and Lipschitz with respect to x. Set $A(t) = f_x(t, \bar{x})$ for all $t \geq 0$; then $\bar{x}$ is exponentially stable if it is an exponentially stable equilibrium of the linear time-variant system $\dot{x} = A(t)(x - \bar{x})$.

Proof Let $Q(t)$ be a symmetric positive definite solution to the Lyapunov differential equation (2.46) for $R(t) \equiv I$ such that $\alpha_1 I \leq Q(t) \leq \alpha_2 I$ for some $\alpha_1, \alpha_2 > 0$, and let $V(t,x) = x'Q(t)x$, for all $t \geq 0$. Then, with the abbreviation $\Delta(t,x) = f(t,x) - A(t)(x - \bar{x})$, and given some $\delta > 0$,

$$\begin{aligned}\dot{V}(t,x) &= (x-\bar{x})'\dot{Q}(x-\bar{x}) + (x-\bar{x})'Q(t)f(t,x) + f'(t,x)Q(t)(x-\bar{x}) \\ &= (x-\bar{x})'\left(\dot{Q}(t) + Q(t)A(t) + A'(t)Q(t)\right)(x-\bar{x}) + 2x'Q(t)\Delta(t,x) \\ &= -(x-\bar{x})'R(t)(x-\bar{x}) + 2(x-\bar{x})'Q(t)\Delta(t,x) \\ &\leq -\|x-\bar{x}\|_2^2 + 2\alpha_2 L\|x-\bar{x}\|_2^3 \\ &\leq -(1 - 2\alpha_2 k\delta)\|x\|_2^2, \end{aligned} \tag{2.47}$$

for all $\|x\| \leq \delta$, as long as $\delta < 1/(2\alpha_2 k)$; realizing that $\Delta(t,x)$ as a truncated Taylor expansion can be majorized by $k\|x-\bar{x}\|_2^2$ for some $k > 0$. Thus, by remark 2.9 the point $\bar{x}$ is (locally) exponentially stable. ■

2.2.8 Limit Cycles and Invariance

In many practically important situations system trajectories do not converge to an equilibrium but instead approach a certain periodic trajectory, which is referred to as a *limit cycle*. For instance, in the Lotka-Volterra predator-prey system of example 2.8 every single trajectory is

---

30. One can show that if $A(t)$ and $R(t)$ are also bounded, then (2.46) has a solution (for $t \geq t_0$), which is of the form $Q(t) = \int_t^\infty \Phi'(s,t)Q(s)\Phi(s,t)\,ds$, where $\Phi(t,t_0)$ is the fundamental matrix of the homogeneous linear system (2.45); see also section 2.3.2.

a limit cycle. For a given initial state x, the set

$$\mathcal{L}_x^+ = \{\xi \in \mathbb{R}^n : \exists \{t_k\}_{k=0}^\infty \subset \mathbb{R} \text{ s.t. } \lim_{k\to\infty} t_k = \infty \text{ and } \lim_{k\to\infty} \|\phi(t_k, x) - \xi\| = 0\}$$

is called the *positive limit set of x*. It is useful to characterize such a limit set, particularly when it describes a limit cycle instead of an isolated equilibrium.

Lemma 2.4 If for a given $x \in \mathcal{D}$ the trajectory $\phi(t, x)$ is bounded and lies in $\mathcal{D}$ for all $t \geq 0$, then the positive limit set $\mathcal{L}_x^+$ is nonempty, compact, and invariant.

Proof (1) *Nonemptiness.* Since the trajectory $\phi(t, x)$ is by assumption bounded, by the Bolzano-Weierstrass theorem (proposition A.2) there is a sequence $\{t_k\}_{k=0}^\infty$ of time instances t_k, with $t_k < t_{k+1}$ for all $k \geq 0$, such that $\lim_{k\to\infty} x(t_k)$ exists, which implies that the positive limit set $\mathcal{L}_x^+$ is nonempty.

(2) *Compactness.* By the previous argument the set $\mathcal{L}_x^+$ is also bounded. Consider now any converging sequence $\{\hat{x}^j\}_{j=0}^\infty \subset \mathcal{L}_x^+$ such that $\lim_{j\to\infty} \|\hat{x} - \hat{x}^j\| = 0$ for some state $\hat{x}$. Then (omitting some of the details) in every neighborhood of $\hat{x}$ there must be infinitely many points of the trajectory, which implies that $\hat{x} \in \mathcal{L}_x^+$, so $\mathcal{L}_x^+$ is also closed and thus compact.

(3) *Invariance.* Let $\hat{x} \in \mathcal{L}_x^+$ and consider the trajectory $\phi(t, \hat{x})$, which needs to stay in $\mathcal{L}_x^+$ for all $t \geq 0$ to obtain invariance. Since $\hat{x}$ is in the positive limit set $\mathcal{L}_x^+$, there exists a sequence $\{t_k\}_{k=0}^\infty$ of time instances t_k, with $t_k < t_{k+1}$ for all $k \geq 0$, such that $\lim_{k\to\infty} \|\hat{x} - \phi(t_k, x)\| = 0$. Hence, by the group laws for the flow (see remark 2.5) it is

$$\phi(t + t_k, x) = \phi(t, \phi(t_k, x)) = \phi(t, x(t_k)),$$

which by continuity of the flow (see proposition 2.5) implies that

$$\lim_{k\to\infty} \phi(t + t_k, x) = \lim_{k\to\infty} \phi(t, x(t_k)) = \phi(t, \hat{x}),$$

which concludes the proof. ∎

Proposition 2.12 (Invariance Principle) (LaSalle 1968) Let $\Omega \subset \mathbb{R}^n$ be a compact invariant set and $V : \Omega \to \mathbb{R}$ be a continuously differentiable function such that $\dot{V}(x) \leq 0$ for all $x \in \Omega$. If $\mathcal{M}$ denotes the largest invariant set in $\{\xi \in \Omega : \dot{V}(\xi) = 0\}$, then[31]

---

31. A *trajectory* $x(t) = \phi(t, x)$ *converges to a* (nonempty) *set* $\mathcal{M}$ *for* $t \to \infty$ *if* $\lim_{t\to\infty} \inf_{\hat{x} \in \mathcal{M}} \|\hat{x} - \phi(t, x)\| = 0$.

$\phi(t, x) \to \mathcal{M}$ (as $t \to \infty$), $\quad \forall x \in \Omega.$

Proof Let $x \in \Omega$. By invariance of Ω, the system trajectory $x(t) = \phi(t, x)$ stays in Ω for all $t \geq 0$. Since $\dot{V}$ is by assumption nonpositive on Ω, the function $V(x(t))$ is nondecreasing in t. In addition, the continuous function V is bounded from below on the compact set Ω, so $V(x(t))$ has a limit V_∞ as $t \to \infty$. Since Ω is closed, it contains the positive limit set $\mathcal{L}_x^+$. Thus, for any $\hat{x} \in \mathcal{L}_x^+$ there exists a sequence $\{t_k\}_{k=0}^\infty$ with $\lim_{k \to \infty} t_k = \infty$ and $\lim_{k \to \infty} \phi(t_k, x) = \hat{x}$. Hence, by continuity of V,

$$V_\infty = \lim_{k \to \infty} V(\phi(t_k, x)) = V\left(\lim_{k \to \infty} \phi(t_k, x)\right) = V(\hat{x}).$$

By lemma 2.4 the set $\mathcal{L}_x^+$ is invariant, so $\dot{V}$ vanishes on $\mathcal{L}_x^+$. As a result, it must be that $\mathcal{L}_x^+ \subset \mathcal{M} \subset \{\xi \in \Omega : \dot{V}(\xi) = 0\} \subset \Omega$. Since Ω is bounded and invariant, it is $x(t) \to \mathcal{L}_x^+$, which implies that $x(t) \to \mathcal{M}$ as $t \to \infty$. ∎

If, for some $T > 0$, a system trajectory $x(t)$ satisfies the relation

$$x(t) = x(t + T), \quad \forall t \geq 0,$$

then it is called a *closed* (or *periodic*) *trajectory*, and the curve $\gamma = x([0, \infty))$ is referred to as a *periodic orbit* (or *limit cycle*).

When the state space is two-dimensional, it is possible to conclude from a lack of equilibria in a positive limit set that this set must be a periodic orbit. The following result is therefore very useful in practice (see e.g., exercise 2.3c).[32]

Proposition 2.13 (Poincaré-Bendixson Theorem) Let $n = 2$. If the positive limit set $\mathcal{L}_x^+$, for some x, does not contain any equilibrium, then it is a periodic orbit (limit cycle).

Proof The proof proceeds along a series of four claims. A *transversal* $\Theta \subset \mathcal{D}$ of the autonomous system $\dot{x} = f(x)$ is a segment of a line defined by $a'x - b = 0$ for some $a \in \mathbb{R}^2$ and $b \in \mathbb{R}$ such that $a'f(x) \neq 0$ for all $x \in \Theta$. Consider now any such transversal, assumed to be open such that it does not contain any endpoints.

---

32. The Poincaré-Bendixson theorem does not generalize to state spaces of dimension greater than 2. Indeed, the well-known *Lorenz oscillator* (Lorenz 1963), described by $\dot{x}_1 = \sigma(x_2 - x_1)$, $\dot{x}_2 = x_1(\rho - x_3) - x_2$, and $\dot{x}_3 = x_1 x_2 - \beta x_3$, where, in the original interpretation as an atmospheric convection system, $\sigma > 0$ is the *Prandtl number*, $\rho > 0$ a (normalized) *Rayleigh number*, and $\beta > 0$ a given constant, produces trajectories that do not converge to a limit cycle (e.g., $(\sigma, \rho, \beta) = (10, 28, 8/3)$, as discussed by Lorenz himself). The positive limit sets of such systems are often referred to as *strange attractors*.

Claim 1 For any $\hat{x} \in \Theta$ and any $\varepsilon > 0$ there exists $\delta > 0$ such that $\|\hat{x} - x(0)\| < \delta$ implies that $x(t) \in \Theta$ for some $t \in (-\varepsilon, \varepsilon)$. If the trajectory is bounded, then the intersection point $x(t)$ tends to $\hat{x}$ as $\varepsilon \to 0^+$. Let $x \in \mathcal{D}$ and set $g(t, x) = a'\phi(t, x) - b$. Thus, $\phi(t, x) \in \Theta$ if and only if $g(t, x) = 0$. Since $\hat{x} \in \Theta$, it is $g(0, \hat{x}) = 0$ and $g_t(0, \hat{x}) = a'f(\phi(0, \hat{x})) = a'f(\hat{x}) \neq 0$. Thus, by the implicit function theorem (proposition A.7) the continuous map $\tau(x)$ can for some $\rho > 0$ be implicitly defined by $g(\tau(x), x) = 0$ for $\|\hat{x} - x\| \leq \rho$. By continuity of τ for any $\varepsilon > 0$ there exists $\delta > 0$ such that

$$\|\hat{x} - x\| < \delta \implies |\tau(\hat{x}) - \tau(x)| = |\tau(x)| < \varepsilon.$$

Furthermore, choose $\delta \in (0, \min\{\varepsilon, \rho\})$ and let $\mu = \sup\{\|f(\phi(t, x))\| : \|\hat{x} - x\| \leq \rho, t \in \mathbb{R}\}$ (which is finite by assumption); then

$$\|\hat{x} - \phi(\tau(x), x)\| \leq \|\hat{x} - x\| + \|x - \phi(\tau(x), x)\| < \varepsilon + \mu|\tau(x)| < (1 + \mu)\varepsilon,$$

because $\|\phi(0, x) - \phi(\tau(x), x)\| \leq \mu|\tau(x)|$.

Claim 2 If there exist $t_0 < t_1 < t_2$ such that $\{t \in [t_0, t_2] : x(t) \in \Theta\} = \{t_0, t_1, t_2\}$, then $x(t_1) = (1 - \lambda)x(t_0) + \lambda x(t_2)$ for some $\lambda \in (0, 1)$. Let $x^k = x(t_k)$ for $k \in \{0, 1, 2\}$. If the claim fails, then without any loss of generality one can assume that $\lambda > 1$, that is, x^2 lies between x^0 and x^1. Let γ denote the closed curve that is obtained by joining the trajectory segment $x([t_0, t_1])$ with the line segment $\overline{x^1 x^0}$ (figure 2.11). By the Jordan curve theorem (proposition A.4) γ has an inside $\mathcal{S}$ and an outside $\mathbb{R}^2 \setminus \mathcal{S}$. Assuming that f points into $\mathcal{S}$ on the transversal Θ,[33] the set $\mathcal{S}$ is invariant and thus contains the trajectory segment $\phi((t_1, t_2), x)$. But since $f(x^2)$ points into $\mathcal{S}$, $\phi(t_2 - \varepsilon, x)$ must lie on the outside of γ, which is a contradiction.

Claim 3 If $\hat{x} \in \mathcal{L}_x^+$ for some x, then $\phi(t, \hat{x})$ can intersect Θ in at most one point. Suppose that for some $\hat{x} \in \mathcal{L}_x^+$ the trajectory $\phi(t, \hat{x})$ intersects the transversal Θ in two points, $x^1 \neq x^2$, that is, $\{x^1, x^2\} \subseteq \phi(\mathbb{R}_+, \hat{x}) \cap \Theta$. Now choose two open subintervals $\Theta_1, \Theta_2 \subset \Theta$ such that $x^j \in \Theta_j$ for $j \in \{1, 2\}$. Given a small $\varepsilon > 0$, define for each $j \in \{1, 2\}$ the flow box $\mathcal{B}_j = \{\phi(t, \xi) : t \in (-\varepsilon, \varepsilon), \xi \in \Theta_j\}$.[34] Since $x^1, x^2 \in \mathcal{L}_x^+$, by definition there exists an increasing sequence $\{t_k\}_{k=1}^\infty$ with $t_k \to \infty$ as $k \to \infty$ such that $\phi(t_{2k-1}, x) \in \mathcal{B}_1$ and $\phi(t_{2k}, x) \in \mathcal{B}_2$ for all $k \geq 1$. Because of the unidirectionality of the

---

33. If f points toward the outside, the same reasoning applies by reversing the orientation of γ and accordingly switching the inside and the outside of the closed curve.
34. These flow boxes exist as a consequence of the *rectifiability theorem* for vector fields (see, e.g., Arnold and Il'yashenko 1988, 14).

Ordinary Differential Equations

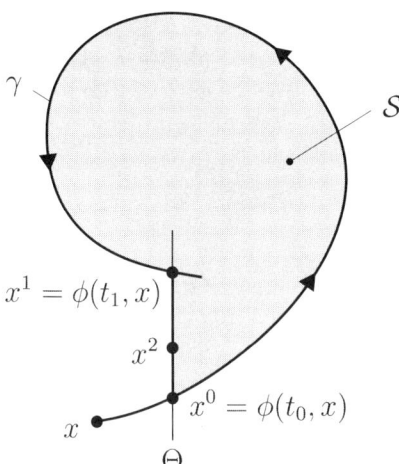

Figure 2.11
Intuition for claim 2 in the proof of the Poincaré-Bendixson theorem.

flow in each flow box, one can without any loss of generality assume that $\phi(t_k, x) \in \Theta$ for all $k \geq 1$. But note that even though $t_k < t_{k+1} < t_{k+2}$, there exists no $\lambda \in (0, 1)$ such that $\phi(t_{k+1}, x) = (1 - \lambda)\phi(t_k, x) + \lambda \phi(t_{k+2}, x)$, a contradiction to claim 2. Hence $x^1 = x^2$, and by claim 1, as $\varepsilon \to 0^+$ the intersection points $\phi(t_k, x) \to x^1 = x^2$ as $k \to \infty$. Thus, the set $\phi(\mathbb{R}_+, \hat{x}) \cap \Theta$ must be a singleton, equal to $\{x^1\}$.

Claim 4 If the positive limit set $\mathcal{L}_x^+$ of a bounded trajectory (starting at some point $x \in \mathcal{D}$) contains a periodic orbit γ, then $\mathcal{L}_x^+ = \gamma$. Let $\gamma \subseteq \mathcal{L}_x^+$ be a periodic orbit. It is enough to show that $\phi(t, x) \to \gamma$ because then necessarily $\mathcal{L}_x^+ \subseteq \gamma$, and thus $\gamma = \mathcal{L}_x^+$. Let $\hat{x} \in \gamma$ and consider a transversal Θ with $\hat{x} \in \Theta$. Thus, since $\hat{x} \in \mathcal{L}_x^+$, there exists an increasing and diverging sequence $\{t_k\}_{k=0}^\infty$ such that

$$\phi(\mathbb{R}_+, x) \cap \Theta = \{\phi(t_k, x)\}_{k=0}^\infty \subset \Theta,$$

and $\lim_{k \to \infty} \phi(t_k, x) = \hat{x}$. By claim 2, the sequence $\{\phi(t_k, x)\}_{k=0}^\infty$ must be monotonic in Θ in the sense that any point $\phi(t_k, x)$ (for $k \geq 2$) cannot lie strictly between any two previous points of the sequence. Moreover, $\phi(t_k, x) \to \hat{x}$ monotonically as $k \to \infty$. Because γ is by assumption a periodic orbit, there is $T > 0$ such that $\phi(T, \hat{x}) = \phi(0, \hat{x}) = \hat{x}$. Fix a small $\varepsilon > 0$ and take δ from claim 1. Then there exists $\hat{k} > 0$ such that

$$k \geq \hat{k} \Rightarrow \|\hat{x} - \phi(t, x^k)\| \leq \delta,$$

where $x^k = \phi(t_k, x)$, so that as a consequence of claim 1, $\phi(t + T, x^k) \in \Theta$ for some $t \in (-\varepsilon, \varepsilon)$. Hence, $t_{k+1} - t_k \leq T + \varepsilon$ for all $k \geq \hat{k}$. Because of the continuous dependence of solutions to ODEs on initial data (proposition 2.5), for any $\hat{\varepsilon} > 0$ there exists $\hat{\delta} > 0$ such that

$$\|\hat{x} - x^k\| \leq \hat{\delta} \Rightarrow \|\phi(t, \hat{x}) - \phi(t, x^k)\| < \hat{\varepsilon}, \qquad \forall t : |t| \leq T + \hat{\varepsilon}.$$

Thus, by choosing $\hat{\varepsilon} = \varepsilon$ and $\hat{\delta} \in [0, \delta]$ it is

$$k \geq k_0 \Rightarrow \|\hat{x} - x^k\| \leq \delta \Rightarrow \|\phi(t, \hat{x}) - \phi(t, x^k)\| < \varepsilon, \quad \forall t : |t| \leq T + \varepsilon,$$

for some $k_0 \geq \hat{k}$. For $t \in [t_k, t_{k+1}]$ and $k \geq k_0$ therefore,

$$\inf_{y \in \gamma} \|y - \phi(t, x)\| \leq \|\phi(t - t_k, \hat{x}) - \phi(t, x)\| = \|\phi(t - t_k, \hat{x}) - \phi(t - t_k, x^k)\| < \varepsilon,$$

since $|t - t_k| \leq T + \varepsilon$ and $\phi(t, x) = \phi(t - t_k, \phi(t_k, x)) = \phi(t - t_k, x^k)$ by the group laws for flows in remark 2.5. Hence, it has been shown that $\phi(t, x) \to \gamma$ as $t \to \infty$, whence $\gamma = \mathcal{L}_x^+$.

Proof of Proposition 2.13 By lemma 2.4, the positive limit set $\mathcal{L}_x^+$ is a nonempty compact invariant set. Let $\hat{x} \in \mathcal{L}_x^+$ and let $y \in \mathcal{L}_{\hat{x}}^+ \subseteq \mathcal{L}_x^+$, and consider a transversal Θ that contains the point y, which by assumption cannot be an equilibrium. Claim 3 implies that $\phi(t, \hat{x})$ can intersect Θ in at most one point. In addition, there is an increasing sequence $\{t_k\}_{k=0}^\infty$ with $t_k \to \infty$ as $k \to \infty$ such that $\lim_{k \to \infty} \phi(t_k, \hat{x}) = y$. The trajectory $\phi(t, \hat{x})$ must therefore cross the transversal Θ infinitely many times, so that (taking into account that $\mathcal{L}_x^+$ does not contain any equilibrium) necessarily $\phi(t, \hat{x}) = \phi(t + T, \hat{x})$ for some $T > 0$. Claim 4 now implies that $\mathcal{L}_x^+$ is a periodic orbit, for it contains a periodic orbit. ∎

In order to exclude the existence of limit cycles for a given two-dimensional system, the following simple result can be used.[35]

Proposition 2.14 (Poincaré-Bendixson Criterion) Let $f : \mathcal{D} \to \mathbb{R}^2$ be continuously differentiable. If on a simply connected domain $\mathcal{D} \subset \mathbb{R}^2$ the function $\operatorname{div} f = \partial f_1 / \partial x_1 + \partial f_2 / \partial x_2$ does not vanish identically and has no sign changes, then the system has no periodic orbit in $\mathcal{D}$.

---

35. In physics, an electrostatic field $E(x) = (E_1(x), E_2(x))$, generated by a distributed charge of density $\rho(x) \geq 0$ (not identically zero), satisfies the Maxwell equation $\operatorname{div}(E(x)) = \rho(x)$ and therefore cannot produce closed trajectories (field lines). By contrast, a magnetostatic field $B(x) = (B_1(x), B_2(x))$ satisfies the Maxwell equation $\operatorname{div} B(x) = 0$ for all x and does always produce closed field lines (limit cycles).

Ordinary Differential Equations

Proof Assume that $\gamma : [0,1] \to \mathcal{D}$ is a periodic orbit in $\mathcal{D}$ (i.e., with $\gamma(0) = \gamma(1)$). Then, by Green's theorem (see J. M. Lee 2003, 363) it is

$$0 = \int_\gamma (f_2(x)\,dx_1 - f_1(x)\,dx_2) = \int_\mathcal{S} \operatorname{div} f(x)\,dx,$$

where the surface $\mathcal{S}$ is the interior of the curve γ such that $\partial \mathcal{S} = \gamma([0,1])$. But the last term cannot vanish, since by assumption $\operatorname{div} f(x)$ either > 0 or < 0 on a nonzero-measure subset of $\mathcal{S}$ (using the continuity of the derivatives of f). This is a contradiction, so there cannot exist a periodic orbit that lies entirely in $\mathcal{D}$. ∎

2.3 Higher-Order ODEs and Solution Techniques

Let $n, r \geq 1$. An rth order ODE is of the form

$$F(t, x(t), \dot{x}(t), \ddot{x}(t), \ldots, x^{(r)}(t)) = 0, \tag{2.48}$$

where $x^{(k)}(t) = d^k x(t)/dt^k$, $k \in \{1,\ldots,r\}$, is the kth total derivative of $x(t)$ with respect to t, using the convenient abbreviations $x^{(1)} = \dot{x}$ and $x^{(2)} = \ddot{x}$. The function $F : \mathbb{R}^{1+n+nr} \to \mathbb{R}^n$ is assumed to be continuously differentiable (at least on a nonempty connected open subset of $\mathbb{R}^{1+n+nr}$). As before, the (real-valued) dependent variable $x(t) = (x_1(t), \ldots, x_n(t))$ varies in $\mathbb{R}^n$ as a function of the independent variable $t \in \mathbb{R}$.

2.3.1 Reducing Higher-Order ODEs to First-Order ODEs

Any rth order ODE in the (implicit) form (2.48) can be transformed to an equivalent first-order ODE; by introducing r functions $y^1, \ldots, y^r$ (setting $x^{(0)} = x$) such that

$$y^k = x^{(k-1)}, \quad k \in \{1,\ldots,r\}, \tag{2.49}$$

one can rewrite (2.48) in the form

$$\hat{F}(t, y, \dot{y}) = \begin{bmatrix} \dot{y}^1 - y^2 \\ \dot{y}^2 - y^3 \\ \vdots \\ \dot{y}^{r-1} - y^r \\ F(t, y^1, \ldots, y^r, \dot{y}^r) \end{bmatrix} = 0,$$

where $y = (y^1, \ldots, y^r) \in \mathbb{R}^{\hat{n}}$ with $\hat{n} = nr$ is the independent variable, and $\hat{F} : \mathbb{R}^{1+2\hat{n}} \to \mathbb{R}^{\hat{n}}$ is a continuously differentiable function.

Remark 2.10 When $F(t, x, \dot{x}, \ddot{x}, \ldots, x^{(r)}) = x^{(r)} - f(t, x, \dot{x}, \ddot{x}, \ldots, x^{(r-1)})$ for some continuous function $f : \mathbb{R}^{1+nr} \to \mathbb{R}^n$, the variables introduced in (2.49) can be used to transform a higher-order ODE into *explicit* form,

$$x^{(r)} = f(t, x, \dot{x}, \ddot{x}, \ldots, x^{(r-1)}),$$

to a first-order ODE in explicit form,

$$\dot{y} = \hat{f}(t, y),$$

where the continuous function $\hat{f} : \mathbb{R}^{1+\hat{n}} \to \mathbb{R}^{\hat{n}}$ is such that

$$\hat{f}(t, y) = \begin{bmatrix} y^2 \\ \vdots \\ y^r \\ f(t, y^1, \ldots, y^r) \end{bmatrix}.$$

□

Since it is possible to reduce higher-order ODEs to first-order ODEs, all techniques described for $n > 1$ in section 2.2 in principle may also be applied to higher-order ODEs.

2.3.2 Solution Techniques

General analytical solution techniques are available for linear systems of the form

$$\dot{x} = A(t)x + b(t), \tag{2.50}$$

where $A(t)$ and $b(t)$ are continuous functions with values in $\mathbb{R}^{n \times n}$ and $\mathbb{R}^n$, respectively. When A, b are constant, the linear ODE (2.50) is said to have constant coefficients (see example 2.11). Using appropriate variable transformations, nonlinear systems sometimes can be reduced to linear systems.

Example 2.14 (Riccati Equation) Consider a matrix Riccati IVP of the form

$$\dot{X} + XA(t) + A'(t)X + XP(t)X = R(t), \quad X(t_0) = K,$$

where $A(t), P(t), R(t) \in \mathbb{R}^{n \times n}$ are continuous, bounded matrix functions such that $P(t)$ and $R(t)$ are symmetric positive definite, for all $t \geq t_0$. The matrix $K \in \mathbb{R}^{n \times n}$ of initial values is assumed to be symmetric and positive definite. Because of the structure of the ODE, a solution $X(t)$ of the matrix Riccati IVP will have symmetric positive definite

Ordinary Differential Equations

values. Moreover, the solution can be written in the form $X = ZY^{-1}$, where $Y(t), Z(t)$, with values in $\mathbb{R}^{n \times n}$, solve the linear IVP

$$\begin{bmatrix} \dot{Y} \\ \dot{Z} \end{bmatrix} = \begin{bmatrix} A(t) & P(t) \\ R(t) & -A'(t) \end{bmatrix} \begin{bmatrix} Y \\ Z \end{bmatrix}, \quad \begin{bmatrix} Y(t_0) \\ Z(t_0) \end{bmatrix} = \begin{bmatrix} I \\ K \end{bmatrix},$$

for all $t \geq t_0$. Indeed, by direct differentiation,

$$\dot{X} = \frac{d}{dt}[ZY^{-1}] = \dot{Z}Y^{-1} - ZY^{-1}\dot{Y}Y^{-1}$$

$$= (R(t)Y - A'(t)Z) Y^{-1} - ZY^{-1}(A(t)Y + P(t)Z) Y^{-1}$$

$$= R(t) - A'(t)X - XA(t) + XP(t)X,$$

for all $t \geq t_0$, and $X(t_0) = KI = Z(t_0)Y^{-1}(t_0)$. □

Linear Systems with Variable Coefficients

As in the solution of the linear ODE for $n = 1$ (see section 2.2.2) the linear system (2.50) is *homogeneous* when $b(t) \equiv 0$ and otherwise *inhomogeneous*.

Homogeneous Linear System Given the initial data $(t_0, x_0) \in \mathbb{R}^{1+n}$, consider the linear homogeneous IVP,

$$\dot{x} = A(t)x, \qquad x(t_0) = x_0. \tag{2.51}$$

By proposition 2.3 a solution to (2.51) exists and is unique. To find the solution, it is convenient to study the solutions of the linear matrix IVP

$$\dot{X} = A(t)X, \qquad X(t_0) = I, \tag{2.52}$$

where $X(t)$ has values in $\mathbb{R}^{n \times n}$ for all $t \in \mathbb{R}$. A solution $\Phi(t, t_0)$ to (2.52) is called *fundamental matrix* (or *state-transition matrix*), and its determinant, $\det \Phi(t, t_0)$, is usually referred to as *Wronskian*.

Lemma 2.5 (Properties of Fundamental Matrix) Let $s, t, t_0 \in \mathbb{R}$. Then the (unique) fundamental matrix $\Phi(t, t_0)$ satisfies the following properties:

1. $\Phi_t(t, t_0) = A(t)\Phi(t, t_0), \qquad \Phi(t_0, t_0) = I.$
2. $\Phi_{t_0}(t, t_0) = -\Phi(t, t_0)A(t_0).$
3. $\Phi(t, t) = I.$
4. $\Phi(t_0, t) = \Phi^{-1}(t, t_0).$
5. $\Phi(t, t_0) = \Phi(t, s)\Phi(s, t_0).$
6. $\det \Phi(t, t_0) = \exp[\int_{t_0}^{t} \text{trace } A(\theta) \, d\theta] > 0.$ (Liouville Formula)

Proof The six properties are best proved in a slightly different order. The statement of property 1 is trivial, since by assumption for any $t, t_0 \in \mathbb{R}$ the fundamental matrix $\Phi(t, t_0)$ solves (2.52). Property 3 follows directly from property 1. Property 5 is a direct consequence of the group laws for flows of nonautonomous systems (see remark 2.4). To prove property 4, note that $I = \Phi(t_0, t_0) = \Phi(t_0, t)\Phi(t, t_0)$ by properties 3 and 5. To prove property 2, note first that by the chain rule of differentiation

$$0 = \frac{d}{dt_0} I = \frac{d}{dt_0} \Phi(t, t_0) \Phi^{-1}(t, t_0)$$

$$= \dot{\Phi}_{t_0}(t, t_0) \Phi^{-1}(t, t_0) + \Phi(t, t_0) \left(\frac{\partial}{\partial t_0} \Phi^{-1}(t, t_0) \right).$$

On the other hand, switching t and t_0 in property 1, multiplying by $\Phi(t, t_0)$ from the left, and using property 4 yields

$$\Phi(t, t_0) \dot{\Phi}_{t_0}(t_0, t) = \Phi(t, t_0) \left(\frac{\partial}{\partial t_0} \Phi^{-1}(t, t_0) \right) = \Phi(t, t_0) A(t_0) \Phi^{-1}(t, t_0),$$

which, together with the preceding relation, implies the result. Finally, to prove property 6: The matrix $\Phi(s, t_0)$ corresponds to the flow of the matrix ODE in (2.52) at time s, starting from the initial value $\Phi(t_0, t_0) = I$ at time t_0. By linearity, the solution to the IVP

$$\dot{X} = A(t)X, \qquad X(s) = \Phi(s, t_0),$$

is $\Phi(t, s)\Phi(s, t_0)$, and by uniqueness of the solution to the IVP (2.52) (using proposition 2.3) this is equal to $\Phi(t, t_0)$.[36] Denote by Φ_i the ith column vector of Φ, so that $\Phi(t, t_0) = [\Phi_1(t, t_0), \ldots, \Phi_n(t, t_0)]$; it is

$$\frac{\partial}{\partial t} \det \Phi(t, t_0) = \sum_{i=1}^{n} \det \left[\Phi_1(t, t_0), \ldots, \Phi_{i-1}(t, t_0), \frac{\partial \Phi_i(t, t_0)}{\partial t}, \right.$$

$$\left. \Phi_{i+1}(t, t_0), \ldots, \Phi_n(t, t_0) \right].$$

Furthermore, using the initial condition $\Phi_i(t_0, t_0) = e_i$ (with e_i the ith Euclidean unit vector) together with property 1 gives

$$\left. \frac{\partial}{\partial t} \right|_{t=t_0} \det \Phi(t, t_0) = \sum_{i=1}^{n} \det [e_1, \ldots, e_{i-1}, A(t_0)e_i, e_{i+1}, \ldots, e_n] = \operatorname{trace} A(t_0).$$

---

36. This is also consistent with the group laws for flows in remark 2.4.

Ordinary Differential Equations

By properties 4 and 5, $\Phi(t_0, t) = \Phi(t_0, s)\Phi(s, t) = \Phi^{-1}(t, s)$, so

$$\text{trace}\, A(t) = \frac{\partial}{\partial t_0}\bigg|_{t_0=t} \det \Phi(t_0, t) = \frac{\frac{\partial}{\partial t_0}\big|_{t_0=t} \det \Phi(t_0, s)}{\det \Phi(t, s)} = \frac{\frac{\partial}{\partial t} \det \Phi(t, s)}{\det \Phi(t, s)}.$$

Thus, setting $s = t_0$ and $\varphi(t) = \det \Phi(t, t_0)$, and using property 3, one obtains that φ solves the linear IVP

$$\dot{\varphi} = \text{trace}\, A(t)\varphi, \qquad \varphi(t_0) = \det \Phi(t_0, t_0) = 1,$$

which, using the Cauchy formula in proposition 2.1, has the unique solution

$$\varphi(t) = \exp\left[\int_{t_0}^{t} \text{trace}\, A(s)\, ds\right],$$

This in turn implies that the Wronskian associated with the fundamental matrix $\Phi(t, t_0)$ is always positive. ■

Remark 2.11 (Peano-Baker Formula) In general, for an arbitrary $A(t)$, it is not possible to directly compute the fundamental matrix $\Phi(t, t_0)$ in closed form. An approximation can be obtained by recursively solving the fixed-point problem

$$X(t) = I + \int_{t_0}^{t} A(s)X(s)\, ds,$$

which is equivalent to the matrix IVP (2.52). This results in the *Peano-Baker formula* for the fundamental matrix,

$$\Phi(t, t_0) = I + \sum_{k=0}^{\infty} \int_{t_0}^{t}\left(\int_{t_0}^{s_1} \cdots \left(\int_{t_0}^{s_{k-1}} \left[\prod_{j=1}^{k} A(s_j)\right] ds_k\right) \cdots ds_2\right) ds_1,$$

for all $t, t_0 \in \mathbb{R}$. Truncating the series on the right-hand side yields an estimate for the fundamental matrix. □

Given the fundamental matrix $\Phi(t, t_0)$ as solution to the matrix IVP (2.52), the homogeneous solution to the linear IVP (2.51) is

$$x_h(t) = \Phi(t, t_0)x_0, \tag{2.53}$$

for all $t \in \mathbb{R}$.

Inhomogeneous Linear System Consider the inhomogeneous linear IVP

$$\dot{x} = A(t)x + b(t), \qquad x(t_0) = x_0. \tag{2.54}$$

As in the case where $n = 1$ (see section 2.2.2), the homogeneous solution in (2.53) is useful to determine the solution to (2.54) by appropriate superposition.

Proposition 2.15 (Generalized Cauchy Formula) The unique solution to the inhomogeneous linear IVP (2.54) is

$$x(t) = \Phi(t, t_0)x_0 + \int_{t_0}^{t} \Phi(t, s)b(s)\, ds,$$

for all $t \in \mathbb{R}$, provided that the fundamental matrix $\Phi(t, t_0)$ solves the matrix IVP (2.52) on $\mathbb{R}$.

Proof This proof uses the same variation-of-constants idea as the proof of the one-dimensional Cauchy formula in section 2.2.2. Let $C = C(t)$, and consider $x_h(t; C) = \Phi(t, t_0)C(t)$ as a candidate for a particular solution $x_p(t)$ of the ODE $\dot{x} = Ax + b$. Substituting $x_p(t) = x_h(t; C)$ into this ODE yields, by virtue of lemma 2.5(4), that

$$\dot{C}(t) = \Phi^{-1}(t, t_0)b(t) = \Phi(t_0, t)b(t),$$

and using the initial condition $C(t_0) = 0$, it is

$$C(t) = \int_{t_0}^{t} \Phi(t_0, s)b(s)\, ds.$$

Thus, taking into account lemma 2.5(5), the particular solution becomes

$$x_p(t) = x_h(t; C(t)) = \int_{t_0}^{t} \Phi(t, t_0)\Phi(t_0, s)b(s)\, ds = \int_{t_0}^{t} \Phi(t, s)b(s)\, ds,$$

whence the generalized Cauchy formula is obtained by superposition,

$$x(t) = x_h(t; x_0) + x_p(t) = \Phi(t, t_0)x_0 + \int_{t_0}^{t} \Phi(t, s)b(s)\, ds,$$

as the (by proposition 2.3, unique) solution to the inhomogeneous linear IVP (2.54). ∎

Linear Systems with Constant Coefficients When the ODE (2.50) has constant coefficients, the fundamental matrix can be determined explicitly by direct integration,

$$\Phi(t,t_0) = e^{At} = \sum_{k=0}^{\infty} A^k \frac{t^k}{k!}, \tag{2.55}$$

where e^{At} is a so-called *matrix exponential*, defined by a power series.[37] Using the generalized Cauchy formula in proposition 2.15, the unique solution to the inhomogeneous linear IVP with constant coefficients,

$$\dot{x} = Ax + b(t), \qquad x(t_0) = x_0, \tag{2.56}$$

becomes

$$x(t) = e^{At} x_0 + \int_{t_0}^{t} e^{A(t-\theta)} b(\theta) \, d\theta, \tag{2.57}$$

for all $t \in \mathbb{R}$.

Remark 2.12 (Nilpotent Linear Systems) A linear system (with constant coefficients) is *nilpotent* if there exist a real number $r \in \mathbb{R}$ and a positive integer κ such that $(A - rI)^\kappa = 0$.[38] It is clear that for nilpotent systems the series expansion for the matrix exponential in (2.55) has only a finite number of terms, since

$$e^{At} = e^{rt} e^{(A-rI)t} = e^{rt}\left(I + (A-rI)t + (A-rI)^2\frac{t^2}{2} + \cdots + (A-rI)^{\kappa-1}\frac{t^{\kappa-1}}{(\kappa-1)!}\right).$$

Note that a matrix is nilpotent if and only if all its eigenvalues are zero. □

Remark 2.13 (Laplace Transform) Instead of solving a system of linear ODEs directly, it is also possible to consider the image of this problem under the *Laplace transform*. Let $x(t)$, $t \geq t_0$, be a real-valued function such that $|x(t)| \leq Ke^{\alpha t}$ for some positive constants K and α. The Laplace transform of $x(t)$ is given by

$$X(s) = \mathcal{L}[x](s) = \int_{t_0}^{\infty} e^{-st} x(t) \, dt, \quad s \in \mathbb{C}, \ \text{Re}(s) > \alpha.$$

The main appeal of the Laplace transform for linear ODEs stems from the fact that the Laplace transform is a linear operator and that $\mathcal{L}[\dot{x}] = sX(s) - x(t_0)$. Thus, an IVP in the original time domain becomes an algebraic equation after applying the Laplace transform. Solving the latter

---

37. Moler and Van Loan (2003) provide a detailed survey of various numerical methods to compute e^{At}.
38. That is, the matrix $A - rI$ is nilpotent.

Table 2.1
Properties of the Laplace Transform ($\alpha, \beta, \tau \in \mathbb{R}$ with $\tau > 0$; $t_0 = 0$)

	Property	t-Domain	s-Domain
1	Linearity	$\alpha x(t) + \beta y(t)$	$\alpha \mathbf{X}(s) + \beta \mathbf{Y}(s)$
2	Similarity	$x(\alpha t)$, $\alpha \neq 0$	$(1/\alpha)\mathbf{X}(s/\alpha)$
3	Right-translation	$x(t - \tau)$	$e^{-\tau s}\mathbf{X}(s)$
4	Left-translation	$x(t + \tau)$	$e^{\tau s}\left(\mathbf{X}(s) - \int_0^\tau e^{-st}f(t)\,dt\right)$
5	Frequency shift	$e^{rt}x(t)$	$\mathbf{X}(s+r)$, $r \in \mathbb{C}$
6	Differentiation	$\dot{x}(t)$	$s\mathbf{X}(s) - x(0^+)$
7	Integration	$\int_0^t f(\theta)\,d\theta$	$(1/s)\mathbf{X}(s)$
8	Convolution	$(x*y)(t) = \int_0^t x(t-\theta)y(\theta)\,d\theta$	$\mathbf{X}(s)\mathbf{Y}(s)$

Table 2.2
Common Laplace Transforms ($\alpha, \tau \in \mathbb{R}$ with $\tau > 0$; $n \in \mathbb{N}$; $t_0 = 0$)

	$x(t)$	$\mathbf{X}(s)$	Domain
1	$\delta(t - \alpha)$	$\exp(-\alpha s)$	$s \in \mathbb{C}$
2	$(t^n/n!)\exp(-\alpha t)\mathbf{1}_{\mathbb{R}_+}$	$1/(s+\alpha)^{n+1}$	$\operatorname{Re}(s) > -\alpha$
3	$t^\alpha \mathbf{1}_{\mathbb{R}_+}$	$s^{-(1+\alpha)}\Gamma(1+\alpha)$	$\operatorname{Re}(s) > 0$
4	$\ln(t/\tau)\mathbf{1}_{\mathbb{R}_+}$	$-(\tau/s)\left(\ln(\tau s) + \gamma\right)$	$\operatorname{Re}(s) > 0$

equation for $\mathbf{X}(s)$, it is possible to obtain the original solution using the *inverse Laplace transform*,

$$x(t) = \mathcal{L}^{-1}[\mathbf{X}](t) = \frac{1}{2\pi i}\int_{c-i\infty}^{c+i\infty} e^{st}\mathbf{X}(s)\,ds,$$

where c is such that $\mathbf{X}(s)$ is analytic for $\operatorname{Re}(s) \geq 0$. The most important properties of the Laplace transform are listed in table 2.1 and common transform pairs in table 2.2.[39] □

Example 2.15 (Linear System with Constant Coefficients) Consider the linear IVP with constant coefficients (2.56) for $t_0 = 0$, which, using the Laplace transform (with properties 1 and 6 in table 2.1), can be written in the s-domain as

$$s\mathbf{X}(s) - x_0 = A\mathbf{X}(s) + \mathbf{B}(s),$$

---

39. More detailed tables are readily available (see, e.g., Bronshtein et al. 2004, 708–721, 1061–1065). The Laplace transform and its inverse can also be applied when $x(t)$ is vector-valued, in which case $\mathbf{X}(s)$ is also vector-valued, as in example 2.15.

where $\mathbf{B}(s) = \mathcal{L}[b](s)$. Solving this algebraic equation for the unknown $\mathbf{X}(s)$ yields

$$\mathbf{X}(s) = -(A - sI)^{-1}\left(\mathbf{B}(s) + x_0\right).$$

By the linearity property and entry 1 in table 2.2, $\mathcal{L}^{-1}[\mathbf{B} + x_0](t) = b(t) + x_0\delta(t)$, where $\delta(t)$ is a Dirac distribution.[40] In addition, the matrix analogue of entry 2 in table 2.2 yields that $\mathcal{L}^{-1}[-(A - sI)^{-1}](t) = \exp(At)$. Hence, properties 1 and 8 in table 2.1, in conjunction with the filter property of the Dirac distribution, immediately imply (2.57). □

2.4 Notes

The theory of ODEs is well developed (see, e.g., Coddington and Levinson 1955; Pontryagin 1962; Hartman 1964; Petrovski[i] 1966). The presentation in this chapter is motivated in part by Arnold (1973), Walter (1998), and Khalil (1992). Arnold (1988) discusses more advanced geometrical methods. Godunov (1994) gives a very detailed account of linear systems with constant coefficients, and Sachdev (1997) provides a compendium of solution methods for nonlinear ODEs.

There are three main situations when the standard theory of ODEs needs to be extended. The first is when the function f on the right-hand side of (2.3) is non-Lipschitz, and thus the problem may not be well-posed in the sense that small changes in initial conditions may lead to arbitrarily large changes in solutions and also to nonuniqueness of solutions. Wiggins (2003) gives an introduction to nonlinear dynamics and chaos.

The second situation occurs when considering the dependence of systems on parameters, which raises questions about the structural stability of solutions. Thom (1972) provided the first coherent account on the matter, which essentially founded the field of catastrophe theory. Arnold (1992) is an excellent nontechnical introduction, while Arnold et al. (1999) is a more rigorous account.[41]

The third situation which requires additional theory is when the right-hand side of (2.34) is discontinuous (Filippov 1988) or set-valued (Smirnov 2002). This third situation is of some importance for the control

---

40. The *Dirac distribution* is defined as the limit $\delta(t) = \lim_{\varepsilon \to 0^+} (1/\varepsilon) \max\{0, 1 - |x/\varepsilon|\}$. Its characteristic properties are (1) (normalization) $\int_{-\varepsilon}^{\varepsilon} \delta(t)\,dt = 1$ for any $\varepsilon > 0$; and (2) (filter property) $\int_{-\varepsilon}^{\varepsilon} \delta(t)y(t)\,dt = y(0)$ for any continuous function y and any $\varepsilon > 0$.
41. Arnold (2000) compares rigid versus flexible models in mathematics with the aid of several interesting examples.

of dynamic systems. Discontinuous controls arise naturally as solutions to standard optimal control problems (see chapter 3). More generally, the relevant dynamic system, when allowing for a choice of controls from a certain set, is described by a differential inclusion (featuring a set-valued right-hand side).

Many of the examples in this chapter have their roots in theoretical biology; see Murray (2007) for further discussion in that context.

2.5 Exercises

2.1 (Growth Models) Given the domain $\mathcal{D} = \mathbb{R}^2_{++}$ and initial data $(t_0, x_0) \in \mathcal{D}$ such that $(t_0, x_0) \gg 0$,[42] solve the initial value problem

$$\dot{x} = f(t, x), \qquad x(t_0) = x_0,$$

where $f(t, x) : \mathcal{D} \to \mathbb{R}$ is one of the continuous functions given in parts a–c. For this, assume that $\bar{x} > x_0$ is a finite carrying capacity and that $\alpha, \beta, \gamma > 0$ are given parameters.

a. (Generalized Logistic Growth) $f(t, x) = \alpha \gamma (1 - \left(\frac{x}{\bar{x}}\right)^{1/\gamma}) x$.

b. (Gompertz Growth); $f(t, x) = \alpha x \ln\left(\frac{\bar{x}}{x}\right)$. Show that Gompertz growth is obtained from generalized logistic growth in part a as $\gamma \to \infty$.

c. (Bass Diffusion) $f(t, x) = \left(1 - \frac{x}{\bar{x}}\right)(\alpha x + \beta)\rho(t)$, where $\rho(t) = 1 + \delta u(t)$. The continuous function $u(t)$ with values in $[0, \bar{u}]$ describes the relative growth of advertising expenditure for some $\bar{u} > 0$, and the constant $\delta \geq 0$ determines the sensitivity of the Bass product-diffusion process to advertising.[43]

d. (Estimation of Growth Models) Let $\delta = 0$. Find some product-diffusion data for your favorite innovation, and estimate the values of the parameters α, β, γ for the models in parts a–c. Plot the three different growth curves together with your data. (There is no need to develop econometrically rigorous estimation procedures; the point of this exercise is to become familiar with some software that can produce graphical output and to experiment with the different growth models.)

---

42. Given two vectors $a, b \in \mathbb{R}^n$, write that $a = (a_1, \ldots, a_n) \gg (b_1, \ldots, b_n) = b$ if and only if $a_i > b_i$ for all $i \in \{1, \ldots, n\}$.

43. This diffusion model was developed by Frank M. Bass in 1969 for $\delta = 0$ (see example 2.1). The original model fits diffusion data surprisingly well, as shown by Bass et al. (1994). It was included in the *Management Science* special issue on "Ten Most Influential Titles of Management Science's First Fifty Years."

Ordinary Differential Equations

2.2 (Population Dynamics with Competitive Exclusion) Consider the evolution of two interacting populations, $\xi_1(\tau)$ and $\xi_2(\tau)$. At time $\tau \geq 0$, the evolution of the populations, with given positive initial sizes of ξ_{10} and ξ_{20}, is described by the following initial value problem:

$$\dot{\xi}_1 = a_1\xi_1\left(1 - \frac{\xi_1}{\bar{\xi}_1} - b_{12}\frac{\xi_2}{\bar{\xi}_1}\right), \qquad \xi_1(0) = \xi_{10},$$

$$\dot{\xi}_2 = a_2\xi_2\left(1 - \frac{\xi_2}{\bar{\xi}_2} - b_{21}\frac{\xi_1}{\bar{\xi}_2}\right), \qquad \xi_2(0) = \xi_{20},$$

where the constants $a_i, \bar{\xi}_i > \xi_{i0}, i \in \{1, 2\}$ are positive and $b_{12}, b_{21} \in \mathbb{R}$.

a. Using a linear variable transformation, de-dimensionalize the model, such that the evolution of the two populations is described by functions $x_1(t)$ and $x_2(t)$ that satisfy

$$\dot{x}_1 = x_1(1 - x_1 - \beta_{12}x_2), \quad x_1(0) = x_{10}, \tag{2.58}$$

$$\dot{x}_2 = \alpha x_2(1 - x_2 - \beta_{21}x_1), \quad x_2(0) = x_{20}, \tag{2.59}$$

where $\alpha, \beta_{12}, \beta_{21}, x_{10}, x_{20}$ are appropriate positive constants. Interpret the meaning of these constants. The description (2.58)–(2.59) is used for parts b–c below.

b. Determine any steady states of the system (2.58)–(2.59). Show that there is at most one positive steady state (i.e., strictly inside the positive quadrant of the (x_1, x_2)-plane). Under what conditions on the parameters does it exist?

c. Determine the stability properties of each steady state determined in part b.

d. Under the assumption that there is a positive steady state, draw a qualitative diagram of the state trajectories in phase space (given any admissible initial conditions).

e. How does the phase diagram in part d support the conclusion (from evolutionary biology) that *when two populations compete for the same limited resources, one population usually becomes extinct*? Discuss your findings using a practical example in economics.

2.3 (Predator-Prey Dynamics with Limit Cycle) Consider two interacting populations of prey $x_1(t)$ and predators $x_2(t)$, which for $t \geq 0$ evolve according to

$$\dot{x}_1 = x_1(1-x_1) - \frac{\delta x_1 x_2}{x_1 + \beta}, \qquad x_1(0) = x_{10}, \qquad (2.60)$$

$$\dot{x}_2 = \alpha x_2 \left(1 - \frac{x_2}{x_1}\right), \qquad x_2(0) = x_{20}, \qquad (2.61)$$

where α, β, δ are positive constants, and $x_0 = (x_{10}, x_{20}) \gg 0$ is a given initial state.

a. Find all steady states of the system (2.60)–(2.61). Show that there is a unique positive steady state x^*.

b. What is the smallest value δ for which the system can exhibit unstable behavior around the positive steady state?

c. Show that if the positive steady state is unstable, there exists a (stable?) limit cycle. For that case, draw a qualitative diagram of solution trajectories in the phase space.

d. Discuss qualitative differences of the diagram obtained in part c from the Lotka-Volterra predator-prey model. Which model do you expect to more robustly fit data? Explain.

2.4 (Predator-Prey Dynamics and Allee Effect) Consider the co-evolution of two related technologies. Let $x_i(t)$ denote the installed base (the number of users) for technology $i \in \{1, 2\}$. The dynamic interaction of the two user populations for times $t \geq 0$ is described by the system of ODEs

$$\dot{x}_1 = x_1 \left(g(x_1) - x_2\right),$$

$$\dot{x}_2 = x_2 \left(x_1 - h(x_2)\right),$$

where the functions $g, h : \mathbb{R}_+ \to \mathbb{R}_+$ have the following properties:

- $g(\cdot)$ has a single peak at $\hat{x}_1$, is concave on $[0, \hat{x}_1]$, $g(\hat{x}_1) > g(0) > 0$, and $\lim_{x_1 \to \infty} g(x_1) = 0$.
- $h(\cdot)$ is concave, increasing, and $h(0) > 0$.

The initial state of the system, $x(0) = x_0 > 0$, is known.

Part 1: System Description

a. Sketch the functions g and h, and briefly discuss how the behavior of this system qualitatively differs from a classical Lotka-Volterra predator-prey system.

b. Describe why this system exhibits the *Allee effect*.[44]

Part 2: Stability Analysis

c. Is the origin a stable or an unstable equilibrium? Prove your statement.

d. Show that there exists a unique positive equilibrium $\bar{x} = (\bar{x}_1, \bar{x}_2)$, and examine its stability properties. Determine conditions on g and h that would guarantee stability or instability. (*Hint:* Write the system in the standard form $\dot{x} = f(x)$, and draw the nullclines in the state space; also distinguish the cases where $\bar{x}_1 < \hat{x}_1$ and where $\bar{x}_1 > \hat{x}_1$.)

Part 3: Equilibrium Perturbations and Threshold Behavior

e. Consider the case where $\bar{x} = (\bar{x}_1, \bar{x}_2) \gg 0$ is an equilibrium such that $\bar{x}_1 > \hat{x}_1$. Assume that the system is perturbed away from $\bar{x}$ so that at time $t = 0$ it starts at the state $x_0 > \bar{x}$. Provide an intuitive proof for the fact that the system behavior becomes qualitatively different as $\|x_0 - \bar{x}\|$ increases and passes a threshold. How does this threshold relate to the Allee effect?

f. Consider the case where $\bar{x} = (\bar{x}_1, \bar{x}_2) \gg 0$ is an equilibrium such that $\bar{x}_1 < \hat{x}_1$. Based on your answer in part d, discuss what happens when the system is perturbed away from the steady state $\bar{x}$. (*Hint:* The Poincaré-Bendixson theorem might be of help.)

Part 4: Interpretation and Verification

g. Interpret the insights obtained from the qualitative analysis in parts 1–3 in the context of technological innovation. Try to find a particular real-world example to illustrate your argument.

h. Provide suitable specifications for the functions g and h to illustrate the system behavior and your earlier conclusions using computer-generated phase diagrams (including trajectories). Describe any interesting observations that the preceding qualitative analysis may have missed.

---

[44]. The effect is named after Allee (1931; 1938); see also the discussion by Stephens et al. (1999). It refers to the observation that (at least below a certain population threshold) the growth rate of a population increases with its size.

3 Optimal Control Theory

Imagination is good.
But it must always be critically
controlled by the available facts.
—Albert Einstein

3.1 Overview

Chapter 2 discussed the evolution of a system from a known initial state $x(t_0) = x_0$ as a function of time $t \geq t_0$, described by the differential equation $\dot{x}(t) = f(t, x(t))$. For well-posed systems the initial data (t_0, x_0) uniquely determine the state trajectory $x(t)$ for all $t \geq t_0$. In actual economic systems, such as the product-diffusion process in example 2.1, the state trajectory may be influenced by a decision maker's actions, in which case the decision maker exerts control over the dynamic process. For example, product diffusion might depend on the price a firm charges and on its marketing effort. The lower the price, or the higher the marketing effort, the faster one expects the product's consumer base to increase.

When all the decision maker's choice variables are collected in a vector-valued control $u(t)$, the evolution of the state is now described by an augmented ordinary differential equation (ODE),

$$\dot{x}(t) = f(t, x(t), u(t)),$$

which includes the impact of the control. Through the choice of the control variable, the decision maker may be able to steer the system from the given initial state x_0 at time t_0 to a desired final state x_T at time T. For example, by charging a very low price and investing in a marketing campaign, a firm might double its consumer base within 1 year.

If a given system can be steered from any x_0 to any x_T in finite time using an admissible control $u(t)$, $t \in [t_0, T]$, it is said to be controllable.

Not every system is controllable, especially if the set of feasible controls is constrained. For example, when a firm's marketing budget is small and a product's elevated marginal cost imposes a high minimum price, it may well be impossible to double the current consume base within any amount of time, especially if consumers are quick to discard the product. Exercise 3.1 discusses this example in greater detail. Section 3.2 provides a characterization of controllability for linear systems, both with bounded and unbounded sets of feasible controls. Controllability properties of general nonlinear systems are more difficult to establish and are therefore examined here using ad hoc techniques.

Beyond the question of the mere possibility of steering a system from one given state to another, a decision maker is often interested in finding an input trajectory $u(t)$, $t \in [t_0, T]$, so as to maximize the objective functional

$$J(u) = \int_{t_0}^{T} h(t, x(t), u(t))\, dt + S(T, x(T)),$$

where $h(t, x(t), u(t))$ is an instantaneous payoff and $S(T, x(T))$ is a terminal value, for instance, of selling the firm at time T (see remark 3.14). Section 3.3 discusses this simplest version of an optimal control problem and presents necessary as well as sufficient conditions for its solution. The maximum principle developed by Lev Pontryagin and his students provides necessary conditions. The main idea of these conditions is that they define the joint evolution of the state $x(t)$ and a co-state $\psi(t)$ (also known as adjoint variable), where the co-state represents the per-unit value (or shadow price) of the current velocity of the system at time t. Sufficient conditions, on the other hand, can be formulated in terms of the so-called Hamilton-Jacobi-Bellman equation, which describes the decrease of the value function $V(t, x)$ with respect to time t as the maximum attainable sum of instantaneous payoff and the value of the current velocity of the system. The value function $V(t, x)$ represents the optimal payoff of the system over the interval $[t, T]$ when started at the time-t state $x(t) = x$. Because this payoff decreases to zero, or more precisely, to the system's terminal value, it makes sense that at the optimum the decrease of this value be as fast as possible. If a (necessarily unique) value function is found that solves the Hamilton-Jacobi-Bellman equation, then it is also possible to derive from it the solution to the optimal control problem.

Section 3.4 considers the general finite-horizon optimal control problem with endpoint constraints and state-control constraints, which can be effectively dealt with using a more general version of the Pontryagin maximum principle (PMP), the proof of which is discussed in section 3.6. Section 3.5 deals with optimal control problems where the decision horizon T goes to infinity. While one might argue that in practice it seems hard to imagine that the decision horizon would actually be infinity, setting $T = \infty$ removes end-of-horizon effects and thus expresses a going concern by the decision maker. For example, when considering the optimal control problem of fishing in a lake over a finite time horizon, then at the end of that horizon the lake would usually contain no fish (provided it is cheap enough to catch the fish) because there is no future value of conservation. An infinite-horizon version of this problem would result in a solution that would most likely approach a steady state (also referred to as a turnpike) where the fish reproduction and the catch balance out so as to guarantee an optimal stream of catch. Determining such a steady state is usually much easier than solving the entire optimal control problem. Section 3.7 discusses the existence of solutions to optimal control problems, notably the Filippov existence theorem.

3.2 Control Systems

A *control system* Σ (figure 3.1) may be viewed as a relation between $m + l$ signals, whereby each *signal* is a real-valued function, defined on a common nonempty time set $\mathcal{T} \subset \mathbb{R}$.[1] The first m signals, denoted by $u_1, \ldots, u_m$, are referred to as *inputs* (or *controls*), and the last l signals, denoted by $y_1, \ldots, y_l$, are called *outputs*. Thus, a system $\Sigma = \{(u_1, \ldots, u_m; y_1, \ldots, y_l)\}$ specifies what $(m + l)$-tuples of signals are admissible. Note that in principle a system need have neither inputs nor outputs.

Based on the theory developed in chapter 2, the focus here is on continuous-time control systems, where the time set $\mathcal{T}$ is a nontrivial interval $\mathcal{I} \subset \mathbb{R}$, and where the relation between the m-dimensional control $u = (u_1, \ldots, u_m)$ (with $m \geq 1$) and the l-dimensional *output* $y = (y_1, \ldots, y_l)$ (with $l \geq 1$) is given in *state-space form*,

$$\dot{x}(t) = f(t, x(t), u(t)), \qquad (3.1)$$

$$y(t) = g(t, x(t), u(t)), \qquad (3.2)$$

for all $t \in \mathcal{I}$. The continuous function $f : \mathcal{I} \times \mathcal{X} \times \mathcal{U} \to \mathbb{R}^n$ is referred to as the *system function*, and the continuous function $g : \mathcal{I} \times \mathcal{X} \times \mathcal{U} \to \mathbb{R}^l$ is

---

1. This definition includes *discrete-time* systems, where $\mathcal{T} \subset \mathbb{Z}$. Such discrete-time systems can be obtained by sampling a continuous-time system at countably many time instants. The word *relation* is to be interpreted as follows: given $m + l \geq 0$ signals, it can be decided (based on the relation) whether they are consistent with the system or not.

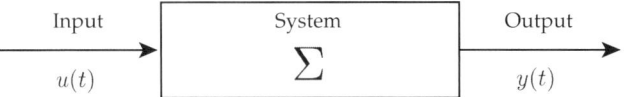

Figure 3.1
A system with input and output.

called the *output function*. The nonempty convex open set $\mathcal{X} \subset \mathbb{R}^n$ is the *state space*, and the nonempty convex compact set $\mathcal{U} \subset \mathbb{R}^m$ is the *control set* (or *control-constraint set*). The special case where $\mathcal{U}$ is a singleton and $g(t, x(t), u(t)) \equiv x(t)$ represents a control system without inputs and with state output (see chapter 2).

The system equation (3.1) determines the evolution of the state $x(t)$ at time t given a control $u(t)$. The output equation (3.2) determines the output $y(t)$ at time t as a function of the state $x(t)$ and the control $u(t)$. Fundamental for the analysis of systems are the notions of *controllability*, *reachability*, and *observability*. A system in state-space form is *controllable* if it can be steered from any state $x \in \mathcal{X}$ to any other state $\hat{x} \in \mathcal{X}$ in finite time. A system is said to be *reachable* if all its states can be reached from a particular state (say, the origin) in finite time, that is, if it is controllable from that state. A controllable system is reachable from *any* state. A system is *observable* if, when its output is recorded over some finite time interval, its initial state can be determined.

Remark 3.1   In this section the controllability and reachability properties are particularly needed because the focus is on systems that can be steered from any given state to a particular optimal state. The observability property plays a lesser role here as it is trivially satisfied in economic systems where the states are readily available and can therefore be considered as output.   □

Example 3.1 (Linear Control System)   A linear time-invariant[2] control system is of the form

$$\dot{x} = Ax + Bu, \tag{3.3}$$

$$y = Cx + Du, \tag{3.4}$$

---

2. To avoid any possible confusion, in the context of control systems the term *time-invariant* is preferred to *autonomous* (which was used in chapter 2 for time-invariant ODEs). The reason is that the term *autonomous system* is sometimes used in the literature for a control system without inputs.

Optimal Control Theory

where $A \in \mathbb{R}^{n \times n}$, $B \in \mathbb{R}^{n \times m}$, $C \in \mathbb{R}^{l \times n}$, and $D \in \mathbb{R}^{l \times m}$ are given matrices. The question naturally arises, Under what conditions is this linear control system controllable, reachable, and observable? Note that the notions of controllability and reachability coincide for linear systems because the set of states that can be reached in finite time, say, from the origin, is a linear subspace of $\mathbb{R}^n$. □

3.2.1 Linear Controllability and Observability

The following result characterizes the controllability and observability of linear systems in terms of a simple algebraic condition.

Proposition 3.1 (Linear Controllability and Observability) Consider a linear control system Σ described by (3.3)–(3.4). (1) Σ is controllable if and only if the *controllability matrix* $R[A, B] = [B, AB, A^2B, \ldots, A^{n-1}B]$ has rank n. (2) Σ is observable if and only if the *observability matrix*

$$O[A, C] = \begin{bmatrix} C \\ CA \\ \vdots \\ CA^{n-1} \end{bmatrix}$$

has rank n.

Proof (1) First observe that for any row vector $v \in \mathbb{R}^n$ it is

$$v e^{At} B = \sum_{k=0}^{\infty} v A^k B \frac{t^k}{k!}. \tag{3.5}$$

Furthermore, by the Cayley-Hamilton theorem[3] it is $A^n = \sum_{k=0}^{n-1} \alpha_k A^k$, where $A^0 = I$, and where $\alpha_0, \ldots, \alpha_{n-1}$ are the coefficients of the characteristic polynomial of A, $\det(sI - A) = s^n - \sum_{k=0}^{n-1} \alpha_k s^k$. Thus, any power A^k for $k \geq n$ can be written as a linear combination of $A^0, \ldots, A^{n-1}$. One can therefore conclude that

$$\text{rank } R[A, B] < n \Leftrightarrow \exists v \in \mathbb{R}^n : v A^k B = 0, \quad \forall k \geq 0. \tag{3.6}$$

Let $\mathcal{R}_T(x) \subset \mathbb{R}^n$ be the set of all states that can be reached by the system Σ in time $T \geq 0$ starting from x, that is, for any $\hat{x} \in \mathcal{R}_T(x)$ there exists

---

3. The *Cayley-Hamilton theorem* states that every square matrix $A \in \mathbb{R}^{n \times n}$ satisfies its own characteristic equation, i.e., if $p(\lambda) = \det(A - \lambda I)$ is its characteristic polynomial as a function of λ, then $p(A) = 0$ (Hungerford 1974, 367).

a bounded measurable control $u : [0,T] \to \mathbb{R}^m$ such that $\phi_u(T,x) = \hat{x}$, where ϕ_u denotes the flow of the system under the control u. Then

$$\Sigma \text{ is controllable} \Leftrightarrow \mathcal{R}_1(0) = \mathbb{R}^n. \tag{3.7}$$

The last statement is true, since for any $T > 0$ the set $\mathcal{R}_T(0)$ is a linear subspace of the set $\mathcal{R}(0)$ of all states that are reachable from the origin in *any* time, and the latter is equivalent to $\mathcal{R}_T(0) = \mathcal{R}(0)$ for all $T > 0$. Now the task is to establish the equivalence of the rank condition and controllability of Σ.

$\Rightarrow$: If Σ is not controllable, then by (3.7) there is a nonzero vector $v \in \mathbb{R}^n$ such that $v \perp \mathcal{R}(0)$, that is, $\langle v, x \rangle = 0$ for all $x \in \mathcal{R}(0)$. Consider the control $u(t) = B'e^{A'(1-t)}v'$ for all $t \in [0,1]$. Then

$$0 = \langle v, x \rangle = \int_0^1 \langle v, e^{A(1-t)} Bu(t) \rangle dt = \int_0^1 \langle v, e^{As} Bu(1-s) \rangle ds$$

$$= \int_0^1 \langle v, e^{As} BB' e^{A's} v' \rangle ds = \int_0^1 \|v e^{As} B\|_2^2 ds,$$

so $ve^{At}B = 0$ for all $t \in [0,1]$, which by virtue of (3.6) implies that rank $R[A,B] < n$, that is, the rank condition fails.

$\Leftarrow$: If rank $R[A,B] < n$, then by (3.6) there exists a nonzero $v \in \mathbb{R}^n$ such that $vA^kB = 0$ for all $k \geq 0$. Then, given any admissible control $u(t)$ for $t \in [0,1]$, it is by (3.5)

$$\langle v, \phi_u(1,0) \rangle = \left\langle v, \int_0^1 e^{At} Bu(1-t)\, dt \right\rangle = \int_0^1 \langle v, e^{At} Bu(1-t) \rangle dt = 0,$$

so $v \perp \mathcal{R}(0) \neq \mathbb{R}^n$, that is, by (3.7) the system is not controllable.

(2) Note first that $(O[A,C])' = R[A',C']$, namely, by part (1) the observability of $\Sigma = (A,B)$ can be understood in terms of the controllability of the dual system $\Sigma' = (A',C')$. Let $T > 0$. Given an $(n-1)$-times continuously differentiable control $u : [0,T] \to \mathbb{R}^m$, one can differentiate the output relation (3.4) successively to obtain that

$$\begin{bmatrix} y \\ \dot{y} \\ \ddot{y} \\ \vdots \\ y^{(n-1)} \end{bmatrix} = O[A,C]x + \begin{bmatrix} D & 0 & & & 0 \\ CB & D & 0 & \cdots & 0 \\ CAB & CB & D & 0 & \cdots & 0 \\ \vdots & \vdots & \vdots & \vdots & & \vdots \\ CA^{n-2}B & CA^{n-3}B & \cdots & CAB & CB & D \end{bmatrix} \begin{bmatrix} u \\ \dot{u} \\ \ddot{u} \\ \vdots \\ u^{(n-1)} \end{bmatrix}$$

on the time interval $[0, T]$. The preceding relation can be used to infer the evolution of the state $x(t)$ on $[0, T]$ if and only if $O[A, C]$ has rank n (by using any left-inverse of $O[A, C]$). ∎

Example 3.2 (1) The linear control system (harmonic oscillator)

$$\dot{x}_1 = x_2 + u,$$

$$\dot{x}_2 = -x_1,$$

with $A = \begin{bmatrix} 0 & 1 \\ -1 & 0 \end{bmatrix}$ and $B = \begin{bmatrix} 1 \\ 0 \end{bmatrix}$ is controllable because $R[A, B] = \begin{bmatrix} 1 & 0 \\ 0 & -1 \end{bmatrix}$ has full rank. (2) The linear control system

$$\dot{x}_1 = x_1 + u,$$

$$\dot{x}_2 = x_2,$$

with $A = I$ and $B = \begin{bmatrix} 1 \\ 0 \end{bmatrix}$ is not controllable because

$$\text{rank } R[A, B] = \text{rank } \begin{bmatrix} 1 & 1 \\ 0 & 0 \end{bmatrix} = 1 < n = 2.$$

The system in fact can be controlled only in the direction of x_1, that is, the set of all states that are reachable from the origin is $\mathcal{R}(0) = \mathbb{R} \times \{0\}$. □

In general, the control $u(t)$ can take on values only in the bounded control set $\mathcal{U} \subset \mathbb{R}^m$, which may severely limit the set of states that can be reached from the origin. The set of all $\mathcal{U}$-*controllable states* is denoted by $\mathcal{R}^{\mathcal{U}}(0) \subset \mathbb{R}^n$.

Proposition 3.2 (Bounded Controllability) Let the control set $\mathcal{U}$ be bounded and such that it contains a neighborhood of the origin. Then the linear control system Σ described by (3.3)–(3.4) is controllable if and only if the controllability matrix $R[A, B]$ has rank n and the system matrix A has no eigenvalues with negative real part.

Proof See Sontag (1998, 117–122). ∎

The intuition for this result is that since $\mathcal{R}^{\mathcal{U}}(0) \subset \mathcal{R}(0)$, the controllability matrix must have full rank, as in proposition 3.1. Furthermore, if A has a negative eigenvalue, then there will be a system trajectory that would tend toward the origin if the applied control is too small, limiting the reachability of states far enough away from the origin. To be able to reach those states, the system matrix cannot have asymptotically stable components. On the other hand, if these two properties are satisfied, then any state in $\mathbb{R}^n$ can be reached using controls with values in the bounded set $\mathcal{U}$.

Remark 3.2 (Output Controllability) A system Σ is called *output controllable* if it is possible to steer it from any point $y \in \mathbb{R}^l$ in finite time to any other point $\hat{y} \in \mathbb{R}^l$. In the case of the linear time-invariant system (3.3)–(3.4) with $D = 0$ it is clear, based on proposition 3.1, that output controllability obtains if and only if the *output controllability matrix* $CR[A, B] = [CB, CAB, \ldots, CA^{n-1}B]$ has rank l. □

3.2.2 Nonlinear Controllability

The question of whether a general nonlinear system Σ described by (3.1)–(3.2) is controllable is difficult. Good answers are available only when the system has a special structure, for example, when it is affine in the control. The tools needed for the corresponding analysis (e.g., Lie algebra and differential geometry) are beyond the scope of this book. Both Isidori (1995) and Sontag (1998) provide good introductory presentations of this advanced topic, establishing local results, at least for time-invariant, control-affine systems.

Remark 3.3 (Controllability of Time-Varying Systems) To extend the notion of controllability to time-varying systems, it is useful to think of *events* (t, x) instead of points in the state space. An event (t, x) can be *controlled to* an event $(\hat{t}, \hat{x})$ (with $t < \hat{t}$) if (using a feasible control input) there is a system trajectory $x(s)$, $s \in [t, \hat{t}]$, such that $x(t) = x$ and $x(\hat{t}) = \hat{x}$. Hence, the state x can be *controlled to* the state $\hat{x}$ if there exist $t, T \geq 0$ such that the event (t, x) can be controlled to the event $(t + T, \hat{x})$. □

3.3 Optimal Control—A Motivating Example

3.3.1 A Simple Optimal Control Problem

To motivate the discussion of optimal control theory, first consider a simple finite-horizon dynamic optimization problem for a system with unconstrained states. Given an interval $\mathcal{I} = [t_0, T]$ with the finite time horizon $T > t_0 \geq 0$, and a system in state-space form with state output, that is, where $y = x$, a decision maker would like to find a bounded (measurable) control $u : [t_0, T] \to \mathcal{U}$, where $\mathcal{U}$ is a nonempty convex

Optimal Control Theory

compact control set, so as to maximize an *objective* (*functional*) of the form

$$J(u) = \int_{t_0}^{T} h(t, x(t), u(t)) \, dt, \tag{3.8}$$

where the continuous function $h : [t_0, T] \times \mathcal{X} \times \mathcal{U} \to \mathbb{R}$ describes a decision maker's instantaneous benefit at time t, given the state $x(t)$ and the control $u(t)$. The latter corresponds to an action taken by the decision maker at time t. The fact that there are no state constraints is taken into account by the assumption that $\mathcal{X} = \mathbb{R}^n$.[4]

Given an initial state $x_0 \in \mathcal{X}$, the decision maker's *optimal control problem* (OCP) can therefore be written in the form

$$J(u) \longrightarrow \max_{u(\cdot)}, \tag{3.9}$$

$$\dot{x}(t) = f(t, x(t), u(t)), \tag{3.10}$$

$$x(t_0) = x_0, \tag{3.11}$$

$$u(t) \in \mathcal{U}, \tag{3.12}$$

for all $t \in [t_0, T]$. In other words, the optimal control problem consists in maximizing the objective functional in (3.9) (defined by (3.8)), subject to the *state equation* (3.10), the initial condition (3.11), the *control constraint* (3.12), and possibly an *endpoint constraint*,

$$x(T) = x_T, \tag{3.13}$$

for a given $x_T \in \mathcal{X}$, which is relevant in many practical situations.

3.3.2 Sufficient Optimality Conditions

To derive sufficient optimality conditions for the optimal control problem (3.9)–(3.12), note first that given any continuously differentiable function $V : [t_0, T] \times \mathcal{X} \to \mathbb{R}$, which satisfies the boundary condition

$$V(T, x) = 0, \quad \forall x \in \mathcal{X}, \tag{3.14}$$

it is possible to replace $h(t, x, u)$ by

$$\hat{h}(t, x, u) = h(t, x, u) + \dot{V}(t, x) = h(t, x, u) + V_t(t, x) + \langle V_x(t, x), f(t, x, u) \rangle,$$

---

[4]. A weaker assumption is to require that the state space $\mathcal{X}$ be invariant given any admissible control input u, which includes the special case where $\mathcal{X} = \mathbb{R}^n$.

without changing the problem solution. This is true because the corresponding objective functional,

$$\hat{J}(u) = \int_{t_0}^{T} \hat{h}(t, x(t), u(t)) \, dt = \int_{t_0}^{T} h(t, x(t), u(t)) \, dt + V(T, x(T)) - V(t_0, x_0)$$
$$= J(u) - V(t_0, x_0), \tag{3.15}$$

is, up to the constant $V(t_0, x_0)$, identical to the original objective functional $J(u)$. Moreover, if the function V is such that, in addition to the boundary condition (3.14), it satisfies the so-called *Hamilton-Jacobi-Bellman* (HJB) *inequality*

$$0 \geq h(t, x, u) + \dot{V}(t, x) = h(t, x, u) + V_t(t, x) + \langle V_x(t, x), f(t, x, u) \rangle$$

for all $(t, x, u) \in [t_0, T] \times \mathcal{X} \times \mathcal{U}$, then by integration

$$0 \geq \hat{J}(u) = J(u) - V(t_0, x_0)$$

for any admissible control u.[5] Hence, the constant $V(t_0, x_0)$ is an upper bound for the objective functional $J(u)$, which, when attained for some admissible control trajectory $u^*(t), t \in [t_0, T]$, would imply the optimality of that trajectory. The optimal control u^* then renders the HJB inequality binding, maximizing its right-hand side. The following sufficient optimality condition has therefore been established.

Proposition 3.3 (HJB Equation) (1) Let $V : [t_0, T] \times \mathcal{X} \to \mathbb{R}$ be a continuously differentiable function such that it satisfies the HJB *equation*

$$-V_t(t, x) = \max_{u \in \mathcal{U}} \{h(t, x, u) + \langle V_x(t, x), f(t, x, u) \rangle\},$$
$$\forall (t, x) \in [t_0, T] \times \mathcal{X}, \tag{3.16}$$

together with the boundary condition (3.14). Take any measurable feedback law $\mu : [t_0, T] \times \mathcal{X}$ such that

$$\mu(t, x) \in \arg\max_{u \in \mathcal{U}} \{h(t, x, u) + \langle V_x(t, x), f(t, x, u) \rangle\},$$
$$\forall (t, x) \in [t_0, T] \times \mathcal{X}, \tag{3.17}$$

and let $x^*(t)$ be a solution to the corresponding initial value problem (IVP)

$$\dot{x}^*(t) = f(t, x^*(t), \mu(t, x^*(t))), \quad x(t_0) = x_0, \tag{3.18}$$

---

5. The term *admissible control* is defined in assumption A2 in section 3.4.

Optimal Control Theory

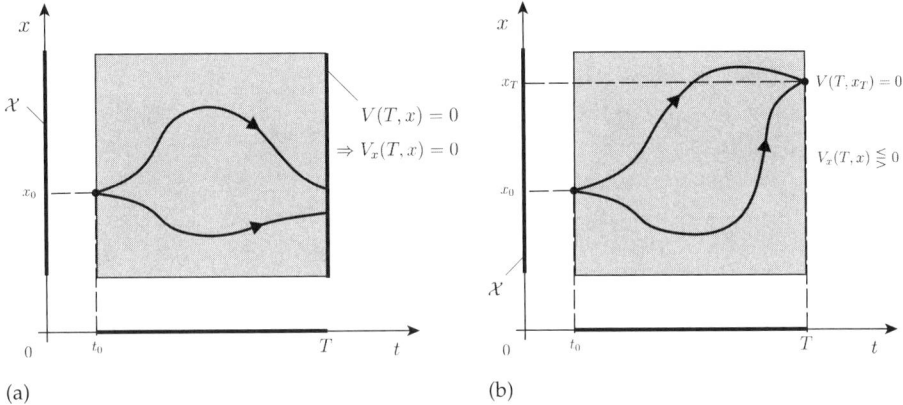

Figure 3.2
The boundary condition (3.14) for the HJB equation can be either (a) active, when the terminal state $x(T)$ is free, or (b) inactive, in the presence of the endpoint constraint (3.13).

for all $t \in [t_0, T]$. Then $u^*(t) = \mu(t, x^*(t))$, $t \in [t_0, T]$, is a solution to the optimal control problem (3.9)–(3.12), and

$$V(t, x^*(t)) = \int_t^T h(s, x^*(s), u^*(s))\, ds, \qquad \forall\, t \in [t_0, T], \tag{3.19}$$

with $V(t_0, x_0) = J(u^*)$ its (unique) optimal value for any initial data $(t_0, x_0) \in [0, T] \times \mathcal{X}$. (2) If a continuously differentiable function $V : [t_0, T] \times \mathcal{X} \to \mathbb{R}$ solves the HJB equation (3.16) (without boundary condition), the state-control trajectory $(x^*(t), u^*(t))$, determined by (3.17)–(3.18) with $u^*(t) = \mu(t, x^*(t))$, subject to (3.13), solves the endpoint-constrained optimal control problem (3.9)–(3.13).

The intuition for why the boundary condition (3.14) becomes inactive in the presence of the endpoint constraint (3.13) is that the endpoint constraint forces all trajectories to the given point x_T at the end of the horizon. This is illustrated in figure 3.2.

Remark 3.4 (Uniqueness of Solution to HJB Equation) Any solution $V(t, x)$ to the HJB equation (3.16) with boundary condition (3.14) is unique on $[t_0, T] \times \mathcal{X}$. To see this, consider without loss of generality the case where $t_0 = 0$. This implies by proposition 3.3 that $V(\tau, \xi)$ is the optimal value of an optimal control problem (of the type (3.9)–(3.12)) with initial data $(\tau, \xi) \in (0, T] \times \mathcal{X}$. Since any such value is unique and V is continuous on $[0, T] \times \mathcal{X}$, the function $V(t, x)$ is uniquely determined on $[0, T] \times \mathcal{X}$. □

Remark 3.5 (Principle of Optimality) The HJB equation (3.16) implies that the optimal policy $u^*(t) = \mu(t, x^*(t))$ does not depend on how a particular state $x^*(t)$ was reached; it requires only that all subsequent decisions be optimal. This means that an optimal policy is fundamentally obtained by backward induction, starting at the end of the time horizon where value of decision vanishes and the boundary condition (3.14) holds. The principle of optimality is usually attributed to Bellman (1957, ch. III.3). The idea of backward induction was mentioned earlier (see chapter 1, footnote 14). □

Remark 3.6 (HJB Equation with Discounting and Salvage Value) Let $t_0 = 0$.[6] If the objective functional in (3.8) is replaced with the seemingly more general

$$J(u) = \int_0^T e^{-rt} h(t, x(t), u(t))\, dt + e^{-rT} S(x(T)),$$

where $r \geq 0$ is a given *discount rate* and $S : \mathcal{X} \to \mathbb{R}$ is a continuously differentiable function that can be interpreted as the *terminal value* (or *salvage value*) of the state at the end of the time horizon, then proposition 3.3 can still be used to obtain a sufficient optimality condition.

Let

$$\hat{h}(t, x, u) = e^{-rt} h(t, x, u) + \langle e^{-rT} S_x(x), f(t, x, u) \rangle;$$

the resulting objective functional $\hat{J}$ differs from the original objective functional J only by a constant:

$$\hat{J}(u) = \int_0^T \hat{h}(t, z(t), u(t))\, dt = \int_0^T \left(e^{-rt} h(t, x(t), u(t)) + e^{-rT} \frac{d}{dt} S(x(t)) \right) dt$$

$$= J(u) - e^{-rT} S(x_0),$$

so the optimization problem remains unchanged when substituting $\hat{h}$ for h. By Proposition 3.3 the HJB equation becomes

$$-\hat{V}_t(t, x) = \max_{u \in \mathcal{U}} \{ \hat{h}(t, x, u) + \langle \hat{V}_x(t, x), f(t, x, u) \rangle \}$$

$$= \max_{u \in \mathcal{U}} \{ e^{-rt} h(t, x, u) + \langle \hat{V}_x(t, x) + e^{-rT} S_x(x), f(t, x, u) \rangle \},$$

for all $(t, x) \in [0, T] \times \mathcal{X}$, with boundary condition $\hat{V}(T, x) = 0$ for all $x \in \mathcal{X}$. Multiplying the HJB equation by e^{rt} on both sides and setting

---

6. In the case where $t_0 \neq 0$ it is enough to multiply all objective values by e^{rt_0}.

Optimal Control Theory

$V(t,x) \equiv e^{rt}(\hat{V}(t,x) + e^{-rT}S(x))$ yields the following HJB equation with discounting and salvage value:

$$rV(t,x) - V_t(t,x) = \max_{u \in \mathcal{U}}\{h(t,x,u) + \langle V_x(t,x), f(t,x,u)\rangle\}, \quad (3.20)$$

for all $(t,x) \in [0,T] \times \mathcal{X}$, with boundary condition

$$V(T,x) = S(x), \quad \forall x \in \mathcal{X}. \quad (3.21)$$

Note that along an optimal state-control trajectory $(x^*(t), u^*(t))$,

$$V(t, x^*(t)) = \int_t^T e^{-r(s-t)} h(s, x^*(s), u^*(s))\,ds + e^{-r(T-t)} S(x^*(T)),$$

for all $t \in [0,T]$, so $V(t, x^*(t))$ is the optimal discounted time-t value. □

Example 3.3 (Linear-Quadratic Regulator) Consider the problem of steering a linear time-variant system,

$$\dot{x} = f(t,x,u) = A(t)x + B(t)u,$$

where $A(t)$, $B(t)$ are continuous bounded matrix functions, with values in $\mathbb{R}^{n \times n}$ and $\mathbb{R}^{n \times m}$, respectively, from a given initial state $x(t_0) = x_0 \in \mathbb{R}^n$ (for some $t_0 \geq 0$) over the finite time horizon $T > t_0$ so as to maximize the discounted (with rate $r \geq 0$) quadratic objective functional

$$e^{-rt_0} J(u) = -\int_{t_0}^T e^{-rt}\left(x'(t)R(t)x(t) + u'(t)S(t)u(t)\right)dt - e^{-rT} x'(T) K x(T),$$

where $R(t)$ and $S(t)$ are continuous bounded matrix functions, with symmetric positive definite values in $\mathbb{R}^{n \times n}$ and $\mathbb{R}^{m \times m}$, respectively. The terminal-cost matrix $K \in \mathbb{R}^{n \times n}$ is symmetric positive definite. Assuming that the convex compact control set $\mathcal{U}$ is large enough (so as to allow for an unconstrained maximization), the corresponding optimal control problem is of the form (3.9)–(3.12), and the HJB equation (3.20) becomes

$$rV(t,x) - V_t(t,x) = \max_{u \in \mathcal{U}}\{-x'R(t)x - u'S(t)u + \langle V_x(t,x), A(t)x + B(t)u\rangle\},$$

for all $(t,x) \in [t_0, T] \times \mathbb{R}^n$. Recall that in the analysis of the stability properties of linear time-variant systems, a quadratic Lyapunov function proved to be useful (see example 2.13). It turns out that essentially the same functional form can be used to solve the HJB equation for the linear-quadratic regulator problem. Let

$$V(t,x) = -x'Q(t)x,$$

for all $(t, x) \in [t_0, T] \times \mathbb{R}^n$, where $Q(t)$ is a continuously differentiable matrix function with symmetric positive definite values in $\mathbb{R}^{n \times n}$. Substituting $V(t, x)$ into the HJB equation yields

$$x'\dot{Q}x = \max_{u \in \mathcal{U}}\{x'\left(rQ - QA(t) - A'(t)Q - R(t)\right)x - u'S(t)u - 2x'QB(t)u\},$$

for all $(t, x) \in [t_0, T] \times \mathbb{R}^n$. Performing the maximization on the right-hand side provides the optimal control in terms of a linear feedback law,

$$\mu(t, x) = -S^{-1}(t)B'(t)Q(t)x, \quad \forall (t, x) \in [t_0, T] \times \mathbb{R}^n,$$

so that the HJB equation is satisfied if and only if $Q(t)$ solves the *Riccati differential equation*

$$\dot{Q} - rQ + QA(t) + A'(t)Q + QB(t)S^{-1}(t)B'(t)Q = -R(t), \tag{3.22}$$

for all $t \in [t_0, T]$. The boundary condition (3.14) is satisfied if $V(T, x) = -x'Q(T)x = -x'Kx$ for all $x \in \mathbb{R}^n$, or equivalently, if

$$Q(T) = K. \tag{3.23}$$

As shown in chapter 2, the solution to this Riccati IVP can be written in the form $Q(t) = Z(t)Y^{-1}(t)$, where the matrices $Y(t), Z(t), t \in [0, T]$, solve the linear IVP

$$\begin{bmatrix} \dot{Y} \\ \dot{Z} \end{bmatrix} = \begin{bmatrix} A(t) - (r/2)I & B(t)S^{-1}(t)B'(t) \\ -R(t) & -A'(t) + (r/2)I \end{bmatrix} \begin{bmatrix} Y \\ Z \end{bmatrix}, \quad \begin{bmatrix} Y(T) \\ Z(T) \end{bmatrix} = \begin{bmatrix} I \\ K \end{bmatrix}.$$

As an illustration consider the special case where the system is one-dimensional (with $m = n = 1$) and time-invariant, and the cost is also time-invariant (except for the discounting), so that $A = \alpha < 0$, $B = \beta \neq 0$, $R = \rho > 0$, $S = \sigma > \rho(\beta/\alpha)^2$, and $K = k = 0$. The corresponding Riccati IVP becomes

$$\dot{q} + 2aq + bq^2 = -\rho, \quad q(T) = k,$$

for all $t \in [t_0, T]$, where $a = \alpha - (r/2) < 0$ and $b = \beta^2/\sigma > 0$. Thus, setting $q(t) = z(t)/y(t)$ leads to the linear IVP

$$\begin{bmatrix} \dot{y} \\ \dot{z} \end{bmatrix} = \begin{bmatrix} a & b \\ -\rho & -a \end{bmatrix} \begin{bmatrix} y \\ z \end{bmatrix}, \quad \begin{bmatrix} y(T) \\ z(T) \end{bmatrix} = \begin{bmatrix} 1 \\ k \end{bmatrix},$$

on the time interval $[t_0, T]$, which has the unique solution

$$\begin{bmatrix} y(t) \\ z(t) \end{bmatrix} = \exp\left(\begin{bmatrix} a & b \\ -\rho & -a \end{bmatrix}(T - t)\right)\begin{bmatrix} 1 \\ k \end{bmatrix}.$$

Using the abbreviation $\kappa = \sqrt{a^2 - \rho b}$ for the absolute value of the eigenvalues of the system and the fact that by assumption $k = 0$ yields

$$q(t) = \frac{z(t)}{y(t)} = \left(\frac{\rho}{\kappa - a}\right) \frac{1 - e^{-2\kappa(T-t)}}{1 + \left(\frac{\kappa+a}{\kappa-a}\right) e^{-2\kappa(T-t)}},$$

for all $t \in [t_0, T]$. Therefore, the optimal state feedback is $\mu(t, x) = -(\beta/\sigma)q(t)x$, and in turn the optimal state trajectory becomes

$$x^*(t) = x_0 \exp\left(\alpha(t - t_0) - \frac{\beta}{\sigma} \int_{t_0}^{t} q(s)\, ds\right),$$

so the optimal control trajectory is

$$u^*(t) = \mu(t, x^*(t)) = -x_0 \left(\frac{\beta\, q(t)}{\sigma}\right) \exp\left(\alpha(t - t_0) - \frac{\beta}{\sigma} \int_{t_0}^{t} q(s)\, ds\right),$$

for all $t \in [t_0, T]$. □

3.3.3 Necessary Optimality Conditions

In principle, the HJB equation in proposition 3.3 requires too much in that it must hold for all $(t, x) \in [t_0, T] \times \mathcal{X}$. But it is clear that for optimality it needs to hold only in a neighborhood of an optimal trajectory $x^*(t)$, $t \in [t_0, T]$.[7] This is illustrated in figure 3.3. This section examines the local properties of the HJB equation in a neighborhood of a solution to the optimal control problem (3.9)–(3.12) in order to derive the necessary optimality conditions.

Differentiating the HJB equation (3.16) with respect to x the envelope theorem (proposition A.15 in appendix A) yields that

$$\begin{aligned}
0 &= h_x(t, x, \mu(t, x)) + V_{tx}(t, x) + V_{xx}(t, x)f(t, x, \mu(t, x)) \\
&\quad + V_x(t, x)f_x(t, x, \mu(t, x)) \\
&= h_x(t, x, \mu(t, x)) + \dot{V}_x(t, x) + V_x(t, x)f_x(t, x, \mu(t, x)),
\end{aligned} \qquad (3.24)$$

for all $(t, x) \in [t_0, T] \times \mathcal{X}$. Similarly, differentiating (3.16) with respect to t and applying the envelope theorem once more gives

$$\begin{aligned}
0 &= h_t(t, x, \mu(t, x)) + V_{tt}(t, x) + \langle V_{tx}(t, x), f(t, x, \mu(t, x))\rangle \\
&\quad + \langle V_x(t, x), f_t(t, x, \mu(t, x))\rangle \\
&= h_t(t, x, \mu(t, x)) + \dot{V}_t(t, x) + \langle V_x(t, x), f_t(t, x, \mu(t, x))\rangle,
\end{aligned} \qquad (3.25)$$

---

7. The fact that the HJB equation needs to hold on an optimal state trajectory $x^*(t)$, $t \in [0, T]$ follows immediately by substituting the optimal value-to-go $V(t, x^*(t))$ as defined in equation (3.19).

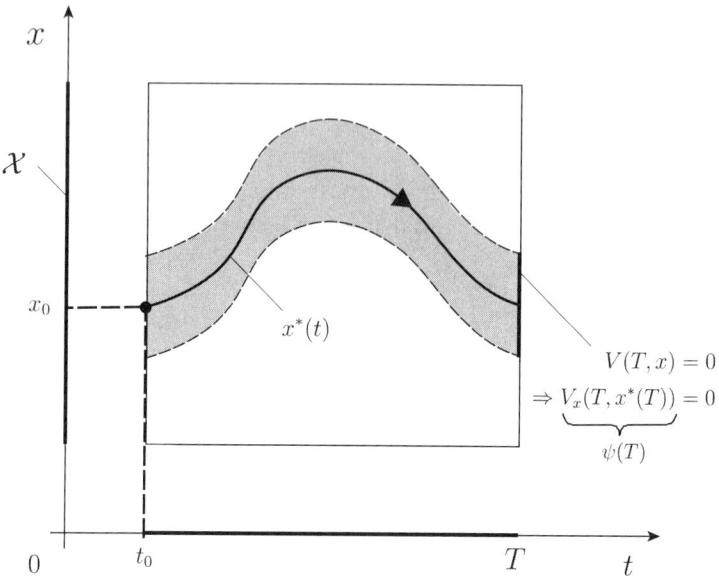

Figure 3.3
Development of necessary optimality conditions, such as transversality, in the neighborhood of an optimal state trajectory.

for all $(t,x) \in [t_0,T] \times \mathcal{X}$. To formulate necessary optimality conditions it is convenient to introduce the so-called *Hamilton-Pontryagin function* (or *Hamiltonian*, for short) $H : [t_0,T] \times \mathcal{X} \times \mathcal{U} \times \mathbb{R}^n \to \mathbb{R}$ such that

$$(t,x,u,\psi) \mapsto H(t,x,u,\psi) = h(t,x,u) + \langle \psi, f(t,x,u) \rangle$$

for all $(t,x,u,\psi) \in [t_0,T] \times \mathcal{X} \times \mathcal{U} \times \mathbb{R}^n$.

Consider now an optimal state-control tuple $(x^*(t), u^*(t))$ on $[t_0,T]$, and introduce the *adjoint variables* $\psi(t) = V_x(t,x^*(t))$ and $\psi^0(t) = V_t(t,x^*(t))$. Then equation (3.24) implies the *adjoint equation*

$$\dot{\psi}(t) = -H_x(t,x^*(t),u^*(t),\psi(t)), \quad \forall t \in [t_0,T], \tag{3.26}$$

and the boundary condition (3.14) gives the *transversality condition*

$$\psi(T) = 0. \tag{3.27}$$

The maximization in (3.17) implies the *maximality condition*

$$u^*(t) \in \arg\max_{u \in \mathcal{U}} H(t,x^*(t),u,\psi(t)), \quad \forall t \in [t_0,T]. \tag{3.28}$$

Finally, equation (3.25), in which $-\psi^0(t)$ by virtue of the HJB equation (3.16) is equal to the *maximized Hamiltonian*

$$H^*(t) = -V_t(t, x^*(t)) = H(t, x^*(t), u^*(t), \psi(t)),$$

gives the *envelope condition*

$$\dot{H}^*(t) = H_t(t, x^*(t), u^*(t), \psi(t)), \qquad \forall t \in [t_0, T]. \tag{3.29}$$

Provided that a continuously differentiable solution to the HJB equation (3.16) with boundary condition (3.14) exists, the following result has been established.[8]

Proposition 3.4 (Pontryagin Maximum Principle) (1) Given an admissible state-control trajectory $(x^*(t), u^*(t))$, $t \in [t_0, T]$, that solves the optimal control problem (3.9)–(3.12), there exists an absolutely continuous function $\psi : [t_0, T] \to \mathbb{R}^n$ such that conditions (3.26)–(3.29) are satisfied. (2) If $(x^*(t), u^*(t))$, $t \in [t_0, T]$, is an admissible state-control trajectory that solves the endpoint-constrained optimal control problem (3.9)–(3.13), then there exists an absolutely continuous function $\psi : [t_0, T] \to \mathbb{R}^n$ such that conditions (3.26) and (3.28)–(3.29) are satisfied.

Remark 3.7 (Endpoint Constraints) Analogous to the brief discussion after proposition 3.3 (see figure 3.2), the endpoint constraint (3.13) affects the transversality condition. Without endpoint constraint, one finds from the HJB boundary condition (3.14) that $\psi(T) = V_x(x^*(T), T) = 0$, which corresponds to the transversality condition (3.27). The endpoint constraint (3.13), however, frees up $V(x, T)$ for all $x \neq x_T$, so that the transversality condition $V_x(x, T) = 0$ is no longer valid, and consequently $\psi(T)$ does not have to vanish. This insight applies to each constrained component $x_i(T)$ of $x(T)$ and the corresponding component $\psi_i(T)$ of $\psi(T)$, for $i \in \{1, \ldots, n\}$. Furthermore, inequality constraints on $x_i(T)$ translate to inequality constraints on $\psi_i(T)$. Section 3.4 discusses the general finite-horizon optimal control problem and formulates the corresponding necessary optimality conditions provided by the PMP. □

Example 3.4 (Optimal Consumption) Consider an investor who, at time $t = 0$, is endowed with an initial capital of $x(0) = x_0 > 0$. At any time $t \in [0, T]$ (where $T > 0$ is given) he decides about his rate of consumption

---

8. The precise proof of the PMP is more complicated; see section 3.6 and some further remarks in section 3.8. The informal derivation given here provides the full intuition but may be technically incorrect if the value function $V(t, x)$ is not continuously differentiable (which is possible, even for problems with smooth primitives).

$c(t) \in [0, \bar{c}]$, where $\bar{c} > 0$ is a large maximum allowable rate of consumption.[9] Thus, his capital stock evolves according to

$$\dot{x} = \alpha x - c(t),$$

where $\alpha > 0$ is a given rate of return. The investor's time-t utility for consuming at a rate $c(t)$ is $U(c(t))$, where $U : \mathbb{R}_+ \to \mathbb{R}$ is his increasing, strictly concave utility function. The investor's problem is to find a consumption plan $c(t)$, $t \in [0, T]$, so as to maximize his discounted utility

$$J(c) = \int_0^T e^{-rt} U(c(t)) \, dt,$$

where $r \geq 0$ is a given discount rate, subject to the solvency constraint that the capital stock $x(t)$ must stay positive for all $t \in [0, T)$.[10] To deal with the solvency constraint, first observe that any optimal state trajectory $x^*(t)$ takes on its smallest value x_T at the end of the horizon, that is, $x_T = \min\{x^*(t) : t \in [0, T]\}$. Consuming at the maximum rate $\bar{c}$ implies that $x(t) = x_0 - (\bar{c}/\alpha)(1 - e^{-\alpha t})$, whence it is possible to reach $x(T) = 0$ within the given time horizon if and only if $\bar{c} \geq \alpha x_0 / (1 - e^{-\alpha T})$. Assume that the last inequality holds (otherwise $c^*(t) \equiv \bar{c}$ is the unique optimal consumption plan and $x^*(t) = x_0 - (\bar{c}/\alpha)(1 - e^{-\alpha t}) \geq 0$ on $[0, T]$) and that the investor maximizes his discounted utility subject to the endpoint constraint $x(T) = 0$. He therefore faces an endpoint-constrained optimal control problem, of the form (3.9)–(3.13). The PMP in proposition 3.4(2) provides the corresponding necessary optimality conditions. Given the Hamiltonian $H(t, x, c, \psi) = e^{-rt} U(c) + \psi (\alpha x - c)$ on an optimal state-control trajectory $(x^*(t), c^*(t))$, $t \in [0, T]$, the adjoint equation (3.26) becomes

$$\dot{\psi}(t) = -H_x(t, x^*(t), c^*(t), \psi(t)) = -\alpha \psi(t), \qquad t \in [0, T],$$

so $\psi(t) = \psi_0 e^{-\alpha t}$, where $\psi_0 = \psi(0) > 0$ is an initial value that will be determined by the endpoint constraint. Positivity of ψ_0 is consistent with the maximality condition (3.28), which yields

$$c^*(t; \psi_0) = \min\{U_c^{-1}(\psi_0 e^{-(\alpha - r)t}), \bar{c}\} \in \arg\max_{c \in [0, \bar{c}]} \{e^{-rt} U(c) - \psi_0 e^{-\alpha t} c\},$$

---

9. One practical reason for such an upper bound may be that authorities tend to investigate persons believed to be living beyond their means, relative to declared income.
10. The solvency constraint is used to simplify the problem so that it can be tackled with the optimality conditions already discussed. When the solvency constraint is relaxed to the nonnegativity constraint $x(t) \geq 0$ for all $t \in [0, T]$, then it may be optimal to consume all available capital by the time $\hat{T} < T$, so the spending horizon $\hat{T} \in [0, T]$ becomes subject to optimization as well.

Optimal Control Theory

for all $t \in [0, T]$. Based on the last optimality condition the Cauchy formula in proposition 2.1 provides the optimal state trajectory,[11]

$$x^*(t; \psi_0) = x_0 e^{\alpha t} - \int_0^t e^{\alpha(t-s)} c^*(s; \psi_0)\, ds, \qquad \forall t \in [0, T].$$

The missing constant ψ_0 is determined by the endpoint constraint $x^*(T; \psi_0) = 0$, namely,

$$x_0 = \int_0^T e^{-\alpha t} \min\{U_c^{-1}(\psi_0 e^{-(\alpha-r)t}), \bar{c}\}\, dt$$

$$= \begin{cases} \dfrac{\bar{c}}{\alpha}(1 - e^{-\alpha \tau}) + \displaystyle\int_\tau^T e^{-\alpha t} U_c^{-1}(\psi_0 e^{-(\alpha-r)t})\, dt & \text{if } \alpha < r, \\ U_c^{-1}(\psi_0)(1 - e^{-\alpha T})/\alpha & \text{if } \alpha = r, \\ \displaystyle\int_0^\tau e^{-\alpha t} U_c^{-1}(\psi_0 e^{-(\alpha-r)t})\, dt + \dfrac{\bar{c}}{\alpha}(e^{-\alpha \tau} - e^{-\alpha T}) & \text{otherwise,} \end{cases} \qquad (3.30)$$

where the switching time for $\alpha \neq r$,

$$\tau = \min\left\{T, \left[\frac{\ln(\psi_0/U_c(\bar{c}))}{\alpha - r}\right]_+\right\},$$

depends on ψ_0 and lies in $(0, T)$ if ψ_0, as a solution to (3.30), is strictly between $U_c(\bar{c})$ and $U_c(\bar{c})e^{(\alpha-r)T}$. If $\tau \in \{0, T\}$, then the solution is interior in the sense that the control constraint $c^*(t) \leq \bar{c}$ is automatically satisfied; in other words, the optimal control remains unchanged when the constraint is relaxed. In the special case where $\alpha = r$, one obtains from (3.30) that $\psi_0 = U_c\left(\alpha x_0/(1 - e^{-\alpha T})\right) > U_c(\bar{c})$, which yields the optimal state-control trajectory (x^*, c^*) with

$$c^*(t) = \frac{\alpha x_0}{1 - e^{-\alpha T}} \quad \text{and} \quad x^*(t) = \frac{1 - e^{-\alpha(T-t)}}{e^{\alpha T} - 1} x_0,$$

---

11. Implicitly assume here that $x^*(t) > 0$ for all $t \in [0, T)$. If, contrary to this assumption, it is best for the investor to consume all his wealth *before* the end of the horizon, then necessarily the utility of consuming at a rate of zero must be finite, i.e., $U(0) > -\infty$. By considering $\hat{U}(c) = U(c) - U(0)$ instead of $U(c)$, and solving the investor's problem over the interval $[0, \hat{T}]$ instead of $[0, T]$, with a *variable* horizon $\hat{T} \in [0, T]$ subject to optimization, it is possible to leave all arguments (and assumptions) of the analysis in place (with T replaced by $\hat{T}$) and to simply add an additional optimality constraint for $\hat{T}$. The corresponding optimal control problem with general endpoint constraints is discussed in section 3.4. The resulting additional optimality condition amounts to requiring that the Hamiltonian vanish at $t = \hat{T}^*$ if $\hat{T}^* \in (0, T)$ is an optimal (interior) value for $\hat{T}$.

for all $t \in [0, T]$. Interior solutions to the investor's optimal consumption problem for $\alpha \neq r$ are discussed in example 3.5. □

Remark 3.8 (Calculus of Variations) The PMP can be used to reestablish well-known classical optimality conditions in the following simplest problem of the *calculus of variations*,

$$\int_0^T F(t, x(t), \dot{x}(t)) \, dt \longrightarrow \max_{x(\cdot)}, \tag{3.31}$$

subject to $x(0) = x_0$ and $x(T) = x_T$, where boundary points $x_0, x_T \in \mathbb{R}^n$ and the time horizon $T > 0$ are given. The function $F : \mathbb{R}^{1+2n} \to \mathbb{R}$ is assumed twice continuously differentiable, bounded, and strictly concave in $\dot{x}$. Introducing the control $u = \dot{x}$ on $[0, T]$, by setting $h(t, x, u) \equiv F(t, x, u)$ and $f(t, x, u) \equiv u$ one obtains an endpoint-constrained optimal control problem of the form (3.9)–(3.13), where the control constraint (3.12) is assumed inactive at the optimum. The Hamilton-Pontryagin function is

$$H(t, x, u, \psi) = F(t, x, u) + \langle \psi, u \rangle,$$

so along an optimal state-control trajectory $(x^*(t), u^*(t))$, $t \in [0, T]$, the adjoint equation (3.26) becomes

$$\dot{\psi}(t) = -F_x(t, x^*(t), u^*(t)), \qquad \forall t \in [0, T],$$

and the maximality condition (3.28) yields

$$F_{\dot{x}}(t, x^*(t), u^*(t)) + \psi(t) = 0, \qquad \forall t \in [0, T].$$

Differentiating the last relation with respect to time gives the *Euler equation* along an optimal solution $x^*(t)$, $t \in [0, T]$, of the initial problem,

$$F_x(t, x^*(t), \dot{x}^*(t)) = \frac{d}{dt} F_{\dot{x}}(t, x^*(t), \dot{x}^*(t)), \qquad \forall t \in [0, T], \tag{3.32}$$

provided the time-derivative $\dot{x}^*(t) = u^*(t)$ is absolutely continuous. The Euler equation in its integral form,

$$F_{\dot{x}}(t, x^*(t), \dot{x}^*(t)) = \int_0^t F_x(s, x^*(s), \dot{x}^*(s)) \, ds + C, \qquad \forall t \in [0, T], \tag{3.33}$$

where C is an appropriate constant, is often referred to as the *DuBois-Reymond equation* and is valid even when $\dot{x}^*(t) = u^*(t)$ exhibits jumps. □

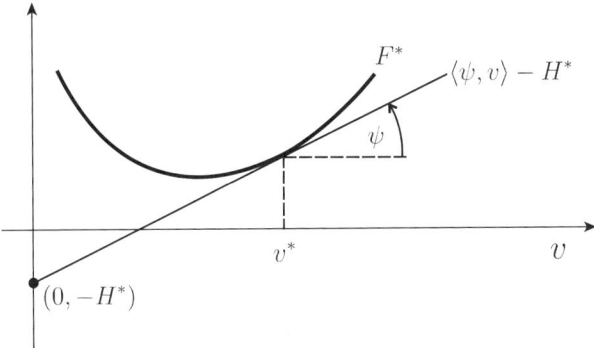

Figure 3.4
Geometric intuition for a (maximized) Hamiltonian (see remark 3.9).

Remark 3.9 (Geometric Interpretation of Hamiltonian) Building on the calculus-of-variations problem (3.31), consider the *Lagrange problem*,

$$-\int_0^T F(t, x(t), \dot{x}(t))\, dt \longrightarrow \max_{x(\cdot)},$$

with the initial condition $x(0) = x_0$, subject to the state equation $\dot{x} = f(t, x, u)$ and the control constraint $\dot{x} = u \in \mathcal{U}$, where $\mathcal{U} \subset \mathbb{R}$ is a nonempty convex compact control set. Denote by

$$F^*(t, x, v) = \inf_{u \in \mathcal{U}} \{F(t, x, u) : v = f(t, x, u)\}$$

the lower envelope of F with respect to the feasible slopes of the state trajectory. The corresponding (maximized) Hamiltonian is then

$$H^*(t, x, \psi) = \max_{u \in \mathcal{U}} \{\langle \psi, f(t, x, u) \rangle - F(t, x, u)\} = \sup_{v \in \mathbb{R}^n} \{\langle \psi, v \rangle - F^*(t, x, v)\}.$$

The right-hand side of the last relation is also called the *dual* (or *Young-Fenchel transform*) of F^* (see, e.g., Luenberger 1969; Magaril-Il'yaev and Tikhomirov 2003). Figure 3.4 provides the geometric intuition (with $v^* = f(t, x, u^*)$). Thus, the maximized Hamiltonian H^* can be interpreted as the dual of F^*.[12] □

Example 3.5 (Optimal Consumption, Revisited) To see the use of the Euler equation introduced in remark 3.8, consider interior solutions to the optimal consumption problem discussed in example 3.4. Indeed, the

---

12. Gamkrelidze (2008) provides another interesting perspective on this duality relation.

investor's endpoint-constrained problem of maximizing his discounted utility can be rewritten in the form (3.31) for

$$F(t, x, \dot{x}) = e^{-rt} U(\alpha x - \dot{x})$$

with the constraints $x(0) = x_0$ and $x(T) = 0$. The Euler equation (3.32) becomes

$$\alpha e^{-rt} U_c(c(t)) = -\frac{d}{dt}[e^{-rt} U_c(c(t))] = re^{-rt} U_c(c(t)) - e^{-rt} U_{cc}(c(t)) \dot{c}(t),$$

for all $t \in [0, T]$, provided that $c(t)$ is interior, namely, that $0 < c(t) < \bar{c}$ almost everywhere. This optimality condition can also be written as an autonomous ODE, in the form

$$\rho_A(c) \dot{c} = -\frac{U_{cc}(c)}{U_c(c)} \dot{c} = \alpha - r,$$

where $\rho_A(c) = -U_{cc}(c)/U_c(c)$ is the investor's *(Arrow-Pratt) coefficient of absolute risk aversion*. This condition means that it is optimal for the investor to consume such that the (absolute) growth, $\dot{c}$, of the optimal consumption path is equal to the ratio of excess return, $\alpha - r$, and the investor's absolute risk aversion, $\rho_A(c) > 0$. Thus, with the strictly increasing first integral of absolute risk aversion,

$$R_A(c; c_0) = \int_{c_0}^{c} \rho_A(\zeta) d\zeta,$$

for $c \in (0, \bar{c})$ and some reference level c_0, the inverse $R_A^{-1}(\cdot; c_0)$ exists, and the optimal consumption path becomes

$$c^*(t; c_0) = R_A^{-1}((\alpha - r)t; c_0),$$

where $c_0 = c^*(0) \in [0, \bar{c}]$ is a constant that can be determined from the zero-capital endpoint constraint,

$$x_0 = \int_0^T e^{-\alpha t} c^*(t; c_0) ds.$$

For example, if the investor has *constant absolute risk aversion* (CARA) so that $\rho_A(c) \equiv \rho > 0$, then $R_A(c; c_0) = \rho(c - c_0)$ and $c^*(t; c_0) = c_0 + (\alpha - r)(t/\rho)$, that is, the optimal consumption path is linearly increasing (resp. decreasing) if the excess return $\alpha - r$ is positive (resp. negative). If the investor has *constant relative risk aversion* (CRRA) so that $c\rho_A(c) \equiv$

Optimal Control Theory

$\rho_R > 0$, then $R_A(c; c_0) = \rho_R \ln(c/c_0)$ and $c^*(t; c_0) = c_0 \exp[(\alpha - r)(t/\rho_R)]$, that is, the optimal consumption path is exponentially increasing (resp. decreasing) if the excess return $\alpha - r$ is positive (resp. negative).[13] □

3.4 Finite-Horizon Optimal Control

Consider the following general finite-horizon optimal control problem:

$$J(u, \omega) = \int_{t_0}^{T} h(t, x(t), u(t))\, dt + K^0(\omega) \longrightarrow \max_{u(\cdot),\, \omega}, \qquad (3.34)$$

$$\dot{x}(t) = f(t, x(t), u(t)), \qquad x(t_0) = x_0, \qquad x(T) = x_T, \qquad (3.35)$$

$$K^1(\omega) \geq 0, \qquad K^2(\omega) = 0, \qquad (3.36)$$

$$R(t, x, u^1) \geq 0, \qquad (3.37)$$

$$u = (u^1, u^2), \qquad u^2(t) \in \mathcal{U}^2(t) \;\forall t, \qquad (3.38)$$

$$t \in [t_0, T], \qquad t_0 < T, \qquad (3.39)$$

where $\omega = (t_0, x_0; T, x_T)$ is the vector of endpoint data, $x = (x_1, \ldots, x_n)$ with values in $\mathbb{R}^n$ is the state variable, and $u = (u_1, \ldots, u_m)$ with values in $\mathbb{R}^m$ is the control variable. The control variable u is represented as a tuple of the form[14]

$$u = (u^1, u^2),$$

where $u^1 = (u_1, \ldots, u_{m_1})$ and $u^2 = (u_{m_1+1}, \ldots, u_m)$ with $m_1 \in \{1, \ldots, m\}$. The first control component, u^1, appears in the state-control constraint (3.37), and the second control component, u^2, satisfies the geometric control constraint (3.38). The general optimal control problem (3.35)–(3.39) is considered under the following assumptions.

A1. The functions $h : \mathbb{R}^{1+n+m} \to \mathbb{R}$, $f : \mathbb{R}^{1+n+m} \to \mathbb{R}^n$, $R : \mathbb{R}^{1+n+m_1} \to \mathbb{R}^{k_R}$ are continuously differentiable with respect to (x, u) (resp. (x, u^1)) for a.a. (almost all) t, and measurable in t for any (x, u), where $k_R = \dim(R)$. On any bounded set these functions and their partial derivatives with

---

13. A CARA investor has a utility function of the form $U(c) = -e^{-\rho c}$, whereas a CRRA investor has a utility function of the form $U(c) = c^{1-\rho_R}/(1-\rho_R)$ for $\rho_R \neq 1$ and $U(c) = \ln(c)$ for $\rho_R = 1$. In either case, the investor's utility function is determined only up to a *positive affine transformation* $\alpha U(c) + \beta$ with $\alpha > 0$ and $\beta \in \mathbb{R}$.
14. This division of the control variable into two vectors goes back to a seminal paper by Dubovitskii and Milyutin (1965).

respect to (x, u) (resp. (x, u^1)) are bounded and continuous, uniformly in (t, x, u).

A2. The set of *admissible* controls is

$$\{u(\cdot) = (u^1(\cdot), u^2(\cdot)) \in \mathbf{L}_\infty : u^2(t) \in \mathcal{U}^2(t) \ \forall t\},$$

where $t \mapsto \mathcal{U}^2(t) \subseteq \mathbb{R}^{m-m_1}$ is a measurable set-valued mapping such that $\mathcal{U}^2(t) \neq \emptyset$ for all t.[15]

A3. The functions $K^j : \mathbb{R}^{2(1+n)} \to \mathbb{R}^{k_j}$, for $j \in \{0, 1, 2\}$, are continuously differentiable, where $k_j = \dim(K^j)$.

A4. The endpoint constraints (3.36) are *regular*, that is, for any $\omega = (t_0, x_0; T, x_T)$ satisfying (3.36) it is $\text{rank}(K_\omega^2(\omega)) = \dim(K^2)$ and there exists $\hat{\omega} = (\hat{t}_0, \hat{x}_0; \hat{T}, \hat{x}_T) \in \mathbb{R}^{2(1+n)}$ such that

$$K_i^1(\omega) = 0 \Rightarrow \left\langle \hat{\omega}, \frac{\partial K_i^1}{\partial \omega}(\omega) \right\rangle > 0, \qquad i \in \{1, \dots, k_1\},$$

and

$$\frac{\partial K_i^2(\omega)}{\partial \omega} \hat{\omega} = 0, \qquad i \in \{1, \dots, k_2\}.$$

A5. The state-control constraints (3.37) are *regular* in the sense that for any $c > 0$ there exists $\delta > 0$ such that for any x, u^1 and a.a. t for which

$$\|x\| \leq c, \qquad \|u^1\| \leq c, \qquad |t| \leq c, \qquad R_j(t, x, u^1) \geq -\delta, \qquad j \in \{1, \dots, k_R\},$$

there exists $v = v(t, x, u^1) \in \mathbb{R}^{m_1}$ with $\|v\| \leq 1$ satisfying

$$R_j(t, x, u^1) \leq \delta \Rightarrow \left\langle v, \frac{\partial R_j}{\partial u^1}(t, x, u^1) \right\rangle \geq \delta, \qquad j \in \{1, \dots, k_R\}.$$

Remark 3.10 If R and the gradient R_{u^1} are continuously differentiable, the condition in A5 can be reformulated as follows: if (t, x, u^1) satisfies (3.37), then the set of vectors

$$\left\{ \frac{\partial R_j}{\partial u^1}(t, x, u^1) : R_j(t, x, u^1) = 0, j \in \{1, \dots, k_R\} \right\}$$

---

15. A measurable function $\sigma(\cdot)$ is called a *measurable selector* of the set-valued mapping $\mathcal{U}^2(\cdot)$ if $\sigma(t) \in \mathcal{U}^2(t)$ for a.a. t. A set-valued mapping $\mathcal{U}(\cdot)$ is called *measurable* if there exists a sequence of measurable selectors $\{\sigma^k(\cdot)\}_{k \in \mathbb{N}}$ such that the set $\{\sigma^k(t)\}_{k \in \mathbb{N}}$ is everywhere dense in $\mathcal{U}(t)$ for a.a. t. A set $\mathcal{S}$ is *dense* in $\mathcal{U}(t)$ if every open set that contains a point of $\mathcal{U}(t)$ has a nonempty intersection with $\mathcal{S}$.

is positive-linearly independent, that is, there exists $v(t) \in \mathbb{R}^{m_1}$ such that

$$R_j(t,x,u^1) = 0 \Rightarrow \left\langle v(t), \frac{\partial R_j}{\partial u^1}(t,x,u^1) \right\rangle > 0, \qquad j \in \{1,\ldots,k_R\},$$

for (almost) all t. □

The optimal control problem (3.34)–(3.39) is very general, except for the possibility of state constraints (see remark 3.12). Assumptions A1–A3 ensure the regularity of the primitives of the problem. Assumptions A4 and A5, also referred to as *Mangasarian-Fromovitz conditions*, guarantee that the endpoint and state-control constraints (3.36)–(3.37) are regular in the sense that whenever a constraint is binding, the gradient of the relevant constraint function with respect to the decision variables is nonzero. As a result, the decision maker is never indifferent about the choice of control or endpoints whenever the constraints are binding. Note that when endpoint constraints are imposed, one implicitly assumes that the underlying dynamic system is sufficiently controllable over the available time horizon (see the discussion of controllability issues in section 3.2).

Remark 3.11 (Lagrange, Mayer, and Bolza Problems) Depending on the form of the objective functional in (3.34), one may distinguish several cases: the *Lagrange problem* when $K^0 = 0$, the *Mayer problem* when $h = 0$, and the *Bolza problem* in the general (mixed) case. □

Remark 3.12 (State Constraints) This book intentionally avoids state constraints because reasonable economic problems can usually be formulated without them. Some state constraints appear because they describe undesirable possibilities. For instance, an investor's account balance may be constrained to be nonnegative in order to ensure liquidity (see example 3.4). In that situation, it is typically possible to tolerate a violation of the constraint at an appropriate cost: the investor may be able to borrow additional funds at higher interest rates. Those rates would realistically escalate as the borrowed amount increases, which effectively replaces the state constraint with a barrier that imposes a cost penalty, which in turn will naturally tend to move the optimal state-control path away from the state constraint. Other state constraints express fundamental impossibilities. For instance, when considering the evolution of the consumer base for a given product (e.g., when designing an optimal dynamic pricing policy), it is impossible for this consumer base to become smaller than zero or larger than 100 percent (see exercise 3.1). Yet, in those cases the system dynamics usually support the set of reasonable states $\mathcal{X}$ as an invariant set. For example, as the consumer base approaches 100 percent, the speed of adoption tends to zero. A similar phenomenon occurs when the consumer base is near zero: the discard rate must tend to zero because there are fewer and fewer consumers who are able to discard the product. Hence, in this example the state space $\mathcal{X} = [0, 1]$ must be invariant under any admissible control.[16] From a theoretical point of view, the PMP with state constraints (formulated for the first time by Gamkrelidze (1959) and more completely by Dubovitskii and Milyutin (1965)) involves the use of measures and can create discontinuities in the adjoint variable. In the presence of state constraints, the optimality conditions provided by the PMP are usually not complete; only rarely can they be used (at least with reasonable effort) to construct an optimal solution.[17] □

Pontryagin et al. (1962) formulated the necessary optimality conditions for the optimal control problem (3.34)–(3.39); the version provided here corresponds to the one given by Arutyunov (2000), which is in part based on Dubovitskii and Milyutin (1965; 1981). The following three additional assumptions simplify its statement.

- *Condition S (smoothness)* The optimal control problem (3.34)–(3.39) is said to satisfy the *smoothness condition* if the functions $f(t, x, u)$, $h(t, x, u)$, and $R(t, x, u^1)$ are continuously differentiable (in all arguments) and the multivalued mapping $\mathcal{U}^2(\cdot)$ is constant.
- *Condition B (boundedness)* An admissible process $(x(\cdot), u(\cdot), \omega)$ is said to satisfy the *boundedness condition* if all sets $\mathcal{U}(t, x) = \mathcal{U}^1(t, x) \times \mathcal{U}^2(t)$, with

$$\mathcal{U}^1(t, x) = \{u^1 \in \mathbb{R}^{m_1} : R(t, x, u^1) \geq 0\},$$

are uniformly bounded relative to all (t, x) that lie in neighborhoods of the points (t_0, x_0) and (T, x_T).
- *Condition C (compactness)* An admissible process $(x(\cdot), u(\cdot), \omega)$ is said to satisfy the *(additional) compactness condition* if there are neighborhoods

---

16. This simple insight is very broad. Because of the inherent finiteness of all human endeavors, any realistic finite-dimensional state space can reasonably be assumed bounded (typically even compact) and invariant.

17. Hartl et al. (1995) review results concerning necessary optimality conditions for state-constrained optimal control problems. Even though the conditions provided by Dubovitskii, Milyutin, and their students for problems with state constraints may appear complete, their application is usually very difficult and impractical.

Optimal Control Theory

of the time endpoints t_0 and T such that for a.a. t belonging to $[t_0, T]$ the sets $\mathcal{U}^2(t)$ are compact. The multivalued mapping $\mathcal{U}^2(\cdot)$ and the mapping $g = (f, h)$ are left-continuous at the point t_0 and right-continuous at the point T, and the sets $g(t_0, x_0, \mathcal{U}(t_0, x_0))$ and $g(T, x_T, \mathcal{U}(T, x_T))$ are convex (see footnote 30).

The *Hamiltonian* (or *Hamilton-Pontryagin function*) is given by

$$H(t, x, u, \psi, \lambda_0) = \lambda_0 h(t, x, u) + \langle \psi, f(t, x, u) \rangle, \quad (3.40)$$

where $\lambda_0 \in \mathbb{R}$ is a constant and $\psi \in \mathbb{R}^n$ is the adjoint variable. The *(small) Lagrangian* is given by

$$L(\omega, \lambda) = \lambda_0 K^0(\omega) + \langle \lambda_1, K^1(\omega) \rangle + \langle \lambda_2, K^2(\omega) \rangle, \quad (3.41)$$

where $\lambda_1 \in \mathbb{R}^{k_1}$, $\lambda_2 \in \mathbb{R}^{k_2}$, $\omega = (t_0, x_0; T, x_T)$. The necessary optimality conditions for the general optimal control problem (3.34)–(3.39) can now be formulated as follows.

Proposition 3.5 (Pontryagin Maximum Principle: General OCP) Let assumptions A1–A5 be satisfied, and let $(x^*(\cdot), u^*(\cdot), \omega^*)$, with the endpoint data $\omega^* = (t_0^*, x_0^*; T^*, x_T^*)$, be a solution to the optimal control problem (3.34)–(3.39), such that conditions B and C hold. (1) There exist a vector $\lambda \in \mathbb{R}^{1+k_1+k_2}$, a measurable, essentially bounded function $\rho : [t_0^*, T^*] \to \mathbb{R}^{k_R}$, and an absolutely continuous function $\psi : [t_0^*, T^*] \to \mathbb{R}^n$ such that the following optimality conditions are satisfied:

- Adjoint equation:

$$-\dot{\psi}(t) = H_x(t, x^*(t), u^*(t), \psi(t), \lambda_0) + \rho(t) R_x(t, x^*(t), u^{1*}(t)), \quad (3.42)$$

for all $t \in [t_0^*, T^*]$.

- Transversality

$$\psi(t_0^*) = -L_{x_0}(\omega^*, \lambda) \quad \text{and} \quad \psi(T^*) = L_{x_T}(\omega^*, \lambda). \quad (3.43)$$

- Maximality

$$u^*(t) \in \arg\max_{u \in \mathcal{U}(t, x^*(t))} H(t, x^*(t), u, \psi(t), \lambda_0), \quad (3.44)$$

$$\rho_j(t) \geq 0 \quad \text{and} \quad \rho_j(t) R_j(t, x^*(t), u^{1*}(t)) = 0, \quad \forall j \in \{1, \ldots, k_R\}, \quad (3.45)$$

$$H_{u^1}(t, x^*(t), u^*(t), \psi(t), \lambda_0) - \rho(t) R_{u^1}(t, x^*(t), u^{1*}(t)) = 0, \quad (3.46)$$

for a.a. $t \in [t_0^*, T^*]$.

- Endpoint optimality

$$\lambda_0 \geq 0, \quad \lambda_1 \geq 0, \quad \forall j \in \{1,\ldots,k_1\}: \lambda_{1,j} K_j^1(\omega^*) = 0, \tag{3.47}$$

$$\sup_{u \in \mathcal{U}(t_0^*, x_0^*)} H\left(t_0^*, x_0^*, u, -L_{x_0}(\omega^*, \lambda), \lambda_0\right) - L_{t_0}(\omega^*, \lambda) = 0, \tag{3.48}$$

$$\sup_{u \in \mathcal{U}(T^*, x_T^*)} H\left(T^*, x_T^*, u, L_{x_T}(\omega^*, \lambda), \lambda_0\right) + L_T(\omega^*, \lambda) = 0. \tag{3.49}$$

- Nontriviality[18]

$$\|\lambda\| + \|\psi(t)\| \neq 0, \quad \forall t \in [t_0^*, T^*]. \tag{3.50}$$

(2) If condition S also holds, then in addition to these optimality conditions, the following obtains:

- Envelope condition

$$\dot{H}(t, x^*(t), u^*(t), \psi(t), \lambda_0)$$
$$= H_t(t, x^*(t), u^*(t), \psi(t), \lambda_0) + \langle \rho(t), R_t(t, x^*(t), u^*(t)) \rangle, \tag{3.51}$$

for a.a. $t \in [t_0^*, T^*]$.

Proof See section 3.6. ∎

The first five optimality conditions are sometimes termed the weakened maximum principle. The envelope condition can also be viewed as an adjoint equation with respect to time, where $\psi^0(t) = -H(t, x^*(t), u^*(t), \psi(t), \lambda_0)$ is the corresponding adjoint variable (see section 3.3.3, where the same adjoint variable appears).

Remark 3.13 (Simplified OCP) In many practical models, the initial data are fixed, there are no state-control constraints, and the control set is constant ($\mathcal{U}(t, x) \equiv \mathcal{U}$), which yields the following simplified optimal control problem:

$$J(u, T, x_T) = \int_{t_0}^{T} h(t, x(t), u(t))\, dt \longrightarrow \max_{u(\cdot), T, x_T}, \tag{3.52}$$

$$\dot{x}(t) = f(t, x(t), u(t)), \quad x(t_0) = x_0, \quad x(T) = x_T, \tag{3.53}$$

$$K^1(T, x_T) \geq 0, \quad K^2(T, x_T) = 0, \tag{3.54}$$

---

18. As shown in section 3.6, if condition S holds, the nontriviality condition (3.50) can be strengthened to $\lambda_0 + \|\psi(t)\| > 0$, for all $t \in (t_0^*, T^*)$.

Optimal Control Theory

$$u(t) \in \mathcal{U}, \quad \forall t, \tag{3.55}$$

$$t \in [t_0, T], \quad t_0 < T, \tag{3.56}$$

given (t_0, x_0).

Note that in the PMP for the simplified optimal control problem (3.52)–(3.56) one can without loss of generality assume that $\lambda_0 = 1$ because the optimality conditions in proposition 3.5 become positively homogeneous with respect to $\psi(\cdot)$ and λ_0. Thus, λ_0 can be dropped from consideration without any loss of generality. □

Remark 3.14 (Salvage Value in the Simplified OCP) The objective functional (3.52) admits a continuously differentiable salvage value. Let

$$\hat{J}(u, T, x_T) = \int_{t_0}^{T} \hat{h}(t, x(t), u(t)) \, dt + S(T, x(T)), \tag{3.57}$$

where $S : \mathbb{R}^{1+n} \to \mathbb{R}$ is a continuously differentiable function describing the terminal value. The objective function (3.57) can be written in the simple integral form (3.52) to obtain

$$\hat{J}(u, T, x_T) = \int_{t_0}^{T} \hat{h}(t, x(t), u(t)) \, dt + S(T, x(T))$$

$$= \int_{t_0}^{T} \hat{h}(t, x(t), u(t)) \, dt + \int_{t_0}^{T} \dot{S}(t, x(t)) \, dt + S(t_0, x_0)$$

$$= \int_{t_0}^{T} [\hat{h}(t, x(t), u(t)) + \langle S_x(t, x(t)), f(t, x(t), u(t)) \rangle + S_t(t, x(t))] \, dt$$

$$+ S(t_0, x_0)$$

$$= \int_{t_0}^{T} h(t, x(t), u(t)) \, dt + S(t_0, x_0)$$

$$= J(u, T, x_T) + S(t_0, x_0),$$

provided one sets $h(t, x, u) \equiv \hat{h}(t, x, u) + \langle S_x(t, x, u), f(t, x, u) \rangle + S_t(t, x)$.[19] □

Proposition 3.6 (Pontryagin Maximum Principle: Simplified OCP)
Let assumptions A1–A5 be satisfied, and let $(x^*(\cdot), u^*(\cdot), T^*, x_T^*)$ be a solution to the optimal control problem (3.52)–(3.56) such that

---
19. In remark 3.6 a similar idea for substitution was used.

Conditions B and C hold. Then there exist a vector $\lambda \in \mathbb{R}^{1+k_1+k_2}$ and an absolutely continuous function $\psi : [t_0, T^*] \to \mathbb{R}^n$ such that the following optimality conditions are satisfied:

- Adjoint equation

$$\dot{\psi}(t) = -H_x(t, x^*(t), u^*(t), \psi(t)), \qquad \forall\, t \in [t_0, T^*]. \tag{3.58}$$

- Transversality

$$\psi(T^*) = L_{x_T}(T^*, x_T^*, \lambda). \tag{3.59}$$

- Maximality

$$u^*(t) \in \arg\max_{u \in \mathcal{U}} H(t, x^*(t), u, \psi(t)), \qquad \dot{\forall}\, t \in [t_0, T^*]. \tag{3.60}$$

- Endpoint optimality

$$\lambda_1 \geq 0, \qquad \lambda_{1,i} K_i^1(T^*, x_T^*) = 0, \qquad \forall\, i \in \{1, \dots, k_1\}, \tag{3.61}$$

$$\sup_{u \in \mathcal{U}} H\left(T^*, x_T^*, u, L_{x_T}(T^*, x_T^*, \lambda)\right) + L_T(T^*, x_T^*, \lambda) = 0. \tag{3.62}$$

(2) If Condition S also holds, then in addition to these optimality conditions, the following condition is satisfied:

- Envelope condition

$$\dot{H}(t, x^*(t), u, \psi(t)) = H_t(t, x^*(t), u^*(t), \psi(t)), \qquad \dot{\forall}\, t \in [t_0, T^*]. \tag{3.63}$$

Remark 3.15 (Discounting and Current-Value Formulation) In many economic applications, the underlying system is time-invariant, that is, described by the IVP $\dot{x} = f(x, u)$, $x(t_0) = x_0$, and the kernel of the objective functional J comprises time only in the form of an exponential *discount factor*, so

$$J(u) = \int_{t_0}^{T} e^{-rt} h(x(t), u(t))\, dt,$$

where $r \geq 0$ is a given discount rate. In that case, the adjoint equation (3.58) (and correspondingly the entire Hamiltonian system (3.53),(3.58) for the variables x and ψ) admits an alternative time-invariant current-value formulation, which may be more convenient. This formulation is often used when considering optimal control problems with an infinite time horizon (see section 3.5).

To obtain a current-value formulation of the (simplified) PMP, first introduce the *current-value adjoint variable*

Optimal Control Theory

$$v(t) \equiv e^{rt}\psi(t).$$

Then the Hamiltonian (3.40) in problem (3.52)–(3.56) can be written in the form

$$H(t,x,u,\psi) = e^{-rt}\hat{H}(x,u,v),$$

where the *current-value Hamiltonian* $\hat{H}$ is given by

$$\hat{H}(x,u,v) = h(x,u) + \langle v, f(x,u) \rangle. \tag{3.64}$$

Clearly, maximizing the Hamiltonian $H(t,x,u,\psi)$ in (3.40) with respect to u is equivalent to maximizing the current-value Hamiltonian $\hat{H}(x,u,v)$ in (3.64). Hence, the maximality condition (3.60) remains essentially unchanged, in the form

$$u^*(t) \in \arg\max_{u \in \mathcal{U}} \hat{H}(x^*(t), u, v(t)), \quad \forall t \in [t_0, T^*]. \tag{3.65}$$

The current-value version of the adjoint equation (3.58) can be obtained as follows:

$$\dot{v}(t) = \frac{d}{dt}\left(e^{rt}\psi(t)\right) = re^{rt}\psi(t) + e^{rt}\dot{\psi}(t) = rv(t) - e^{rt}H_x(t,x(t),u(t),\psi(t))$$

$$= rv(t) - \hat{H}_x(x(t), u(t), v(t)). \tag{3.66}$$

The transversality condition (3.59) becomes

$$v(T^*) = e^{rT^*} L_{x_T}(T^*, x_T^*, \lambda). \tag{3.67}$$

Condition (3.62) translates to

$$\sup_{u \in \mathcal{U}} \hat{H}\left(x_T^*, u, -L_{x_T}(T^*, x_T^*, \lambda)\right) = -e^{rT^*} L_T(T^*, x_T^*, \lambda). \tag{3.68}$$

To summarize, the necessary optimality conditions (3.58)–(3.62) are equivalent to conditions (3.66), (3.67), (3.65), (3.61), (3.68), respectively. □

Remark 3.16 (Fixed Time Horizon) In the case where the time horizon T is fixed, the optimality conditions in proposition 3.6 specialize to the version of the PMP formulated in proposition 3.4. □

The PMP can be used to identify solution candidates for a given optimal control problem. If a solution exists (see the brief discussion in section 3.8), then uniqueness of the solution candidate also implies its global optimality. The following result by Mangasarian (1966) provides sufficient optimality conditions in the framework of the PMP (rather than in the context of the HJB equation).

Proposition 3.7 (Mangasarian Sufficiency Theorem) Consider the admissible state-control trajectory $(x^*(t), u^*(t))$, $t \in [t_0, T]$, for the optimal control problem (3.52)–(3.56), with fixed finite horizon $T > t_0$ and free endpoint x_T.

(1) If $H(t, x, u, \psi)$ is concave in (x, u), and there exists an absolutely continuous function $\psi : [t_0, T] \to \mathbb{R}^n$ such that

$$\dot{\psi}(t) = -H_x(t, x^*(t), u^*(t), \psi(t)), \quad \forall t \in [t_0, T], \quad \psi(T) = 0,$$

and

$$u^*(t) \in \arg\max_{u \in \mathcal{U}} H(t, x^*(t), u, \psi(t)), \quad \forall t \in [t_0, T],$$

then $(x^*(t), u^*(t))$, $t \in [t_0, T]$, is an optimal state-control trajectory.

(2) If, in addition to the hypotheses in (1), $H(t, x, u, \psi)$ is strictly concave in (x, u), then the optimal state-control trajectory $(x^*(t), u^*(t))$, $t \in [t_0, T]$, is unique.

Proof (1) The state-control trajectory $(x^*(t), u^*(t))$, $t \in [t_0, T]$, is optimal if and only if for any admissible state-control trajectory $(x(t), u(t))$, $t \in [t_0, T]$,

$$\Delta = \int_{t_0}^{T} h(t, x^*(t), u^*(t))\, dt - \int_{t_0}^{T} h(t, x(t), u(t))\, dt \geq 0.$$

But since $H(t, x, u, \psi) = h(t, x, u) + \langle \psi, f(t, x, u) \rangle = h(t, x, u) + \langle \psi, \dot{x} \rangle$, it follows that

$$\Delta = \int_{t_0}^{T} \left[H(t, x^*(t), u^*(t), \psi(t)) - H(t, x(t), u(t), \psi(t)) + \langle \psi(t), \dot{x}(t) - \dot{x}^*(t) \rangle \right] dt.$$

By assumption, the Hamiltonian $H(t, x, u, \psi)$ is concave in (x, u), so

$$\left\langle \frac{\partial H(t, x^*, u^*, \psi)}{\partial (x, u)}, (x^*, u^*) - (x, u) \right\rangle \leq H(t, x^*, u^*, \psi) - H(t, x, u, \psi);$$

the maximality condition implies further that

$$0 \leq \langle H_u(t, x^*, u^*, \psi), u^* - u \rangle.$$

Combining the last two inequalities with the adjoint equation yields that

$$\langle \dot{\psi}, x - x^* \rangle = \langle H_x(t, x^*, u^*, \psi), x^* - x \rangle \leq H(t, x^*, u^*, \psi) - H(t, x, u, \psi).$$

Optimal Control Theory

Hence, using the transversality condition $\psi(T) = 0$ and the fact that $x(t_0) = x^*(t_0) = x_0$, it is[20]

$$\Delta \geq \int_{t_0}^{T} [\langle \dot{\psi}(t), x(t) - x^*(t) \rangle + \langle \psi(t), \dot{x}(t) - \dot{x}^*(t) \rangle] dt$$

$$= \langle \psi(T), x(T) - x^*(T) \rangle = 0,$$

which in turn implies that $(x^*(t), u^*(t))$, $t \in [0, T]$, is an optimal state-control trajectory.

(2) Because of the strict concavity of $H(t, x, u, \psi)$, if there are two optimal state-control trajectories $(x^*(t), u^*(t))$ and $(\hat{x}^*(t), \hat{u}^*(t))$ (for $t \in [0, T]$), then any convex combination $\lambda(x^*(t), u^*(t)) + (1 - \lambda)(\hat{x}^*(t), \hat{u}^*(t))$ for $\lambda \in (0, 1)$ yields a strictly higher value of the objective functional unless $(x^*(t), u^*(t))$ and $(\hat{x}^*(t), \hat{u}^*(t))$ are the same (possibly up to a measure-zero set of time instances in $[0, T]$), which concludes the proof. ∎

3.5 Infinite-Horizon Optimal Control

In practical decision problems, it may be either difficult or impossible to specify a plausible planning horizon T. For example, an electric power utility may not operate with any finite time horizon in mind, reflecting a going concern for preserving its economic viability indefinitely. Similarly, when considering the evolution of an economy in terms of its capital growth, for example, it is natural to assume an infinite time horizon. Accordingly, consider the following (simple) infinite-horizon optimal control problem:

$$J(u) = \int_{t_0}^{\infty} e^{-rt} h(x(t), u(t)) \, dt \longrightarrow \max_{u(\cdot)}, \qquad (3.69)$$

$$\dot{x}(t) = f(x(t), u(t)), \quad x(t_0) = x_0, \qquad (3.70)$$

$$u(t) \in \mathcal{U}, \quad \forall t, \qquad (3.71)$$

$$t \in [t_0, \infty), \qquad (3.72)$$

given (t_0, x_0), where $r > 0$ is a given discount rate, and the control set $\mathcal{U} \subset \mathbb{R}^m$ is nonempty, convex, and compact. Also assume that

---

20. Note that this inequality, and with it the Mangasarian sufficiency theorem, remains valid if the problem is in part endpoint-constrained.

the primitives of the problem satisfy assumptions A1 and A2 (see section 3.4). In addition, assume either that h is bounded, or that the state space $\mathcal{X} \subset \mathbb{R}^n$ is a compact invariant set. In either case the image $h(\mathcal{X}, \mathcal{U})$ is bounded, so the objective functional $J(u)$ must be bounded as well.

An infinite-horizon optimal control problem of this kind was considered by Pontryagin et al. (1962, ch. 4). With the exception of the transversality condition (3.27), the optimality conditions (3.26)–(3.29) of the finite-horizon PMP, as formulated in proposition 3.4, carry over to the infinite-horizon problem (3.69)–(3.72) by taking the limit for $T \to \infty$. Intuitively, one might expect that

$$\lim_{T \to \infty} \psi(T) = \lim_{T \to \infty} e^{-rT} v(T) = 0 \qquad (3.73)$$

would be a natural transversality condition for the infinite-horizon optimal control problem (3.69)–(3.72). Yet, the following example demonstrates that the natural transversality condition (3.73) should not be expected to hold in general.

Example 3.6 (Halkin 1974) Assume that the discount rate r is zero, and consider the infinite-horizon optimal control problem

$$J(u) = \int_0^\infty (1 - x(t))u(t)\, dt \longrightarrow \max_{u(\cdot)},$$

$$\dot{x}(t) = (1 - x(t))u(t), \quad x(0) = 0,$$

$$u(t) \in [0, 1], \quad \forall t,$$

$$t \in [0, \infty).$$

Using the state equation and the initial condition, it is clear that

$$J(u) = \lim_{T \to \infty} \int_0^T \dot{x}(t)\, dt = \lim_{T \to \infty} x(T) \le 1.$$

Indeed, integrating the state equation directly yields

$$x(t) = 1 - \exp\left[-\int_0^t u(s)\, ds\right],$$

so that any $u(t) \in [0, 1]$, $t \ge 0$, with $\int_0^t u(s)\, ds \to \infty$ as $t \to \infty$ is optimal, and for any such optimal control $u^*(t)$, the optimal value of the objective function is $J^* = J(u^*) = 1$. For example, if one chooses the constant optimal control $u^*(t) \equiv u_0 \in (0,1)$, then by the maximality condition of the PMP

$$u_0 \in \arg\max_{u \in [0,1]} \{(1 - x^*(t))(1 + \psi(t))u\},$$

so $\psi(t) \equiv -1$. Hence, $\psi(t)$ cannot converge to zero as $t \to \infty$. □

The available transversality condition in infinite-horizon problems is often weaker (and sometimes stronger) than the natural transversality in (3.73).[21] From a practical point of view it is often useful to first try natural transversality, and then to prove the optimality of the resulting trajectory by a different argument, using the specific structure of the problem at hand. The most likely solution scenario is that the candidate solution determined by the conditions of the PMP converges to a (dynamically optimal) steady state $\bar{x}$ as $T \to \infty$. The latter is determined as part of a *turnpike* $(\bar{x}, \bar{v}; \bar{u})$, which consists of an equilibrium state/co-state $(\bar{x}, \bar{v})$ of the Hamiltonian system and an equilibrium control $\bar{u}$, so that

$$0 = f(\bar{x}, \bar{u}), \tag{3.74}$$

$$0 = r\bar{v} - \hat{H}_x(\bar{x}, \bar{u}, \bar{v}), \tag{3.75}$$

$$\bar{u} \in \arg\max_{u \in \mathcal{U}} \hat{H}(\bar{x}, u, \bar{v}). \tag{3.76}$$

Intuitively, a turnpike is an optimal equilibrium state, which would be maintained at the optimum should the system be started at that state.[22] The following remark provides a different perspective on how the turnpike can be obtained computationally.

Remark 3.17 (Implicit Programming Problem) Feinstein and Luenberger (1981) developed an alternative approach to determining a turnpike.

---

21. For example, from the Mangasarian sufficiency theorem (proposition 3.7) one can obtain the transversality condition $\lim\inf_{T \to \infty} \langle \psi(T), x(T) - x^*(T) \rangle = 0$, given any admissible trajectory $x(t)$, $t \geq 0$. Other transversality conditions can be obtained by successive approximation of the infinite-horizon problem by a sequence of finite-horizon problems (Aseev 1999). The corresponding results are, because of their technical complexity, at present available only in specialized monographs (Aseev and Kryazhimskii 2007; Aseev 2009).

22. It is important to note the difference between a turnpike and an *optimal steady state* tuple $(\hat{x}_0, \hat{u}_0)$. The optimal steady state tuple $(\hat{x}_0, \hat{u}_0)$ maximizes $h(x,u)$ subject to $f(x,u) = 0$ and $u \in \mathcal{U}$, and is therefore independent of the discount rate. In economics, this is often referred to as the golden rule. At the turnpike, on the other hand, the system evolves optimally over time, taking into account that moving to a different state (such as $\hat{x}_0$, implied by the golden rule) may be too costly (or that moving away from a state such as $\hat{x}_0$ using a nonstationary control may increase the decision maker's discounted payoff).

Instead of thinking of the turnpike $(\bar{x}, \bar{u})$ as an equilibrium of the Hamiltonian system, they showed that any solution to the so-called *implicit programming problem*

$$\bar{x} \in \arg\max_{(x,u) \in \mathcal{X} \times \mathcal{U}} h(x, u), \tag{3.77}$$

subject to

$$f(x, u) = r(x - \bar{x}), \quad u \in \mathcal{U}, \tag{3.78}$$

constitutes in fact a turnpike. The problem is termed an implicit programming problem because it essentially determines $\bar{x}$ as a fixed point, featuring its solution $\bar{x}$ in (3.77) and in the constraint (3.78). Indeed, the optimality conditions for the constrained optimization problem (3.77)–(3.78) are equivalent to finding an equilibrium of the Hamiltonian system and using the maximality condition, that is, equivalent to relations (3.74)–(3.76). □

Example 3.7 (Infinite-Horizon Optimal Consumption) Consider an infinite-horizon version of the optimal consumption problem in example 3.4. To keep things as simple as possible, the present example concentrates on the special case where the investor's utility function $U(c) = \ln(c)$, and there is no excess return, so that $\alpha = r$. Then the investor solves the infinite-horizon optimal control problem

$$J(c) = \int_0^\infty e^{-rt} \ln(c(t)) \, dt \longrightarrow \max_{c(\cdot)},$$

$$\dot{x}(t) = rx(t) - c(t), \quad x(0) = x_0,$$

$$c(t) \in [0, c_{\max}], \quad \forall t \geq 0,$$

$$t \in [0, \infty),$$

where the initial capital $x_0 > 0$ and the spending limit $c_{\max} \geq rx_0$ are given constants. The current-value Hamiltonian for this problem is given by

$$\hat{H}(x, c, \nu) = \ln(c) + \nu(rx - c).$$

From the PMP, one obtains the adjoint equation $\dot{\nu}(t) = r\nu(t) - r\nu(t) \equiv 0$, so the current-value adjoint variable is constant, $\nu(t) \equiv \nu(0)$. By the maximality condition, $c(t) = 1/\nu(t) \equiv 1/\nu(0)$. Thus, integrating the state equation over any finite time horizon $T > 0$, it is

$$x(T) = x_0 e^{rT} - \int_0^T \frac{e^{rt}}{v(0)} dt = x_0 e^{rT} - \frac{e^{rT}-1}{rv(0)} \equiv x_0,$$

for all $T > 0$, provided that $v(0) = 1/(rx_0)$. Hence, the unique turnpike is $(\bar{x}, \bar{c}) = (x_0, rx_0)$. The golden rule of optimal consumption for this problem is that when there is no excess return, it is best to maintain a constant capital and to consume at a rate equal to the interest revenue rx_0. □

Example 3.8 (Optimal Advertising) Consider the system

$$\dot{x} = (1-x)a^\kappa - \beta x \quad (3.79)$$

with some constants $\beta > 0$ and $\kappa \in (0,1)$. The control variable $a \in [0, \bar{a}]$ (with some upper bound $\bar{a} > 0$) denotes the intensity of a firm's advertising activity and the state variable $x \in [0,1]$ the installed base, namely, the percentage of the entire population of potential customers who have already bought the firm's product. The problem is, given an initial state $x(0) = x_0 \in (0,1)$, to maximize the firm's discounted profits[23]

$$J(a) = \int_0^\infty e^{-rt} \left((1-x(t))a^\kappa(t) - ca(t) \right) dt,$$

where $c > 0$ is the cost of advertising. Note first that a simple integration by parts using (3.79) yields $J(a) = c\hat{J}(u) - x_0$, where $u = a^\kappa$ and

$$\hat{J}(u) = \int_0^\infty e^{-rt} \left(\gamma x(t) - u^{1/\kappa}(t) \right) dt,$$

with $\gamma = (r+\beta)/c$. Given an optimal state-control trajectory $(x^*(t), u^*(t))$, $t \geq 0$, by the maximality condition of the PMP it is

$$u^*(t) = \min\{\bar{a}^\kappa, (\kappa(1-x(t))v(t))^{\frac{\kappa}{1-\kappa}}\} > 0, \quad \forall t \geq 0,$$

and for $\bar{a} > (\kappa \bar{v})^{1/\kappa}$, the Hamiltonian system becomes

$$\dot{x}^* = (1-x^*)^{\frac{1}{1-\kappa}} (\kappa v)^{\frac{\kappa}{1-\kappa}} - \beta x^*,$$

$$\dot{v} = -\gamma + (r+\beta)v + (\kappa(1-x^*))^{\frac{\kappa}{1-\kappa}} v^{\frac{1}{1-\kappa}}.$$

This system possesses a unique equilibrium $(\bar{x}, \bar{v})$, characterized by $(\dot{x}, \dot{v})|_{(\bar{x}, \bar{v})} = 0$, or equivalently

---

23. The demand is equal to the positive inflow to the installed base, $(1-x)a^\kappa$. The firm is assumed to be a price taker in a market for durable goods, where the price has been normalized to 1. The products have a characteristic lifetime of $1/\beta$ before being discarded. The exponent κ models the effect of decreasing returns to investment in advertising.

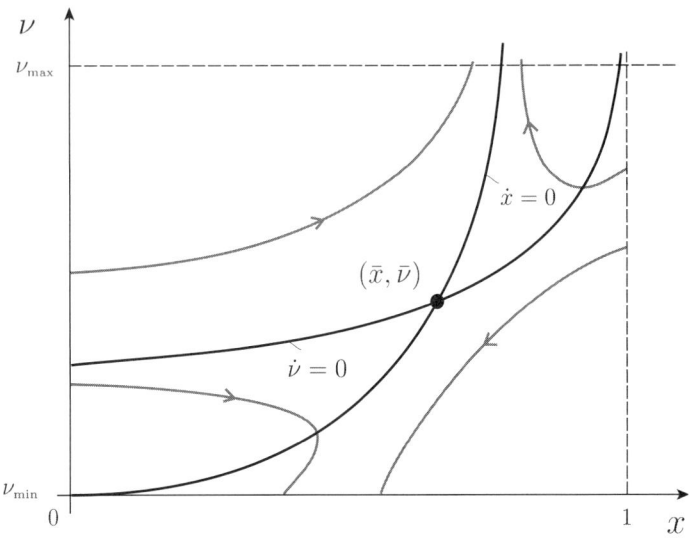

Figure 3.5
Trajectories of a Hamiltonian system (see example 3.8).

$$\kappa \gamma = \left(\frac{r+\beta}{(1-\bar{x})^{\frac{1}{\kappa}}} + \frac{\beta \bar{x}}{(1-\bar{x})^{\frac{1-\kappa^2}{\kappa(1-\kappa)}}} \right) (\beta \bar{x})^{\frac{1-\kappa}{\kappa}},$$

$$\bar{\nu} = \frac{(\beta \bar{x})^{\frac{1-\kappa}{\kappa}}}{\kappa(1-\bar{x})^{\frac{1}{\kappa}}}.$$

For $\bar{u}$ large enough there is a (unique) turnpike; a corresponding qualitative phase diagram is given in figure 3.5. Weber (2006) showed that the adjoint variable ν is bounded. But the only trajectories that can satisfy such bounds are such that they asymptotically converge to the turnpike. Because the equilibrium $(\bar{x}, \bar{\nu})$ is a saddle point, all other trajectories of the Hamiltonian system cannot satisfy the bounds on the adjoint variable. Thus, uniqueness, asymptotic convergence, and the precise form of the optimal solution are essentially implied when the adjoint variable is bounded.[24] □

---

24. This example is based on Weber's (2006) discussion of a classic model for optimal advertising spending by Vidale and Wolfe (1957), as an application of a stronger version of the maximum principle for a class of infinite-horizon problems, which guarantees (instead of a standard transversality condition) that the current-value adjoint variable ν is bounded.

Remark 3.18 (Bellman Equation) Because of the time invariance of the infinite-horizon optimal control problem (3.69)–(3.72), its value function $V(x)$ does not depend on time. Consequently, the relevant HJB equation (3.20) for $T \to \infty$ simplifies to

$$rV(x) = \max_{u \in \mathcal{U}} \{h(x,u) + \langle V_x(x), f(x,u) \rangle\}, \qquad \forall x \in \mathcal{X}, \tag{3.80}$$

which is commonly referred to as the *Bellman equation*. □

Example 3.9 (Infinite-Horizon Linear-Quadratic Regulator) Consider a time-invariant infinite-horizon version of the linear-quadratic regulator problem in example 3.3, where all the matrices are constant, and the time horizon $T \to \infty$. If one sets $V(x) = -x'Qx$ for some positive definite matrix $Q \in \mathbb{R}^{n \times n}$, then the Bellman equation (3.80) becomes

$$0 = \max_{u \in \mathcal{U}} \{-x' \left(R - rQ + QA + A'Q\right) x - u'Su\}, \qquad \forall x \in \mathbb{R}^n,$$

which, as in the finite-horizon case, yields a linear optimal feedback law,

$$\mu(x) = -S^{-1}B'Qx, \qquad \forall x \in \mathbb{R}^n,$$

where Q solves the algebraic matrix Riccati equation

$$-rQ + QA + A'Q + QBS^{-1}B'Q = -R.$$

Hence, the optimal control becomes $u^*(t) = \mu(x^*(t))$, for all $t \geq 0$, which, together with the IVP $\dot{x}^*(t) = Ax^*(t) + Bu^*(t)$, $x^*(0) = x_0$, determines the optimal state-control trajectory. □

3.6 Supplement 1: A Proof of the Pontryagin Maximum Principle

This section provides an (almost complete) rigorous proof of the PMP as formulated in proposition 3.5 for the finite-horizon optimal control problem (3.34)–(3.39) with state-control constraints. The proof has two major parts. The first part establishes a simplified version of the maximum principle for a problem without state-control constraints and without endpoint constraints; the second part considers the missing constraints in the linear-concave case, where (strictly speaking) the system function f is linear in the control and the cost function h is concave in the control. In that case it is further assumed that the control-constraint set is always convex and compact.

Both parts of the proof are very instructive. The first part shows how to construct appropriate variations in the endpoint data and the control, which leads to the adjoint equation (including transversality), the maximality condition, and the endpoint-optimality conditions. The common intuition for these optimality conditions is that at the optimum the variations vanish, quite similar to the standard interpretation of Fermat's lemma in calculus (see proposition A.11). The techniques in the second part of the proof are geared toward constructing a sequence of simplified optimal control problems that converges to the constrained optimal control problem (3.34)–(3.39). In the simplified optimal control problems, the constraints are relaxed and violations of these constraints are increasingly penalized. One can then show that the sequence of optimality conditions (corresponding to the sequence of simplified problems) converges toward optimality conditions of the constrained problem.

Some of the techniques used in the second part of the proof are beyond the scope of this book. The reader can skip that part without any consequence. Restricting attention to a somewhat simpler case clarifies the main intuition and required tools. Note also that all applications with state-control constraints in this book (see chapter 5) are linear-concave. A complete proof of the PMP for the optimal control problem (3.34)–(3.39) with additional pure state constraints (see remark 3.12) is given, for example, by Arutyunov (2000, ch. 2).

3.6.1 Problems without State-Control Constraints

Consider a simplified version of the finite-horizon optimal control problem (3.34)–(3.39), where the functions K^1, K^2, and R vanish, and the control $u(t)$ lies in the uniformly bounded, measurable control-constraint set $\mathcal{U}(t)$. The resulting problem is therefore free of constraints on the endpoint data ω and does not exhibit any state-control constraints:

$$J(u,\omega) = \int_{t_0}^{T} h(t,x(t),u(t))\,dt + K^0(\omega) \longrightarrow \max_{u(\cdot),\omega}, \qquad (3.81)$$

$$\dot{x}(t) = f(t,x(t),u(t)), \qquad x(t_0) = x_0, \qquad x(T) = x_T, \qquad (3.82)$$

$$u \in \mathcal{U}(t), \qquad \forall\, t, \qquad (3.83)$$

$$t \in [t_0, T], \qquad t_0 < T. \qquad (3.84)$$

Let (x^*, u^*, ω^*), with $\omega^* = (t_0^*, x_0^*; T^*, x_T^*)$, be a solution to the initial value problem (3.82), so that

$$\dot{x}^*(t) = f(t, x^*(t), u^*(t)), \qquad x^*(t_0^*) = x_0^*, \qquad \forall\, t \in [t_0^*, T^*],$$

Optimal Control Theory

and by compatibility of the endpoint data with the state trajectory, it is $x^*(T^*) = x_T^*$. Consider now the adjoint equation (3.42),

$$\dot{\psi}(t) = -\lambda_0 h_x(t, x^*(t), u^*(t)) - \psi'(t) f_x(t, x^*(t), u^*(t)). \tag{3.85}$$

By proposition 2.15 there exists a solution $\psi(t)$, $t \in [t_0^*, T^*]$, to this linear time-varying ODE, with initial condition

$$\psi(t_0^*) = -\lambda_0 K_{x_0}^0(\omega^*). \tag{3.86}$$

Let $\lambda_0 = 1$. The following three steps show that the adjoint variable ψ, which by construction satisfies the adjoint equation, is also compatible with the transversality, maximality, and endpoint-optimality conditions of the PMP in proposition 3.5.

Step 1: Transversality Fix a vector $\hat{x} \in \mathbb{R}^n$. Then, by continuous dependence of the solutions to a well-posed ODE with respect to initial conditions,[25] for any $\alpha \geq 0$ the IVP

$$\dot{x}(t) = f(t, x(t), u^*(t)), \qquad x(t_0^*) = x_0^* + \alpha \hat{x} \tag{3.87}$$

has a solution $x(t, \alpha)$, $t \in [t_0^*, T^*]$. Let

$$\omega(\alpha) = (t_0^*, x(t_0^*, \alpha); T^*, x(T^*, \alpha))$$

be the corresponding vector of endpoint data, and let $J^* = J(u^*, \omega^*)$ be the maximized objective. Then

$$\frac{J(u^*, \omega(\alpha)) - J^*}{\alpha} \leq 0, \qquad \forall \alpha > 0.$$

Taking the limit for $\alpha \to 0^+$ implies that $\left.\frac{dJ(u^*, \omega(\alpha))}{d\alpha}\right|_{\alpha=0^+} \leq 0$, that is,

$$\langle K_{x_0}^0(\omega^*), \hat{x} \rangle + \langle K_{x_T}^0(\omega^*), x_\alpha(T^*, 0) \rangle$$

$$+ \int_{t_0^*}^{T^*} \langle h_x(t, x(t, 0), u^*(t)), x_\alpha(t, 0) \rangle dt \leq 0. \tag{3.88}$$

Recall the discussion in chapter 2 (remark 2.3) on the evolution of the sensitivity matrix $S(t) = x_\alpha(t, 0)$,[26] which satisfies the linear time-variant IVP

---

25. If one sets $y = x - \alpha \hat{x}$, $y_0 = x_0^* + \alpha \hat{x}$, and $\hat{f}(t, y, \alpha) \equiv f(t, y + \alpha \hat{x}, u^*(t))$, then the IVP $\dot{y} = \hat{f}(t, y, \alpha)$, $y(t_0) = y_0$, is equivalent to the IVP (3.87), and proposition 2.5 guarantees continuous dependence.
26. The sensitivity "matrix" here is just a vector.

$$\dot{S} = f_x(t, x(t, 0), u^*(t))S, \qquad S(t_0^*) = \hat{x}. \qquad (3.89)$$

Since $x^*(t) = x(t, 0)$ and $\psi(t)$ satisfies the adjoint equation (3.85),

$$-\frac{d}{dt}\langle \psi(t), S(t) \rangle = \langle H_x(t, x^*(t), u^*(t), \psi(t)), S(t) \rangle - \langle \psi(t), f_x(t, x^*(t), u^*(t))S(t) \rangle$$
$$= \langle h_x(t, x^*(t), u^*(t)), S(t) \rangle,$$

for all $t \in [t_0^*, T^*]$. Integrating the last identity between $t = t_0^*$ and $t = T^*$, and combining the result with inequality (3.88), yields that

$$\langle \psi(t_0^*), S(t_0^*) \rangle - \langle \psi(T^*), S(T^*) \rangle = \int_{t_0^*}^{T^*} \langle h_x(t, x^*(t), u^*(t)), S(t) \rangle dt$$
$$\leq -\langle K_{x_0}^0(\omega^*), \hat{x} \rangle - \langle K_{x_T}^0(\omega^*), S(T^*) \rangle.$$

Let $\Phi(t, t_0^*)$ be the fundamental matrix (see section 2.3.2) so that $\Phi(t_0^*, t_0^*) = I$, and $S(T^*) = \Phi(T^*, t_0^*)\hat{x}$. Using the initial condition (3.86) and the initial condition in (3.89) gives

$$0 \leq \langle \psi(T^*) - K_{x_T}^0(\omega^*), \Phi(T^*, t_0^*)\hat{x} \rangle,$$

independent of the chosen $\hat{x}$. Hence, the transversality condition,

$$\psi(T^*) = K_{x_T}^0(\omega^*),$$

must necessarily hold because $\Phi(T^*, t_0^*)$ is a nonsingular matrix by lemma 2.5(6), which states that the Wronksian, $\det \Phi(t, t_0^*)$, is positive.

Step 2: Maximality Since (in the absence of state-control constraints) the set-valued mapping $\mathcal{U}(t)$ is measurable, there exists a sequence of measurable selectors $\{\sigma^k(\cdot)\}_{k \in \mathbb{N}}$ such that the set $\{\sigma^k(t)\}_{k \in \mathbb{N}}$ is everywhere dense in $\mathcal{U}(t)$ for a.a. $t \in [t_0^*, T^*]$ (see footnote 15). It can be shown that the maximality condition holds for a given Lebesgue point $\hat{t} \in (t_0^*, T^*)$ of the function $f(t, x^*(t), u^*(t))$.[27] For a given integer $k \geq 0$ and real number $\alpha > 0$, let

$$u^k(t, \alpha) = \begin{cases} \sigma^k(t) & \text{if } t \in (\hat{t} - \alpha, \hat{t}), \\ u^*(t), & \text{otherwise,} \end{cases}$$

---

27. The point $\hat{t}$ is a *Lebesgue point* of a measurable function $\varphi(t)$ if

$$\lim_{\varepsilon \to 0} (1/\varepsilon) \int_{\hat{t}-\varepsilon}^{\hat{t}} \|\varphi(\hat{t}) - \varphi(t)\| \, dt = 0,$$

that is, if φ does not vary too much at that point. Because the control is essentially bounded, almost all points are Lebesgue points.

Optimal Control Theory

be a *needle variation* of the optimal control u^*. Let $x^k(t, \alpha)$ be the solution to the corresponding IVP,

$$\dot{x}(t) = f(t, x(t), u^k(t, \alpha)), \qquad x(t_0^*) = x_0^*;$$

this solution exists on $[t_0^*, T^*]$, as long as α is sufficiently small. Then, because $\hat{t}$ is a Lebesgue point, it follows that

$$\frac{x^*(\hat{t}) - x^*(\hat{t} - \alpha)}{\alpha} = f(\hat{t}, x^*(\hat{t}), u^*(\hat{t})) + O(\alpha),$$

and

$$\frac{x^k(\hat{t}, \alpha) - x^*(\hat{t} - \alpha)}{\alpha} = f(\hat{t}, x^k(\hat{t}, \alpha), \sigma^k(\hat{t})) + O(\alpha),$$

where $O(\cdot)$ is the familiar Landau notation (see footnote 16). Hence, the limit

$$\Delta(\hat{t}) \equiv \lim_{\alpha \to 0^+} \frac{x^k(\hat{t}, \alpha) - x^*(\hat{t})}{\alpha} = f(\hat{t}, x^*(\hat{t}), \sigma^k(\hat{t})) - f(\hat{t}, x^*(\hat{t}), u^*(\hat{t})) \qquad (3.90)$$

is well-defined. Since the optimal control $u^*(t)$ is applied for $t > \hat{t}$, the state trajectories $x^*(t)$ and $x^k(t, \alpha)$ satisfy the same ODE for those t. By the same logic as in remark 2.3 (on sensitivity analysis),

$$\dot{\Delta}(t) = f_x(t, x^*(t), u^*(t)) \Delta(t), \qquad \forall t \in (\hat{t}, T^*),$$

where $\Delta(t) = x_\alpha^k(t, 0)$. Thus, using the adjoint equation (3.85), it is

$$\frac{d}{dt} \langle \psi(t), \Delta(t) \rangle = -\langle h_x(t, x^*(t), u^*(t)), \Delta(t) \rangle, \qquad \forall t \in (\hat{t}, T^*).$$

Integrating the last equation between $t = \hat{t}$ and $t = T^*$ and using the transversality condition (3.86) at the right endpoint (see step 1) yields

$$\langle \psi(\hat{t}), \Delta(\hat{t}) \rangle = \int_{\hat{t}}^{T^*} \langle h_x(s, x^*(s), u^*(s)), \Delta(s) \rangle \, ds + \langle K_{x_T}^0(\omega^*), \Delta(T^*) \rangle. \qquad (3.91)$$

If one sets $J^* = J(u^*, \omega^*)$, then

$$0 \leq \lim_{\alpha \to 0^+} \frac{J^* - J(u^k, \omega(\alpha))}{\alpha}$$

$$= \lim_{\alpha \to 0^+} \int_{\hat{t} - \alpha}^{\hat{t}} \frac{h(t, x^*(t), u^*(t)) - h(t, x^k(t, \alpha), \sigma^k(t))}{\alpha} \, dt$$

$$+ \lim_{\alpha \to 0^+} \int_{\hat{t}}^{T^*} \frac{h(t, x^*(t), u^*(t)) - h(t, x^k(t, \alpha), u^*(t))}{\alpha} dt$$

$$+ \lim_{\alpha \to 0^+} \frac{K^0(\omega^*) - K^0(\omega(\alpha))}{\alpha}$$

$$= h(\hat{t}, x^*(\hat{t}), u^*(\hat{t})) - h(\hat{t}, x^*(\hat{t}), \sigma^k(\hat{t})) - \int_{\hat{t}}^{T^*} \langle h_x(t, x^*(t), u^*(t)), \Delta(t) \rangle dt$$

$$- \langle K_{x_T}^0(\omega^*), \Delta(T^*) \rangle$$

$$= h(\hat{t}, x^*(\hat{t}), u^*(\hat{t})) - h(\hat{t}, x^*(\hat{t}), \sigma^k(\hat{t})) - \langle \psi(\hat{t}), \Delta(\hat{t}) \rangle,$$

where the last equal sign is due to (3.91). Thus substituting the expression for $\Delta(\hat{t})$ in (3.90) gives

$$H(\hat{t}, x^*(\hat{t}), \sigma^k(\hat{t}), \psi(\hat{t})) \leq H(\hat{t}, x^*(\hat{t}), u^*(\hat{t}), \psi(\hat{t})),$$

namely, maximality at $t = \hat{t}$. The previous inequality holds for any $k \geq 0$. The maximality condition (3.44) is therefore established at $\hat{t}$, because the sequence $\{\sigma^k(\hat{t})\}_{k=0}^{\infty}$ is by assumption everywhere dense in $\mathcal{U}(\hat{t})$. This implies that maximality holds a.e. (almost everywhere) on $[t_0^*, T^*]$.

Step 3: Endpoint Optimality Now consider the transversality with respect to the endpoint constraint. If for $\alpha > 0$ one sets $\omega(\alpha) = (t_0^*, x_0^*; T^* - \alpha, x^*(T^* - \alpha))$, then

$$0 \leq \frac{J^* - J(u^*, \omega(\alpha))}{\alpha}$$

$$= \frac{1}{\alpha} \int_{T^*-\alpha}^{T^*} h(t, x^*(t), u^*(t)) \, dt + \frac{K^0(t_0^*, x_0^*; T^*, x_T^*)}{\alpha}$$

$$- \frac{K^0(t_0^*, x_0^*; T^* - \alpha, x^*(T^* - \alpha))}{\alpha}$$

$$= \frac{1}{\alpha} \int_{T^*-\alpha}^{T^*} \left(h(t, x^*(t), u^*(t)) + \frac{dK^0(t_0^*, x_0^*; t, x^*(t))}{dt} \right) dt$$

$$= \int_{T^*-\alpha}^{T^*} \frac{H(t, x^*(t), u^*(t), K_{x_T}^0(t_0^*, x_0^*; t, x^*(t))) + K_T^0(t_0^*, x_0^*; t, x^*(t))}{\alpha} dt$$

$$= \sup_{u \in \mathcal{U}(T^*, x_T^*)} H(T^*, x_T^*, u, K_{x_T}^0(\omega^*)) + K_T^0(\omega^*) + O(\alpha),$$

Optimal Control Theory

where $J^* = J(u^*, \omega^*)$. Recall the small Lagrangian in (3.41), $L = \lambda_0 K^0 = K^0$, so that taking the limit for $\alpha \to 0^+$ implies, via Fermat's lemma, that the endpoint-optimality condition (3.49) holds. Condition (3.48) obtains in a completely analogous manner.

3.6.2 Problems with State-Control Constraints

Now consider the general finite-horizon optimal control problem (3.34)–(3.39), referred to here as problem (P). The proof outline proceeds in six steps (and omits some technical details).

Step 1: Approximate Problem (P) by a Sequence of Problems $\{(\bar{P}_k)\}_{k=1}^{\infty}$
Choose positive numbers ε, δ, and[28]

$$\bar{u} \geq 2 + \operatorname{ess\,sup}_{t \in [t_0^*, T^*]} \|u^*(t)\|,$$

relative to which the sequence $\{(\bar{P}_k)\}_{k=1}^{\infty}$ of relaxed problems will be defined. For any $k \geq 1$, let

$$h^k(t, x, u) = h(t, x, u) - \delta \|u(t) - u^*(t)\|^2 - k\|R_-(t, x, u^1)\|^2, \qquad (3.92)$$

where $R_- = \min\{0, R\}$ has nonzero (negative) components whenever the state-control constraint (3.37) in (P) is violated. Similarly, for any feasible boundary data $\omega \in \mathbb{R}^{2(1+n)}$ set

$$K^{0,k}(\omega) = K^0(\omega) - \|\omega - \omega^*\|^2 - k\left(\|K_-^1(\omega)\|^2 + \|K^2(\omega)\|^2\right), \qquad (3.93)$$

where $K_-^1 = \min\{0, K^1\}$, to penalize deviations from the endpoint constraints (3.36) in (P). For any $k \geq 1$ the relaxed problem $(\bar{P}_k)$ can now be formulated, given an optimal solution (x^*, u^*, ω^*) to the original problem (P):

$$\begin{cases} J^k(u, \omega) = \int_{t_0}^T h^k(t, x(t), u(t))\, dt + K^{0,k}(\omega) \longrightarrow \max_{u(\cdot), \omega} \\ \text{s.t.} \\ \dot{x}(t) = f(t, x(t), u(t)), \quad x(t_0) = x_0, \quad x(T) = x_T, \\ \varepsilon \geq \|x(t) - x^*(t)\|_\infty + \|\omega - \omega^*\|^2, \quad (*) \\ u(t) \in \mathcal{U}_{\hat{\varepsilon}, \bar{u}}(t, x(t)), \quad \forall\, t, \\ t \in [t_0, T], \quad t_0 < T, \end{cases} \qquad (\bar{P}_k)$$

---

28. The *essential supremum* of a nonempty set $\mathcal{S} \subset \mathbb{R}$, denoted by ess sup $\mathcal{S}$, is the smallest upper bound M such that the set of elements in $\mathcal{S}$ greater than M is of measure zero.

where (with $e = (1, \ldots, 1) \in \mathbb{R}_+^{k_R}$ and $\hat{\varepsilon} > 0$ small) the relaxed constraint set is given by

$$\mathcal{U}_{\hat{\varepsilon}, \bar{u}}(t, x) = \{(u^1, u^2) \in \mathbb{R}^{m_1} \times \mathcal{U}^2(t) : \|(u^1, u^2)\| \leq \bar{u}, R(t, x, u^1) \geq -\hat{\varepsilon} e\}.$$

Denote a solution to the relaxed problem $(\bar{P}_k)$ by (x^k, u^k, ω^k), where $\omega^k = (t_0^k, x_0^k; T^k, x_T^k)$ is the vector of endpoint data. For all $t \notin [t_0^k, T^k]$ extend any such solution continuously in such a way that the state trajectory x^k is constant outside $[t_0^k, T^k]$.

Step 2: Show That the Relaxed Problem $(\bar{P}_k)$ Has a Solution for All k ≥ 1 Let $\{(x^{k,j}, u^{k,j}, \omega^{k,j})\}_{j=1}^{\infty}$ be an admissible *maximizing sequence* for the problem $(\bar{P}_k)$.[29] Since $u^{k,j}$ takes values in the closed ball of $\mathbb{R}^n$ at 0 of radius $\bar{u}$, and $x(t)$ lies, by virtue of the constraint (*) in $(\bar{P}_k)$, in a neighborhood of the (uniformly bounded) $x^*(t)$ for all t, this maximizing sequence is uniformly bounded, which allows the following three conclusions for an appropriate subsequence (for simplicity the original maximizing sequence is identified with this subsequence by relabeling the indices if necessary). First, from the definition of an admissible sequence $\{x^{k,j}\}_{j=1}^{\infty} \subset \mathbf{W}_{1,\infty}$ (see remark A.2) and the uniform boundedness of $\{\dot{x}^{k,j}\}_{j=1}^{\infty}$ this sequence of state trajectories is equicontinuous,[30] so by the Arzelà-Ascoli theorem (proposition A.5) it converges uniformly to $\hat{x}^k$. Second, one obtains pointwise convergence of $\omega^{k,j}$ to $\hat{\omega}^k$ as $j \to \infty$. Third, since the space of admissible controls $\mathbf{L}_{\infty}$ is a subset of the space $\mathbf{L}_2$ (see example A.2), $u^{k,j}$ converges weakly to $\hat{u}^k$ as $j \to \infty$.[31]

Now one can show that in fact the above limits coincide with the solution to $(\bar{P}_k)$, that is,

$$(x^k, u^k, \omega^k) = (\hat{x}^k, \hat{u}^k, \hat{\omega}^k). \tag{3.94}$$

For any $t \in [t_0^k, T^k]$ it is

$$x^{k,j}(t) = x^{k,j}(t_0^{k,j}) + \int_{t_0^{k,j}}^{t} f(\vartheta, x^{k,j}(\vartheta), u^{k,j}(\vartheta))\, d\vartheta,$$

so taking the limit for $j \to \infty$ gives

---

29. The *maximizing sequence* is such that the corresponding sequence of objective values $J(u^{k,j}, \omega^{k,j})$ converges to the optimal value (Gelfand and Fomin, 1963, 193).
30. Equicontinuity is defined in appendix A, footnote 10.
31. By the Banach-Alaoglu theorem the unit ball in $\mathbf{L}_2$ is weakly* (and therefore weakly) compact, so that by the Eberlein-Šmulian theorem it is also weakly sequentially compact (Megginson 1998, 229,248). This property of reflexive Banach spaces can also be deduced from the uniform boundedness principle (Dunford and Schwartz 1958, ch. 2).

Optimal Control Theory

$$\hat{x}^k(t) = \hat{x}^k(\hat{t}_0^k) + \int_{\hat{t}_0^k}^t f(\vartheta, \hat{x}^k(\vartheta), \hat{u}^k(\vartheta))\, d\vartheta.$$

The limiting tuple $(\hat{x}^k, \hat{u}^k, \hat{\omega}^k)$, with $\hat{\omega}^k = (\hat{t}_0^k, \hat{x}_0^k; \hat{T}^k, \hat{x}_T^0)$, solves the IVP

$$\dot{\hat{x}}^k(t) = f(t, \hat{x}(t), \hat{u}^k(t)), \qquad \hat{x}^k(\hat{t}_0^k) = \hat{x}_0^k,$$

for all $t \in [\hat{t}_0^k, \hat{T}^k]$. The state constraint $\varepsilon \geq \|\hat{x}^k - x^*\|_\infty + \|\hat{\omega}^k - \omega^*\|^2$ is satisfied by uniform convergence of the maximizing sequence. Last, the control constraint $\hat{u}^k \in \mathcal{U}_{\hat{\varepsilon},\bar{u}}$ is a.a. satisfied because each $u^{k,j}$, $j = 1, 2, \ldots$, is feasible ($\bar{u}$ has been chosen appropriately large). The weak convergence $u^{k,j} \xrightarrow{w} \hat{u}^k$ as $j \to \infty$ implies, by Mazur's compactness theorem (Megginson 1998, 254), that there exists a sequence $\{v^{k,j}\}_{j=1}^\infty$ with elements in the convex hull co $\{u^{k,j}\}_{j=1}^\infty$, which converges strongly to $\hat{u}^k$ in $\mathbf{L}_2^m[t_0^*, T^*]$.[32] Therefore, equation (3.94) holds, that is, the limit point $(\hat{x}^k, \hat{u}^k, \hat{\omega}^k)$ of the maximizing sequence describes an admissible solution to the relaxed problem $(\bar{P}_k)$.

Step 3: Show That the Solutions of $(\bar{P}_k)_{k \geq 1}$ Converge to the Solution of (P) As before, there exists an admissible tuple $(\hat{x}, \hat{u}, \hat{\omega})$ such that $x^k \rightrightarrows \hat{x}$, $u^k \to \hat{u}$ (a.e.), and $\omega^k \to \hat{\omega}$. Now one can show that

$$(\hat{x}, \hat{u}, \hat{\omega}) = (x^*, u^*, \omega^*), \tag{3.95}$$

in particular that $x^k \rightrightarrows x^*$, $u^k \to u^*$ (a.e.), and $\omega^k \to \omega^*$.

Let $J^k(u, \omega) = \int_{t_0}^T h^k(t, x(t), u(t))\, dt + K^{0,k}(\omega)$, as in step 1. The uniform boundedness of the state-control trajectories implies that there exists a constant $M > 0$ such that $M \geq J(u^k, \omega^k) - J(u^*, \omega^*)$ for all k. Since $J^k(u^k, \omega^k) \geq J^k(u^*, \omega^*) = J(u^*, \omega^*)$, it is

$$\frac{M}{k} \geq \int_{t_0^k}^{T^k} \left(\frac{\delta \|u^k - u^*\|^2}{k} + \|R_-\|^2 \right) dt + \frac{\|\omega^k - \omega^*\|^2}{k} + \|K_-^1(\omega^k)\|^2 + \|K^2(\omega^k)\|^2$$

$$\geq 0.$$

Taking the limit for $k \to \infty$ we obtain by continuity of $K_-^1(\cdot)$ and $K^2(\cdot)$ that $\hat{\omega}$ satisfies the endpoint constraints $K^1(\hat{\omega}) \geq 0$ and $K^2(\hat{\omega}) = 0$. Moreover,

$$\lim_{k \to \infty} \int_{t_0^k}^{T^k} \|R_-(t, x^k(t), u^{1,k}(t))\|^2 dt = 0,$$

---

32. The *convex hull* of a set of points $\xi^1, \ldots, \xi^l$ of a real vector space, denoted by co $\{\xi^1, \ldots, \xi^l\}$, is the minimal convex set containing these points.

whence $R_-(t, \hat{x}(t), \hat{u}^1(t)) = 0$ a.e. on $[t_0^*, T^*]$. Hence, it has been shown that the limit $(\hat{x}, \hat{u}, \hat{\omega})$ is admissible in problem (P). This implies

$$J(u^*, \omega^*) \geq J(\hat{u}, \hat{\omega}). \tag{3.96}$$

On the other hand, $J^k(u^k, \omega^k) \geq J^k(u^*, \omega^*) = J(u^*, \omega^*)$, so

$$J(u^k, \omega^k) - \|\omega^k - \omega^*\|^2 - \delta \int_{t_0^k}^{T^k} \|u^k(t) - u^*(t)\|^2 \, dt \geq J(u^*, \omega^*),$$

for all $k \geq 1$. Taking the limit for $k \to \infty$ yields

$$J(\hat{u}, \hat{\omega}) - \|\hat{\omega} - \omega^*\|^2 - \delta \lim_{k \to \infty} \int_{t_0^k}^{T^k} \|u^k(t) - u^*(t)\|^2 \, dt \geq J(u^*, \omega^*),$$

which together with (3.96) implies that $\hat{\omega} = \omega^*$, $\hat{u} = u^*$, and

$$\lim_{k \to \infty} \int_{t_0^*}^{T^*} \|u^k(t) - u^*(t)\|^2 \, dt = 0,$$

so the sequence $\{u^k\}_{k=1}^{\infty}$ converges to u^* a.e. on $[t_0^*, T^*]$.

Step 4: Show That the Problem ($\bar{P}_k$) becomes a Standard Optimal Control Problem ($\bar{P}'_k$) for Large k Because of the uniform convergence of the optimal state trajectories x^k and the pointwise convergence of the boundary data ω^k (as $k \to \infty$) to the corresponding optimal state trajectory x^* and optimal boundary data ω^* of the original problem (P), respectively, the state constraint in the relaxed problem ($\bar{P}_k$) is not binding, namely,

$$\varepsilon > \|x^k - x^*\|_\infty + \|\omega^k - \omega^*\|^2,$$

as long as k is sufficiently large. Hence, for fixed constants ε, δ, and $\bar{u}$ (see step 1) there exists a $k_0 = k_0(\varepsilon, \delta, \bar{u}) \geq 1$ such that for all $k \geq k_0$ the problem ($\bar{P}_k$) can be rewritten equivalently in the form

$$\begin{cases} J^k(u, \omega) = \int_{t_0}^T h^k(t, x(t), u(t)) \, dt + K^{0,k}(\omega) \longrightarrow \max_{u(\cdot), \omega} \\ \text{s.t.} \\ \dot{x}(t) = f(t, x(t), u(t)), \quad x(t_0) = x_0, \quad x(T) = x_T, \\ u(t) \in \mathcal{U}_{\hat{\varepsilon}, \bar{u}}(t, x(t)), \quad \forall t, \\ t \in [t_0, T], \quad t_0 < T. \end{cases} \tag{$\bar{P}'_k$}$$

Optimal Control Theory

Necessary optimality conditions for this type of optimal control problem without state-control constraints were proved in section 3.6.1.

Step 5: Obtain Necessary Optimality Conditions for ($\bar{P}'_k$) Let $H^k(t, x, u, \psi^k, \lambda_0^k) = \lambda_0^k h^k(t, x, u) + \langle \psi^k, f(t, x, u) \rangle$ be the Hamiltonian associated with problem ($\bar{P}'_k$), where $\lambda_0^k \in \mathbb{R}$ is a constant multiplier, and $\psi^k \in \mathbb{R}^n$ is the adjoint variable. The Hamiltonian represents the instantaneous payoff to the decision maker, including the current benefit of the state velocities. The shadow price of the instantaneous payoff is λ_0^k, and the shadow price of the state velocity is given by the adjoint variable ψ^k. Before formulating the necessary optimality conditions for problem ($\bar{P}'_k$), note that because of the definition of the constraint set $\mathcal{U}_{\hat{\varepsilon}, \bar{u}}(t, x^k)$, which relaxes the state-control constraints by a finite increment $\hat{\varepsilon}$, this set is in fact independent of the state for k large enough. In what follows it is therefore assumed that the sequence of relaxed problems has progressed sufficiently.

Maximum Principle for Problem ($\bar{P}'_k$) If (x^k, u^k, ω^k) is an optimal solution for the problem ($\bar{P}'_k$), then there exist an absolutely continuous function $\psi^k : [t_0^k, T^k] \to \mathbb{R}^n$ and a constant $\lambda_0^k > 0$ such that the following relations hold:

- Adjoint equation

$$-\dot{\psi}^k(t) = H_x^k(t, x^k(t), u^k(t), \psi^k(t), \lambda_0^k). \tag{3.97}$$

- Transversality

$$\psi^k(t_0^k) = -\lambda_0^k K_{x_0}^{0,k}(\omega^k), \tag{3.98}$$

$$\psi^k(T^k) = \lambda_0^k K_{x_T}^{0,k}(\omega^k). \tag{3.99}$$

- Maximality

$$u^k(t) \in \arg\max_{u \in \mathcal{U}_{\hat{\varepsilon}, \bar{u}}(t, x^k(t))} H^k(t, x^k(t), u, \psi^k(t), \lambda_0^k), \tag{3.100}$$

a.e. on $[t_0^k, T^k]$.

- Endpoint optimality

$$\sup_{u \in \mathcal{U}_{\hat{\varepsilon}, \bar{u}}(t_0, x_0^k)} H^k(t_0^k, x_0^k, u, -\lambda_0^k K_{x_0}^{0,k}(\omega^k), \lambda_0^k) - \lambda_0^k K_{t_0}^{0,k}(\omega^k) = 0, \tag{3.101}$$

$$\sup_{u \in \mathcal{U}_{\hat{\varepsilon}, \bar{u}}(T, x_T^k)} H^k(T^k, x_0^k, u, \lambda_0^k K_T^{0,k}(\omega^k), \lambda_0^k) + \lambda_0^k K_T^{0,k}(\omega^k) = 0. \tag{3.102}$$

Applying the necessary optimality conditions (3.97)–(3.102) to the relaxed problem $(\bar{P}'_k)$ yields the adjoint equation

$$-\dot{\psi}^k = \lambda_0^k h_x(t, x^k(t), u^k(t)) + (\psi^k)' f_x(t, x^k(t), u^k(t))$$
$$+ \rho^k(t) R_x(t, x^k(t), u^{1,k}(t)), \tag{3.103}$$

for all $t \in [t_0^*, T^*]$, where

$$\rho^k(t) = -2k\lambda_0^k R_-(t, x^k(t), u^{1,k}(t)) \in \mathbb{R}_+^{k_R}. \tag{3.104}$$

The maximality condition (3.100) contains a constrained optimization problem, for which there exist Lagrange multipliers $\zeta^k(t) \in \mathbb{R}_+^{k_R}$ and $\varsigma^k(t) \in \mathbb{R}_+$ such that

$$\lambda_0^k H_u^k(t, x^k(t), u^k(t), \psi^k(t), \lambda_0^k) + \zeta^k(t) R_u(t, x^*(t), u^{1,k}(t)) + \varsigma^k(t) u^k(t) = 0,$$

with complementary slackness conditions

$$\zeta_j^k(t)(R_j(t, x^k(t), u^{1,k}(t)) - \hat{\varepsilon}) = 0, \qquad j \in \{1, \ldots, k_R\},$$

and

$$\varsigma^k(t)(\|u^k(t)\| - \bar{u}) = 0,$$

for all $t \in [t_0^k, T^k]$ (with the usual extension if needed). By step 3, $u^k \to u^*$ a.e. on $[t_0^*, T^*]$. Hence, by Egorov's theorem (Kirillov and Gvishiani 1982, 24), for any $\bar{\delta} > 0$, there is a subset $\Theta_{\bar{\delta}}$ of $[t_0^*, T^*]$ such that $\int_{[t_0^*, T^*] \setminus \Theta_{\bar{\delta}}} dt < \bar{\delta}$ and $u^k \rightrightarrows u^*$ uniformly on $\Theta_{\bar{\delta}}$. Since u^* is feasible, this uniform convergence implies that $\|u^k\| < \bar{u}$ on $\Theta_{\bar{\delta}}$ for k large enough. By virtue of complementary slackness, the corresponding Lagrange multipliers ζ^k and ς^k therefore vanish on $\Theta_{\bar{\delta}}$ as long as k is large enough. In other words,

$$\lambda_0^k H_u^k(t, x^k(t), u^k(t), \psi^k(t), \lambda_0^k) = 0, \tag{3.105}$$

a.e. on $[t_0^*, T^*]$ as long as k is large enough.

Step 6: Derive Necessary Optimality Conditions for (P) The sequence $\{\lambda_0^k\}_{k=1}^\infty$ is uniformly bounded and $\{\psi^k\}_{k=1}^\infty$ is also equicontinuous. Hence, as in step 2, there exist $\psi_{\delta,\bar{u}}$ and $\lambda_{\delta,\bar{u}}$ such that

$$\psi^k \rightrightarrows \psi_{\delta,\bar{u}}, \qquad \lambda_0^k \to \lambda_{\delta,\bar{u}}.$$

As already indicated through the notation, the limits $\psi_{\delta,\bar{u}}$ and $\lambda_{\delta,\bar{u}}$ generically depend on the constants δ and $\bar{u}$. More specifically, these limits correspond to the optimal solution to (P) if h is replaced by $h - \delta\|u - u^*\|^2$ and the additional constraint $\|u\| \leq \bar{u}$ is introduced.

Optimal Control Theory

Adjoint Equation Since by the maximum principle for problem $(\bar{P}'_k)$ (see step 5) it is $\lambda_0^k > 0$, relations (3.97)–(3.99) are positively homogeneous of degree 1 in ψ^k/λ_0^k, and relation (3.100) is positively homogeneous of degree zero, it is possible to multiply equations (3.97)–(3.99) with positive numbers (and relabel the variables λ_0^k and ψ^k back) such that

$$0 < \lambda_0^k + \max_{t \in [t_0^k, T^k]} \|\psi^k(t)\|^2 \leq 1. \tag{3.106}$$

Integrating the components of the adjoint equation (3.103) yields, using the transversality condition (3.99),

$$\psi^k(t) = \int_t^{T^*} (\lambda_0^k h_x(s, x^k(s), u^k(s)) + (\psi^k(s))' f_x(s, x^k(s), u^k(s))) \, ds$$

$$+ \int_t^{T^*} \rho^k(s) R_x(s, x^k(s), u^{1,k}(s)) \, ds \tag{3.107}$$

for all $t \in [t_0^*, T^*]$ (using the standard extension from $[t_0^k, T^k]$ to $[t_0^*, T^*]$ explained at the end of step 1).

Since the total variation of ψ^k on $[t_0^*, T^*]$ is uniformly bounded for all k by (3.106) (every absolutely continuous function is of bounded variation on a compact interval (Taylor 1965, 412)), and the sequence $\{\psi^k\}$ is also uniformly bounded as a consequence of (3.107), by Helly's selection theorem (Taylor 1965, 398) there exists a function $\hat{\psi}$ such that (a subsequence of) the sequence $\{\psi^k\}$ converges to $\hat{\psi}$. By taking the limit in (3.107) for $k \to \infty$, with $\rho^k \to \rho$, one obtains

$$\hat{\psi}(t) = \int_t^{T^*} (H_x(s, x^*(s), u^*(s), \hat{\psi}(s), \lambda_0) + \rho(s) R_x(s, x^*(s), u^{1*}(s))) \, ds,$$

for all $t \in [t_0^*, T^*]$, where $\hat{\psi} = \psi_{\delta, \bar{u}}$. The adjoint equation (3.42) then follows.

Transversality Since $\omega^k \to \omega^*$ as $k \to \infty$ (see step 3), by setting $\lambda_1 = -\lim_{k \to \infty} 2k\lambda_0^k K_-^1(\omega^k) \in \mathbb{R}_+^{n_1}$ and $\lambda_2 = -\lim_{k \to \infty} 2k\lambda_0^k K^2(\omega^k) \in \mathbb{R}^{n_2}$, one obtains from the transversality condition (3.98) for $k \to \infty$ that

$$\hat{\psi}(t_0^*) = -\sum_{j=0}^{2} \lambda_j K_{x_0}^j(\omega^*) = -L_{x_0}(\omega^*, \lambda),$$

where L is the small Lagrangian, and $\lambda = (\lambda_0, \lambda_1, \lambda_2)$. Similarly, for $k \to \infty$ the transversality condition (3.99) yields that

$$\hat{\psi}(T^*) = \sum_{j=0}^{2} \lambda_j K_{x_T}^j(\omega^*) = L_{x_T}(\omega^*, \lambda).$$

This establishes the transversality conditions (3.43) in proposition 3.5.

Maximality Consider the maximality condition (3.105) for problem $\bar{P}_k$ which holds for k large enough. Since $x^k \rightrightarrows x^*$ and $u^k \to u^*$ (a.e. on $[t_0^*, T^*]$) for $k \to \infty$, one obtains from (3.97) for $k \to \infty$ (with $u = (u^1, u^2)$) that

$$H_{u^1}(t, x^*(t), u^*(t), \hat{\psi}(t), \lambda_0) + \rho(t) R_{u^1}(t, x^*(t), u^{1*}(t)) = 0.$$

Using the definition (3.104) of $\rho^k(t) \geq 0$, which implies that

$$\rho_j^k(t) R_j(t, x^k(t), u^{1,k}(t)) = 0,$$

and taking the limit for $k \to \infty$ yields the complementary slackness condition

$$\rho_j(t) R_j(t, x^*(t), u^{1*}(t)) = 0, \qquad \forall j \in \{1, \ldots, k_R\}.$$

This complementary slackness condition and the maximality condition (3.100) together imply, for $k \to \infty$, and then $\hat{\varepsilon} \to 0^+$ and $\bar{u} \to \infty$, that

$$u^*(t) \in \arg\max_{u \in \mathcal{U}(t, x^*(t))} H(t, x^*(t), u, \hat{\psi}(t), \lambda_0), \qquad \forall t \in [t_0^*, T^*].$$

The maximality conditions (3.44)–(3.46) have thus been established a.e. on $[t_0^*, T^*]$.

Endpoint Optimality The inequalities and complementary slackness condition in (3.47) follows immediately from the definition of the multipliers λ_0 and λ_1. Using the endpoint-optimality conditions (3.101)–(3.102) together with the definitions (3.92)–(3.93) yields

$$\sup_{u \in \mathcal{U}(t_0^*, x_0^*)} H\left(t_0^*, x_0^*, u, -\sum_{j=0}^{2} \lambda_j K_{t_0}^j(\omega^*), \lambda_0\right) - \sum_{j=0}^{2} \lambda_j K_{t_0}^j(\omega^*) = 0$$

and

Optimal Control Theory   133

$$\sup_{u \in \mathcal{U}(T^*, x_T^*)} H\left(T^*, x_T^*, u, \sum_{j=0}^{2} \lambda_j K_T^j(\omega^*), \lambda_0\right) + \sum_{j=0}^{2} \lambda_j K_T^j(\omega^*) = 0,$$

that is, the endpoint-optimality conditions (3.48) and (3.49).

Nontriviality If λ and ψ are trivial, then (λ, ψ) must vanish identically on $[t_0^*, T^*]$. In particular this means that $\lambda_j^k \to 0, j \in \{0, 1, 2\}$ and $\psi^k(t) \rightrightarrows 0$ as $k \to \infty$. For each problem k, $\lambda_0^k > 0$, so all relations of the maximum principle (multiplying with the same positive constant) for $(\bar{P}_k')$ can be renormalized such that

$$\|\lambda^k\| + \sup_{t \in [t_0^k, T^k]} \|\psi^k(t)\| + \int_{t_0^k}^{T^k} \|\rho^k(t)\|^2 dt = 1, \qquad \forall k \geq 1.$$

Thus, taking the limit for $k \to \infty$ yields

$$\|\lambda\| + \sup_{t \in [t_0^*, T^*]} \|\hat{\psi}(t)\| + \int_{t_0^*}^{T^*} \|\rho(t)\|^2 dt = 1. \tag{3.108}$$

From the maximality condition for problem $(\bar{P}_k')$,

$$\lambda_0^k h_{u^1}^k + (\psi^k)' f_{u^1} + \sum_{j=1}^{k_R} \rho_j^k R_{j, u^1} = 0.$$

By assumption A5, for any $j \in \{1, \ldots, k_R\}$ there exists a vector $v \in \mathbb{R}^{m_1}$ with $\|v\| \leq 1$ such that $\langle v, R_{j, u^1} \rangle \geq \hat{\varepsilon}$ whenever $R_j \leq \hat{\varepsilon}$. Hence, there is a positive constant $\kappa > 0$ (independent of j) such that $\|R_{j, u^1}\| \geq \kappa$ whenever $R_j \leq \min\{\hat{\varepsilon}, \kappa\}$. Omitting a few technical details (see, e.g., Arutyunov 2000), the fact that $\lambda_0^k \to 0$ and $\psi^k \rightrightarrows 0$ as $k \to \infty$ therefore implies that $\lim_{k \to \infty} \|\rho^k(t)\| = 0$ a.e. on $[t_0^*, T^*]$. But this is a contradiction to (3.108), which in turn establishes the nontriviality condition (3.50).

Remark 3.19 (Strengthening Nontriviality) If Condition S holds, it is possible to strengthen the nontriviality relation (3.50) to

$$\lambda_0 + \|\psi(t)\| > 0, \qquad \forall t \in (t_0^*, T^*) \tag{3.109}$$

(see footnote 18). Indeed, if (3.109) is violated, then there exists $\tau \in (t_0^*, T^*)$ such that $\lambda_0 = 0$ and $\|\psi(\tau)\| = 0$. By assumption A5 (regularity of the state-control constraint) there exists a function $v(t)$ with values in $\mathbb{R}^{m_1}$ such that for some $\delta > 0$,

$$R_j(t,x,u^1) = 0 \Rightarrow \left\langle v(t), \frac{\partial R_j}{\partial u^1}(t,x,u^1) \right\rangle \geq \delta, \qquad j \in \{1,\ldots,k_R\},$$

a.e. on $[t_0^*, T^*]$. Hence, if the maximality condition (3.46) is scalar-multiplied with $v(t)$, than (analogous to the earlier analysis) there is a positive constant μ such that $0 \leq \rho(t) \leq \mu \|\psi(t)\|$ a.e. on $[t_0^*, T^*]$. Hence, invoking the adjoint equation (3.42), there is a constant $\hat{\mu} > 0$ such that $\|\dot{\psi}(t)\| \leq \hat{\mu}\|\psi(t)\|$ a.e. on $[t_0^*, T^*]$. By the (simplified) Gronwall-Bellman inequality (proposition A.9 and remark A.4), it is $\dot{\psi}(t) \equiv 0$, so that with initial condition $\psi(\tau) = 0$, it is $\psi(t) \equiv 0$. Using the endpoint-regularity assumption A4 together with the transversality conditions in (3.43) therefore yields that λ_1 and λ_2 both vanish. But this yields a contradiction to the nontriviality condition (3.50), which establishes the stronger condition (3.109). □

Envelope Condition To prove the envelope condition, consider the following autonomous (time-invariant) optimal control problem over the fixed time interval $[t_0^*, T^*]$, which features the control variable $\hat{u} = (u,v) \in \mathbb{R}^{m+1}$, the state $\hat{x} = (\xi, x) \in \mathbb{R}^{1+n}$, and the endpoint data $\hat{\omega} = (\xi_0, x_0; \xi_T, x_T)$:

$$\hat{J}^k(\hat{u}, \hat{\omega}) = \int_{t_0^*}^{T^*} (1 + v(t))h(\xi(t), x(t), u(t))\,dt + K^0(\hat{\omega}) \longrightarrow \max_{\hat{u}(\cdot), \hat{\omega}}$$

s.t.

$$\dot{x}(t) = f(\xi(t), x(t), u(t)), \qquad x(t_0) = x_0, \qquad x(T) = x_T,$$

$$\dot{\xi}(t) = 1 + v(t), \qquad \xi(t_0) = \xi_0, \qquad \xi(T) = \xi_T, \qquad (\hat{P})$$

$$0 \leq K^1(\hat{\omega}),$$

$$0 = K^2(\hat{\omega}),$$

$$0 \leq R(\xi(t), x(t), u^1(t)),$$

$$u = (u^1, u^2), \qquad u^2(t) \in \mathcal{U}(t), \qquad \forall t,$$

$$t \in [t_0, T], \qquad t_0 < T.$$

Note first that any admissible $(\hat{x}, \hat{u}, \hat{\omega})$ for $(\hat{P})$ is such that the corresponding (x, u, ω) with

$$\hat{x}(\theta) = (t, x(t)), \qquad \hat{u}(\theta) = (u(t), 0), \qquad \hat{\omega} = (t_0, x_0; T, x_T), \qquad (3.110)$$

where $\theta = \xi^{-1}(t)$, is admissible for (P) (i.e., the general finite-horizon optimal control problem (3.34)–(3.39)). The converse also holds;

Optimal Control Theory

therefore (x^*, u^*, ω^*) solves (P) if and only if the corresponding $(\hat{x}^*, \hat{u}^*, \hat{\omega}^*)$ solves $(\hat{P})$.

As a result, the already established conditions of the PMP can be applied to $(\hat{P})$. In particular, by proposition 3.5 there exist an absolute continuous adjoint variable $\hat{\psi} = (\psi^0, \psi) : [t_0^*, T^*] \to \mathbb{R}^{1+n}$, an essentially bounded function $\rho : [t_0^*, T^*] \to \mathbb{R}^{k_R}$, and a nonnegative constant λ_0 such that (restricting attention to its first component) the adjoint equation

$$-\dot{\psi}^0(t) = (1 + v^*(t))\left(\lambda_0 h_\xi(\xi^*(t), x^*(t), u^*(t)) + \langle \psi(t), f_\xi(\xi^*(t), x^*(t), u^*(t))\rangle\right)$$
$$+ \rho(t) R_\xi(\xi^*(t), x^*(t), u^*(t))$$

holds for all $t \in [t_0^*, T^*]$. In addition, maximality with respect to v implies that

$$\lambda_0 h(\xi^*(t), x^*(t), u^*(t)) + \langle \psi(t), f(\xi^*(t), x^*(t), u^*(t))\rangle + \psi^0(t) = 0.$$

Thus, using the variable transform in (3.110), one obtains

$$-\psi^0(t) = H(t, x^*(t), u^*(t), \psi(t), \lambda_0), \qquad \forall t \in [t_0^*, T^*],$$

and

$$-\dot{\psi}^0(t) = H_t(t, x^*(t), u^*(t), \psi(t), \lambda_0)$$
$$+ \langle \rho(t), R_t(t, x^*(t), u^*(t))\rangle, \qquad \forall t \in [t_0^*, T^*],$$

which establishes the envelope condition (3.51) in proposition 3.5.

This completes the proof of the PMP in proposition 3.5. ∎

3.7 Supplement 2: The Filippov Existence Theorem

Consider the existence of solutions to the general finite-horizon optimal control problem (3.34)–(3.39) with $\mathcal{U}(t, x)$ as in condition B of section 3.4. Let $\mathcal{D} \subset \mathbb{R}^{1+n}$ be a nonempty connected open set, termed domain as in section 2.2.1, and denote its closure by $\bar{\mathcal{D}}$. As before, define

$$\bar{\mathcal{D}}_0 = \{t \in \mathbb{R} : \exists (t, x) \in \bar{\mathcal{D}}\}$$

as the projection of $\bar{\mathcal{D}}$ onto the t-axis, and set

$$\bar{\mathcal{D}}(t) = \{x \in \mathbb{R}^n : (t, x) \in \bar{\mathcal{D}}\},$$

for all $t \in \mathbb{R}$. For any $(t, x) \in \bar{\mathcal{D}}$, let $\mathcal{U}(t, x) \subset \mathbb{R}^m$ be a nonempty control-constraint set, and let $\mathcal{M} = \bigcup_{(t,x) \in \bar{\mathcal{D}}} \{(t, x)\} \times \mathcal{U}(t, x)$ be the set of all feasible (t, x, u) in $\mathbb{R}^{1+n+m}$. Last, let

$$\Omega = \{\omega \in \mathbb{R}^{2(1+n)} : K^1(\omega) \geq 0,\ K^2(\omega) = 0\}$$

be the compact set of possible vectors of endpoint data $\omega = (t_0, x_0; T, x_T)$, for which always $t_0 < T$. The following result by Filippov (1962) guarantees the existence of solutions to the finite-horizon optimal control problem under a few additional assumptions.

Proposition 3.8 (Filippov Existence Theorem) Let $\mathcal{D}$ be bounded, $\Omega \subset \bar{\mathcal{D}} \times \bar{\mathcal{D}}$ be closed. Assume that assumptions A1–A5 are satisfied and that conditions B and C hold. Suppose further there exist an admissible state-control trajectory $(x(t), u(t))$, $t \in [t_0, T]$, and endpoint data $\omega = (t_0, x_0; T, x_T)$ such that $(t_0, x(t_0); T, x(T)) \in \Omega$. If for almost all t the vectograms $\mathcal{V}(t, x) = f(t, x, \mathcal{U}(t, x))$, $x \in \bar{\mathcal{D}}(t)$, are convex, then the optimal control problem (3.34)–(3.39) possesses an admissible solution (x^*, u^*, ω^*).[33]

Proof By condition B the constraint set $\mathcal{U}(t, x)$ is uniformly bounded for $(t, x) \in \bar{\mathcal{D}}$. Thus, the set $\mathcal{M}$ is compact, for $\bar{\mathcal{D}}$ is compact. By continuity of f the vectograms $\mathcal{V}(t, x)$ are therefore also compact, and they are all contained in a certain ball in $\mathbb{R}^n$.

For any admissible $(x(\cdot), u(\cdot), \omega)$, the endpoint vector $(t_0, x(t_0); T, x(T))$ lies in Ω. By the Weierstrass theorem (proposition A.10), the continuous function $K^0(\omega)$ attains its maximum m^0 on Ω. Furthermore, since $\mathcal{M}$ is compact, there exists $M > 0$ such that $|t|$, $\|x\|$, $\|u\|$, $\|f(t, x, u)\|$, $\|h(t, x, u)\| \leq M$ for all $(t, x, u) \in \mathcal{M}$. Thus, in particular $\bar{\mathcal{D}}_0 \subset [-M, M]$. Consider now the augmented vectogram

$$\hat{\mathcal{V}}(t, x) = \{(y_0, y) \in \mathbb{R}^{1+n} : y_0 \leq h(t, x, u),\ y = f(t, x, u),\ u \in \mathcal{U}(t, x)\}.$$

Note that $y \in \mathcal{V}(t, x)$ implies that $(-M, y) \in \hat{\mathcal{V}}(t, x)$. In addition, if $(y_0, y) \in \hat{\mathcal{V}}(t, x)$, then necessarily $y_0 \leq M$. The following problem is equivalent to the optimal control problem (3.34)–(3.39):

$$\hat{J}(\eta, \omega) = \int_{t_0}^T \eta(t)\,dt + K^0(\omega) \longrightarrow \max_{\eta(\cdot),\omega}, \tag{3.111}$$

$$(\eta(t), \dot{x}(t)) \in \hat{\mathcal{V}}(t, x(t)), \qquad x(t_0) = x_0, \qquad x(T) = x_T, \tag{3.112}$$

$$\omega = (t_0, x_0; T, x_T) \in \Omega, \tag{3.113}$$

$$t \in [t_0, T], \qquad t_0 < T. \tag{3.114}$$

---

33. Given a control set $\mathcal{U}$, the *vectogram* $f(t, x, \mathcal{U}) = \{f(t, x, u) : u \in \mathcal{U}\}$ corresponds to the set of all directions in which the system trajectory can proceed from (t, x).

Optimal Control Theory

For any endpoint vector $\omega \in \Omega$, the choice $(\eta(t), \omega)$ with $\eta(t) \equiv -M$ is feasible, that is, it satisfies the constraints (3.111)–(3.114). Note further that any feasible $(\eta(t), \omega)$ satisfies

$$\eta(t) \leq h(t, x(t), u(t)) \leq M, \qquad \forall t \in [t_0, T].$$

By the choice of the constants M, m^0 it is $T - t_0 \leq 2M$ and $K^0(\omega) \leq m^0$, so that

$$\hat{J}(\eta, \omega) \leq J(u, \omega) \leq 2M^2 + m^0.$$

If $\eta(t) = h(t, x(t), u(t))$ for almost all $t \in [t_0, T]$, then $\hat{J}(\eta, \omega) = J(u, \omega)$.

Let $J^* = \sup_{(u,\omega)} J(u, \omega)$ and $\hat{J}^* = \sup_{(\eta,\omega)} \hat{J}(\eta, \omega)$ be the smallest upper bounds for the attainable values of the objective functional in the equivalent problems (3.34)–(3.39) and (3.111)–(3.114), respectively. Both of these bounds are finite; they can both be realized by a feasible solution and are in fact equal.

Let $\{(\eta^k(t), \omega^k)\}_{k=0}^{\infty}$, $t \in \mathcal{I}^k = [t_0^k, T^k]$, be a maximizing sequence (with $\omega^k = (t_0^k, x_0^k; T^k, x_T^k)$ for $k \geq 0$), in the sense that $\hat{J}(\eta^k, \omega^k) \to \hat{J}^*$ as $k \to \infty$, and let $\{x^k(t)\}_{k=0}^{\infty}$, $t \in \mathcal{I}^k$, be the corresponding sequence of state trajectories. Since $\mathcal{V}(t, x^k(t)) = f(t, x^k(t), \mathcal{U}(t, x^k(t)))$, it is $\|\dot{x}^k(t)\| \leq M$, for all $t \in \mathcal{I}^k$ and all $k \geq 0$. As a result, the x^k are Lipschitz on $\mathcal{I}^k$ with the same Lipschitz constant, and therefore also equicontinuous on $\mathcal{I}^k$. In addition, $(t, x^k(t)) \in \bar{\mathcal{D}}$ and $\omega^k \in \Omega$.

The Arzelà-Ascoli theorem (proposition A.5) implies that there exist a subsequence $\{k_j\}_{j=0}^{\infty}$, a point $\omega = (t_0, x_0; T, x_T) \in \Omega$, and a state trajectory $x(t)$, $t \in [t_0, T]$, such that $x^{k_j}(t) \to x(t)$ as $j \to \infty$ uniformly on $[t_0, T]$, namely,

$$\lim_{j \to \infty} (|t_0^{k_j} - t_0| + |T^{k_j} - T| + \sup_{t \in \mathbb{R}} \|x^{k_j}(t) - x(t)\|) = 0,$$

where $x^{k_j}(t)$ and $x(t)$ are extended outside their domains by setting the functions constant (e.g., for $t \geq T$ it is $x(t) = x_T$). Since the sets $\bar{\mathcal{D}}$ and Ω are closed, it is $(t, x(t)) \in \bar{\mathcal{D}}$ and $\omega \in \Omega$ for all $t \in [t_0, T]$. In addition, x is Lipschitz and therefore absolutely continuous. It follows from an appropriate closure theorem (Cesari 1973) that there exists a Lebesgue-integrable function $\eta(t)$, $t \in [t_0, T]$, such that

$$(\eta(t), \dot{x}(t)) \in \hat{\mathcal{V}}(t, x(t)), \qquad \forall t \in [t_0, T],$$

and

$$\int_{t_0}^{T} \eta(t) \, dt \geq \limsup_{j \to \infty} \int_{t_0^{k_j}}^{T^{k_j}} \eta^{k_j}(t) \, dt. \tag{3.115}$$

Note also that by continuity of K^0,
$$\lim_{j\to\infty} K^0(\omega^{k_j}) = K^0(\omega). \tag{3.116}$$

By combining (3.115) and (3.116) one can conclude that $\hat{J}(\eta,\omega) = \hat{J}^*$. Now represent the augmented vectogram $\hat{\mathcal{V}}(t,x)$ in the form
$$\hat{\mathcal{V}}(t,x) = \{(h(t,x,u) - v, f(t,x,u)) : (u,v) \in \mathcal{U}(t,x) \times \mathbb{R}_+\}.$$

By the implicit function theorem (proposition A.7), the system of equations
$$\eta(t) = h(t, x(t), u(t)) - v(t),$$
$$\dot{x}(t) = f(t, x(t), u(t)),$$

has a solution $(u(t), v(t)) \in \mathcal{U}(t, x(t)) \times \mathbb{R}_+$, for almost all $t \in [t_0, T]$. The tuple $(u(t), \omega)$ and the associated state trajectory $x(t)$ are admissible for the optimal control problem (3.34)–(3.39). Since $\hat{J}^*$ is the optimal value of problem (3.111)–(3.114), the function $v(t)$ vanishes a.e. on $[t_0, T]$. But this implies that $\hat{J}^* = \hat{J}(\eta,\omega) = J(u,\omega) = J^*$, completing the proof. ∎

When the vectograms $\mathcal{V}(t,x)$ are nonconvex for some $(t,x) \in \bar{\mathcal{D}}$, there may be no solution to the optimal control problem (3.34)–(3.39). To illustrate this point, Filippov (1962) considered the following example.

Example 3.10 (Nonconvexity of Vectograms) Let $f = (f_1, f_2)$ with $f_1(t,x,u) \equiv u^2 - (x_2)^2$ and $f_2(t,x,u) \equiv u$, $h(t,x,u) \equiv -1$, $K^0(\omega) \equiv 0$,[34] and $\mathcal{U}(t,x) \equiv [0,1]$. Furthermore, assume that $\bar{\mathcal{D}} = [0,2] \times [0,1]^2$ and $\Omega = \{(0,(0,0);T,(1,0)) : T \in [1,2]\}$. All assumptions of proposition 3.8 are satisfied, except for the fact that $\hat{\mathcal{V}}(t,x)$ is nonconvex in the right half-plane where $\hat{x}$ is nonnegative (figure 3.6). The optimal control problem (3.34)–(3.39) with the above primitives has no solution. Note first that $T > 1$ for any solution and that any sequence $\{(u^k(t), \omega^k)\}_{k=2}^\infty$ with $|u^k(t)| = 1$ a.e. and $1 < T^k < 1 + \frac{1}{k^2-1}$ must be a maximizing sequence such that $J(u^k, \omega^k)$ approaches the optimal value $J^* = \sup_{(u,\omega)} J(u,\omega)$ as $k \to \infty$. Yet, any such maximizing sequence implies a sequence of state trajectories, $x^k(t)$, for $t \in [0, T^k]$ and $k \geq 2$, which converges to the trajectory $x(t) = (t,0)$, $t \in [0,1]$, which is not feasible for any $u(t)$ with values in $[0,1]$ because $\dot{x}_2(t) \equiv 0 \neq u(t)$ when $|u(t)| = 1$. □

---

34. The Mayer problem with $h(t,x,u) \equiv 0$ and $K^0(\omega) = T$ is equivalent.

Optimal Control Theory

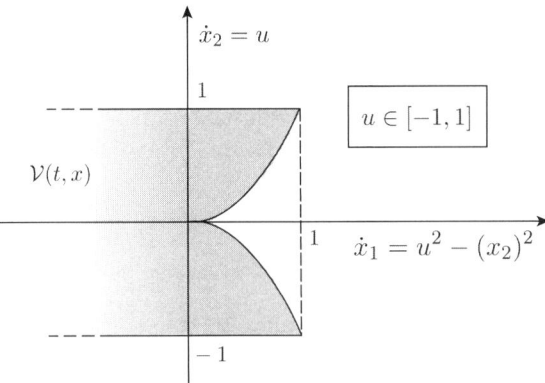

Figure 3.6
System with nonconvex vectogram (see example 3.10).

Remark 3.20 (Sliding-Mode Solutions) When the vectograms $\hat{\mathcal{V}}(t,x)$ are not convex, it is still possible to guarantee the existence of solutions via proposition 3.8, provided that *generalized solutions* (or *sliding-mode solutions*) of the optimal control problem (3.34)–(3.39) are introduced as solutions to the following modified optimal control problem:

$$\check{J}(p, v, \omega) = \int_{t_0}^T \sum_{l=1}^{n+2} p_l(t) h(t, x(t), v^l(t)) \, dt + K^0(\omega) \longrightarrow \max_{p(\cdot), v(\cdot), \omega},$$

$$\dot{x}(t) = \sum_{l=1}^{n+2} p_l(t) f(t, x(t), v^l(t)), \qquad x(t_0) = x_0, \qquad x(T) = x_T,$$

$$K^1(\omega) \geq 0, \qquad K^2(\omega) = 0,$$

$$(p(t), v^l(t)) \in \Delta_{n+2} \times \mathcal{U}(t, x(t)), \qquad l \in \{1, \ldots, n+2\}, \qquad \forall t,$$

$$t \in [t_0, T], \qquad t_0 < T,$$

where $\Delta_{n+2} = \{\pi = (\pi_1, \ldots, \pi_{n+2}) \in \mathbb{R}_+^{n+2} : \sum_{l=1}^{n+2} \pi_l = 1\}$ denotes an $(n+2)$-simplex, and where $p = (p_1, \ldots, p_{n+2})$ and $v = (v^1, \ldots, v^{n+2})$ are the control variables, with values in $\mathbb{R}^{n+2}$ and $\mathbb{R}^{(n+2)m}$, respectively. The modified optimal control problem, with control $u = (p, v)$, is in fact a general finite-horizon optimal control problem, as in section 3.4. Since its vectograms (see the proof of proposition 3.8) are convex (as convex hulls of the original $\mathcal{V}$), the Filippov existence theorem guarantees that there is a generalized solution to the original problem. The intuition for this sliding-mode solution is that it effectively partitions the system payoffs and system dynamics into $n+2$ pieces, which could be considered as random realizations of systems with different controls (in which case the p_l are viewed as probabilities). This allows the (original) system to be steered in the direction of any point in the convex hull of its vectogram.[35] □

3.8 Notes

The presentation of control systems is inspired by Anderson and Moore (1971) and Sontag (1998). Good introductory textbooks on optimal control theory are Warga (1972), Ioffe and Tikhomirov (1979), Seierstad and Sydsæter (1987), Kamien and Schwartz (1991), Sethi and Thompson (2000), and Vinter (2000). Milyutin and Osmolovskii (1998) relate the finite-horizon optimal control problem to the classical calculus of variations and also discuss sufficient optimality conditions. The infinite-horizon optimal control problem was discussed by Carlson et al. (1991), and more recently, with much additional insight, by Aseev and Kryazhimskii (2007) and Aseev (2009). Weber (2005a) considers an optimal advertising model (somewhat similar to the one in example 3.8) and shows asymptotic convergence of an optimal state trajectory through explicit considerations.

The HJB equation was formulated by Bellman (1957), who introduced the method of dynamic programming, which, in its discretized version, is very useful for computational purposes (Bertsekas 2007). For continuous-time systems, the HJB equation is a partial differential equation that is often difficult (or effectively impossible) to solve, even numerically. In addition, very simple, smooth optimal control problems may have value functions that are not continuously differentiable, so for the HJB equation to remain valid it is necessary to use generalized derivatives, giving rise to nonsmooth analysis (Clarke 1983; Clarke et al. 1998). Vinter (1988) provides a related discussion about the link between the HJB equation and the PMP.

L. S. Pontryagin developed the maximum principle together with his students and assistants, V. G. Boltyanskii, R. V. Gamkrelidze, and E. F. Mishchenko (Pontryagin et al. 1962). Boltyanskii (1994) and Gamkrelidze (1999) provide separate accounts of how the maximum

---

35. *Carathéodory's theorem* states that for any subset $\mathcal{S}$ of $\mathbb{R}^n$, any point in its convex hull can be represented as a convex combination of $n+1$ suitable points of $\mathcal{S}$.

principle was discovered. A key issue in its proof was resolved by Boltyanskii (1958), who adapted Weierstrass's idea of needle variations (which take their name from the shape of the corresponding graphs in the limit). The formulation here is adapted from the version for a problem with state constraints given by Arutyunov (1999; 2000) based on earlier work by Dubovitskii and Milyutin (1965; 1981), Dikusar and Milyutin (1989), Afanas'ev et al. (1990), and Dmitruk (1993). Dorfman (1969) provides an early economic interpretation of optimal control theory in the context of capital investment and profit maximization. Comprehensive accounts of the neighboring field of variational analysis are given by Giaquinta and Hildebrandt (1996), Rockafellar and Wets (2004), and Mordukhovich (2006).

The existence of solutions to a general time-optimal control problem was proved by Filippov (1962) and extended to the more general problem by Cesari (1983, 313).[36] As Boltyanski[i] et al. (1998) point out, with a suitable change of variables, a standard optimal control problem can actually be formulated as a time-optimal control problem.

3.9 Exercises

3.1 (Controllability and Golden Rule) (Weber 1997) Consider a model for determining a firm's dynamic policy about what amount $u_1(t)$ to spend on advertising and what price $u_2(t)$ to charge for its homogeneous product at any time $t \geq 0$. The advertising effect x_1 tracks the advertising expenditure, and the firm's installed base x_2 increases when demand $D(x, u_2)$ is larger than the number of products βx_2 that fail due to obsolescence. The evolution of the state variable $x = (x_1, x_2)$ is described by a system of ODEs,

$$\dot{x}_1 = -\alpha_1 x_1 + u_1,$$

$$\dot{x}_2 = D(x, u_2) - \beta x_2,$$

where[37]

$$D(x, u_2) = [1 - x_2 - \gamma u_2]_+ (\alpha_2 x_1 + \alpha_3 x_2)$$

denotes demand, and $\alpha_1, \alpha_2, \alpha_3, \beta, \gamma$ are given positive constants. The control at time $t \geq 0$ is $u(t) = (u_1(t), u_2(t)) \in \mathcal{U} = [0, \bar{u}_1] \times [0, 1/\gamma]$.

---

36. A *time-optimal control problem* is an OCP of the form (3.34)–(3.39), where $h = -1$ and $K^0 = 0$.
37. For any $z \in \mathbb{R}$, the nonnegative part of z is denoted by $[z]_+ = \max\{0, z\}$.

Assume that the initial state $x(0) = (x_{10}, x_{20}) \in (0, \bar{u}_1/\alpha_1) \times (0, 1)$ is known. The constant $\bar{u}_1 > 0$ is a given upper limit on advertising expenditure.

a. Show that, without any loss of generality, one can restrict attention to the case where $\gamma = 1$.

b. Sketch phase diagrams for the cases where $\beta < \alpha_3$ and $\beta \geq \alpha_3$. (*Hint:* The interesting controls to consider are those that drive the system at (or close to) either maximum or minimum velocity (in terms of the right-hand side of the system equation) in the different directions.)

c. Determine a nontrivial compact set $\mathcal{C} \subset (0, 1) \times (0, \bar{u})$ of controllable states, which are such that they can be reached from any other state in that set in finite time. Be sure to show how one could steer the system from x to $\hat{x}$ for any $\hat{x}, x \in \mathcal{C}$. Explain what happens when $x(0) \notin \mathcal{C}$.

d. Consider the problem of maximizing the firm's discounted infinite-horizon profit,

$$J(u) = \int_0^\infty e^{-rt}(u_2(t)D(x(t), u_2(t)) - cu_1(t))\,dt,$$

where $r > 0$ is a given discount rate and $c > 0$ is the unit cost of advertising, with respect to bounded measurable controls $u = (u_1, u_2)$ defined a.e. on $\mathbb{R}_+$, with values in the compact control set $\mathcal{U}$. Can you determine a state $\hat{x} = (\hat{x}_1, \hat{x}_2)$ such that if $x(0) = \hat{x}$, it would be optimal for the firm to stay at that state forever? (*Hint:* Compare the system's turnpike with the golden rule; see footnote 21.)

e. Check if the equilibrium state $\hat{x}$ of part d is contained in set $\mathcal{C}$ of part c. Based on this, explain *intuitively* how to find and implement an optimal policy. Try to verify your policy numerically for an example with $(\alpha_1, \alpha_2, \alpha_3, \beta, \gamma, r, c) = (1, .05, .1, .6, 100, .1, .05)$.

3.2 (Exploitation of an Exhaustible Resource) Let $x(0) = x_0 > 0$ be the initial stock of an exhaustible (also known as nonrenewable or depletable) resource. The utility (to society) of consuming the resource at the nonnegative rate $c(t)$ at time $t \in [0, T]$ (for a given time horizon $T > 0$) is $U(c(t))$, where $U : \mathbb{R}_+ \to \mathbb{R}$ is a utility function that is twice continuously differentiable, increasing, and strictly concave on $\mathbb{R}_{++}$. For any bounded, measurable consumption path $c : [0, T] \to [0, \bar{c}]$, bounded by the maximum extraction rate $\bar{c} > 0$, the *social welfare* is

$$W(c) = \int_0^T e^{-rt} U(c(t))\, dt,$$

where $r > 0$ is the social discount rate. The stock of the resource evolves according to

$$\dot{x} = -c,$$

provided that the feasibility constraint

$$c(t) \in [0, \mathbf{1}_{\{x(t) \geq 0\}} \bar{c}\,]$$

is satisfied a.e. on $[0, T]$, where $\mathbf{1}$ is the indicator function.

a. Formulate the social planner's dynamic welfare maximization problem as an optimal control problem.[38]

b. Using the PMP, provide necessary optimality conditions that need to hold on an optimal state-control path (x^*, c^*). If $\psi(t)$ is the (absolutely continuous) adjoint variable in the PMP, denote by $v(t) = e^{rt}\psi(t)$, for all $t \in [0, T]$, the current-value adjoint variable.

c. Let $\eta = -c\, U_{cc}(c)/U_c(c) > 0$ be the relative risk aversion (or the elasticity of the marginal utility of consumption). Using the conditions in part b, prove the *Hotelling rule*[39] that $\dot{v} = rv$ on $[0, T]$. Explain its economic significance using intuitive arguments. Show also that

$$\frac{\dot{c}}{c} = -\frac{r}{\eta},$$

that is, the relative growth rate of consumption on an optimal resource extraction path is proportional to the ratio of the discount rate and the relative risk aversion.

d. Find a welfare-maximizing policy $c^*(t)$, $t \in [0, T]$, when $U(c) = \ln(c)$. Compute the corresponding optimal state trajectory $x^*(t)$, $t \in [0, T]$.

e. Is it possible to write the optimal policy in part (d) in terms of a feedback law μ, in the form $c^*(t) = \mu(t, x^*(t))$, $t \in [0, T]$?

f. Redo parts a–e when $T \to \infty$. (For each one it is enough to note and discuss the key changes)

3.3 (Exploitation of a Renewable Resource) Let $x(t)$ represent the size of an animal population that produces a useful by-product $y(t)$ (e.g.,

---

38. Instead of letting the control constraint depend on the state, it may be convenient to introduce a constraint on the state endpoint $x(T)$ because $x(T) \leq x(t)$ for all $t \in [0, T]$.
39. See Hotelling (1931).

cows produce milk, bees produce honey) at time $t \in [0, T]$, where $T > 0$ is a given time horizon. The production of the by-product is governed by the production function $y = F(x)$, where $F : \mathbb{R}_+ \to \mathbb{R}$ is continuously differentiable, increasing, strictly concave, and such that $F(0) = 0$. A fraction of $u(t) \in [0, 1]$ of the by-product is extracted at time t, and the remaining fraction $1 - u(t)$ is left with the animals, so that their population evolves according to

$$\dot{x} = \alpha(x - \bar{x}) + (1 - u(t))F(x),$$

where $\bar{x} \geq 0$ is a given critical mass for the population to be able to grow, and $\alpha \geq r$ is a given growth rate. Each unit of the by-product that is extracted can be sold at a profit of 1. A firm is trying to maximize its profit,

$$J(u) = \int_0^T e^{-rt} u(t) F(x(t)) \, dt,$$

where $r > 0$ is a given discount rate, subject to the sustainability constraint

$$x(0) = x(T) = x_0,$$

where $x_0 \leq \bar{x}$, with $\alpha x_0 + F(x_0) > \alpha \bar{x}$, is the given initial size of the animal population.

a. Formulate the firm's profit-maximization problem as an optimal control problem.

b. Use the PMP to provide necessary optimality conditions. Provide a phase diagram of the Hamiltonian system of ODEs.

c. Characterize the optimal policy $u^*(t)$, $t \in [0, T]$, and show that in general it is discontinuous.

d. Describe the optimal policy in words. How is this policy influenced by r and α?

e. For $\alpha = 1$, $r = .1$, $x_0 = 10$, $\bar{x} = 12$, $T = 2$, and $F(x) = \sqrt{x}$, provide an approximate numerical solution to the optimal control problem in part a, that is, plot the optimal state trajectory $x^*(t)$ and the optimal control trajectory $u^*(t)$ for $t \in [0, T]$.

3.4 (Control of a Pandemic) Consider the outbreak of an infectious disease, which poses a public health hazard.[40] At time $t \in [0, T]$, where

---

40. This exercise is related to Sethi (1977); see also Sethi and Thompson (2000, 295–298).

Optimal Control Theory

$T > 0$ is a given intervention horizon, the percentage of infected people in a given population is $x(t) \in [0,1]$. Given a public treatment policy $u(t) \in [0, \bar{u}]$ (with $\bar{u} > \alpha$ a given maximum intervention level defined by the capacity of treatment facilities), the disease dynamics are described by the initial value problem

$$\dot{x} = \alpha(1-x)x - ux, \quad x(0) = x_0,$$

where $\alpha > 0$ is a known infectivity parameter, and the initial spread of the disease $x_0 \in (0,1)$ is known. A social planner would like to maximize the social-welfare functional

$$J(u) = -\int_0^T e^{-rt}\left(x(t) + cu^\kappa(t)\right) dt,$$

where $r > 0$ is the social discount rate, $c > 0$ denotes the intervention cost, and $\kappa \geq 1$ describes the diseconomies when scaling up public treatment efforts.

a. Formulate the social planner's welfare maximization problem as an optimal control problem.

b. Provide a set of necessary optimality conditions for $\kappa \in \{1,2\}$.

c. Characterize the optimal solutions for $\kappa \in \{1,2\}$, and discuss the qualitative difference between the two solutions.

d. Discuss your findings and provide an intuitive description of the optimal policy that a public official in charge of the health care system would understand.

e. What happens as the intervention horizon T goes to infinity?

f. Choose reasonable numerical values for $\alpha, c, r, \bar{u}, x_0$, and plot the optimal state and control trajectories, for $\kappa \in \{1,2\}$.

3.5 (Behavioral Investment Strategies) Consider an investor, who at time $t \geq 0$ consumes at the rate $c(t) \in (0, \bar{c}]$, as long as his capital (bank balance) $x(t)$ is positive, where $\bar{c} > 0$ is a fairly large spending limit. If the investor's capital becomes zero, consumption $c(t)$ must be zero as well. Given a discount rate $r > 0$, the investor's policy is to maximize his discounted utility

$$J_T(c) = \int_0^T e^{-rt} U(c(t)) dt,$$

where $U(c(t)) = \ln(c(t))$ is the investor's time-t utility of consuming at the rate $c(t)$, and $T > 0$ is a given *planning horizon*. The return on invested capital is $\alpha > r$, so the initial value problem

$$\dot{x} = \alpha x - c, \qquad x(0) = x_0,$$

where $x_0 > 0$ is his initial capital, describes the evolution of the investor's bank balance.

Part 1: Optimal Consumption Plan

a. Assuming that the investor's bank balance stays positive for all $t \in [0, T)$, formulate the investor's optimal control problem and determine his optimal *T-horizon consumption plan* $c_T^*(t; x_0)$, $t \in [0, T]$.

b. Determine the investor's optimal *infinite-horizon consumption plan* $c_\infty^*(t; x_0)$, $t \in \mathbb{R}_+$, when the planning horizon $T \to \infty$.

Part 2: Myopic Receding-Horizon Policy

Assume that the investor is myopic, so that, given a finite *planning horizon* $T > 0$ and an *implementation horizon* $\tau \in (0, T)$, he proceeds as follows. For any implementation period $k \geq 0$, the investor implements the consumption plan $c_T^*(t - k\tau; x_k)$, $t \in \mathcal{I}_k = [k\tau, (k+1)\tau]$, where x_k is the amount of capital available at time $t = k\tau$. Let $\hat{c}_{T,\tau}^*(t)$, $t \in \mathbb{R}_+$, be the investor's resulting (T, τ)-*receding-horizon consumption plan*.

c. Discuss what practical reason or circumstance might be causing the investor to choose a receding-horizon consumption plan $\hat{c}_{T,\tau}^*$ over an optimal infinite-horizon consumption plan c_∞^*. Draw a picture that shows how the receding-horizon consumption plan is obtained from the infinite-horizon consumption plan.

d. For a given implementation horizon $\tau > 0$, does $T \to \infty$ imply that $\hat{c}_{T,\tau}^* \to c_\infty^*$ pointwise? Explain.

e. Is it possible that $\hat{c}_{T,\tau}^*$ becomes periodic (but nonconstant)? If yes, try to provide an example. If no, explain.

Part 3: Prescriptive Measures

f. If $x_\infty^*(t; x_0)$, $t \in \mathbb{R}_+$, denotes the state trajectory under the optimal infinite-horizon consumption plan $c_\infty^*(t; x_0)$, $t \in \mathbb{R}_+$, find the long-run steady state $\bar{x}_\infty^* = \lim_{t \to \infty} x_\infty^*(t; x_0)$.

g. Consider the following *modified* (T, τ)-*receding-horizon consumption plan* $\tilde{c}_{T,\tau}^*(t)$, $t \in \mathbb{R}_+$, which is such that on each time interval $\mathcal{I}_k$, $k \geq 0$, the investor implements an optimal endpoint-constrained

Optimal Control Theory

consumption plan $\tilde{c}_T^*(t - k\tau; \tilde{x}_k)$ where (for any $x_0 > 0$) the plan $\tilde{c}_T^*(\cdot; x_0)$ solves the finite-horizon optimal control problem formulated in part a, subject to the additional state-endpoint constraint $x(T) = \bar{x}_\infty^*$,[41] and where $\tilde{x}_k$ is the amount of capital available at time $t = \kappa\tau$. Compare, in words, the receding-horizon consumption plans $\tilde{c}_{T,\tau}^*$ and $\hat{c}_{T,\tau}^*$. Does $\tilde{c}_{T,\tau}^*$ fix some of the weaknesses of $\hat{c}_{T,\tau}^*$ (e.g., those identified in part 2)? Explain.

3.6 (Optimal Consumption with Stochastic Lifetime) Consider an investor, who at time $t \geq 0$ consumes at the rate $c(t) \in [0, \bar{c}]$, as long as his capital (bank balance) $x(t)$ is positive, where $\bar{c} > 0$ is a fairly large spending limit. If the investor's capital becomes zero, consumption $c(t)$ must be zero as well. Given a discount rate $r > 0$, the investor's policy is to maximize his expected discounted utility

$$\bar{J}(c) = E\left[\int_0^{\tilde{T}} e^{-rt} U(c(t))\, dt\right],$$

where $U(c(t)) = \ln(c(t))$ is the investor's time-t utility of consuming at the rate $c(t)$, and $\tilde{T} \geq 0$ represents the investor's *random* remaining lifetime. The latter is exponentially distributed with probability density function $g(T) = \lambda e^{-\lambda T}$ for all $T \geq 0$, where $\lambda > 0$ is a given constant. The return on invested capital is $\alpha > r$, so the initial value problem

$$\dot{x} = \alpha x - c, \qquad x(0) = x_0,$$

where $x_0 > 0$ is his initial capital, describes the evolution of the investor's bank balance.

a. Formulate the investor's optimal consumption problem as a deterministic infinite-horizon optimal control problem.

b. Determine the investor's optimal consumption plan $c^*(t)$, $t \geq 0$.

c. Compare your solution in part b to the optimal consumption plan $c_T^*(t)$, $t \in [0, T]$, when $T > 0$ is perfectly known. How much is the information about T worth?

d. Compare your solution in part b to the optimal (deterministic) infinite-horizon consumption plan $c_\infty^*(t)$, $t \geq 0$.

e. What can you learn from parts b–d for your own financial management?

---

41. Under which conditions on the parameters is this feasible? $\bar{x}_\infty^*$ is the steady state obtained in exercise 3.5f; it remains fixed.

4 Game Theory

Of supreme importance in war is
to attack the enemy's strategy.

—Sun Tzu

4.1 Overview

The strategic interaction generated by the choices available to different agents is modeled in the form of a *game*. A game that evolves over several time periods is called a *dynamic game*, whereas a game that takes place in one single period is termed a *static game*. Depending on the information available to each agent, a game may be either of *complete* or *incomplete information*. Figure 4.1 provides an overview of these main types of games, which are employed for the exposition of the fundamental concepts of game theory in section 4.2.

Every game features a set of players, together with their action sets and their payoff (or utility) functions. The vector of all players' actions is called a strategy profile or an outcome. A given player's *payoff function* (or *utility function*) maps strategy profiles to real numbers (called payoffs). These payoffs represent this player's preferences over the outcomes.[1]

Game theory aims at providing predictions about the possible outcomes of a given game. A *Nash equilibrium* is a strategy profile (i.e., a

---

1. In economics, a player's preferences over the set of outcomes $\mathcal{A}$ would define a *preorder* on that set, i.e., a binary relation $\succeq$ ("is preferred to") which for any $\hat{a}, a, \check{a} \in \mathcal{A}$ satisfies the following two properties: (1) $\hat{a} \succeq a$ or $a \succeq \hat{a}$ (completeness); and (2) $\hat{a} \succeq a$ and $a \succeq \check{a}$ implies that $\hat{a} \succeq \check{a}$ (transitivity). If $\hat{a} \succeq a$ and $a \succeq \hat{a}$, then the player is *indifferent* between $\hat{a}$ and a, which is denoted by $\hat{a} \sim a$. A *utility function* $U : \mathcal{A} \to \mathbb{R}$ represents these preferences if for all $\hat{a}, a \in \mathcal{A}$: $U(\hat{a}) \geq U(a) \Leftrightarrow \hat{a} \succeq a$. One can show that as long as the upper contour set $\mathcal{U}(a) = \{\hat{a} \in \mathcal{A} : \hat{a} \succeq a\}$ and the lower contour set $\mathcal{L}(a) = \{\hat{a} \in \mathcal{A} : a \succeq \hat{a}\}$ are closed for all $a \in \mathcal{A}$, there exists a continuous utility function that represents the preferences.

	Information	
	Complete	Incomplete
Static	Section 4.2.1	Section 4.2.2
Dynamic	Section 4.2.3	Section 4.2.4

Figure 4.1
Classification of games.

vector of actions by the different players) such that no player wants to alter his own action unilaterally, given that all other players play according to this strategy profile. Reasonable outcomes of a game are generally expected to at least satisfy the Nash-equilibrium requirement. Yet, as games become more complex, perhaps because the players' actions are implemented dynamically over time, there may exist many Nash equilibria, some of which are quite implausible. To see this, consider the following simple two-player bargaining game.

Example 4.1 (Ultimatum Bargaining) Two players, Ann and Bert, try to decide how to split a dollar. Ann proposes a fraction of the dollar to Bert. Bert then decides whether to accept or reject Ann's offer. If the offer is accepted, then the proposed money split is implemented and the players' payoffs realize accordingly. Otherwise both players receive a payoff of zero. For example, if Ann proposes an amount of $0.30 to Bert, then if Bert accepts the offer, Ann obtains $0.70 and Bert gets $0.30. If Bert rejects the offer, then both players get a zero payoff (and the dollar disappears). To apply the concept of Nash equilibrium, one needs to look for strategy profiles that would not provoke a unilateral deviation by either player. Note first that because Bert moves after Ann, he can observe her action, that is, her offer $a \in [0, 1]$. Bert's strategy may be to accept the offer if and only if it reaches at least a certain threshold $\alpha \in [0, 1]$. With Bert's choice to accept denoted by $b = 1$ and to reject by $b = 0$, his strategy can be summarized by

$$b(a, \alpha) = \begin{cases} 1 & \text{if } a \geq \alpha, \\ 0 & \text{otherwise.} \end{cases}$$

If Ann believes that Bert will implement this threshold strategy, then her best strategy is to propose $a = \alpha$. Now, checking for a possible deviation by Bert, note that given Ann's proposed amount α, it is best for

Bert to accept (because he would otherwise get a zero payoff, which is never strictly better than α). Therefore, for any $\alpha \in [0,1]$, the strategy profile consisting of $a = \alpha$ and $b(\cdot, \alpha)$ constitutes a Nash equilibrium. Thus, even for this fairly simple dynamic game there exists a *continuum* of Nash equilibria, one for each $\alpha \in [0,1]$. This analysis is not particularly useful for generating a prediction for the game's outcome because any split of the dollar between Ann and Bert can be justified by one of the Nash equilibria. A way out of this dilemma is to refine the concept of Nash equilibrium by imposing an additional requirement. For this, note that because Ann moves before Bert, it is clearly optimal for Bert to accept any amount that Ann offers, even if it is zero, in which case Bert is indifferent between accepting or not.[2] The reason Ann was willing to offer a positive amount (corresponding to positive α) is that she believed Bert's threat of following through with his threshold strategy. But this threat is not credible, or as game theorists like to say, Bert's threshold strategy is not subgame-perfect, because in the subgame that Bert plays (with himself) after Ann has made her offer, it is best for Bert to accept *any* offer. The concept of a subgame-perfect Nash equilibrium requires that a strategy profile induce a Nash equilibrium in each subgame. Provided that the game ends in finite time, one can obtain an equilibrium path by backward induction, starting at the end of the horizon.[3] The only subgame-perfect Nash equilibrium of this bargaining game is therefore for Ann to propose zero and for Bert to accept her ruthless offer. □

The preceding example illustrates the need for *equilibrium refinements* in dynamic games, even when both players have perfect knowledge about each other. But the information that players have about each other might be quite incomplete. For example, when a seller faces multiple potential buyers (or agents) with different valuations for an item, then these valuations are generally not known to her. They belong to the agents' private information that the seller may try to extract. Assume that each buyer i's valuation for the item is given by a nonnegative number θ^i, and suppose that the seller uses some type of auction mechanism to sell the item. Then buyer i's bid b^i for the item will depend on his private valuation θ^i. In other words, from the seller's perspective and

---

2. A standard assumption in game theory is that in the case of indifference, a player does what the game theorist wants him to do, which is usually to play an equilibrium strategy.
3. The intuition is similar to the logic of the Hamilton-Jacobi-Bellman equation (see chapter 3), which contains its boundary condition at the end of the horizon and is therefore naturally solved from the end of the horizon, especially when discretizing the optimal control problem to a finite number of periods (see, e.g., Bertsekas 2007).

the perspective of any other bidder $j \neq i$, buyer i's strategy becomes a function of θ^i, so the corresponding *Bayes-Nash equilibrium* (BNE) of games with incomplete information requires the specification of each player's actions as a function of his private information. The following example illustrates this notion.

Example 4.2 (Second-Price Auction) A seller tries to auction off an item to one of $N \geq 2$ agents with unknown private valuations $\theta^1, \ldots, \theta^N \in [0, 1]$.[4] Assume that the agents' valuations are, from the perspective of both the seller and the agents, independently and identically distributed. Agents submit their bids simultaneously and the highest bidder wins the item.[5] The winning bidder then pays an amount to the seller that is equal to the second-highest bid (corresponding to the highest losing bid). All other bidders obtain a zero payoff. The question is now, What would be a symmetric Bayes-Nash-equilibrium bidding strategy such that each bidder $i \in \{1, \ldots, N\}$ submits a bid $b^i = \beta(\theta^i)$, where $\beta(\cdot)$ is the bidding function to be determined? If bidder i submits a bid b^i strictly less than his private value θ^i, then there is a chance that some bidder j submits a bid $b^j \in (b^i, \theta^i)$, in which case bidder i would have been better off bidding his true valuation. Similarly, if bidder i submits a bid b^i strictly greater than θ^i, then there is a chance that some agent j submits the highest competing bid $b^j \in (\theta^i, b^i)$, which would lead bidder i to win with a payment above his valuation, resulting in a negative payoff. Again, bidder i prefers to bid his true valuation θ^i. Consider now the situation in which all bidders other than bidder i use the bidding function $\beta(\theta) \equiv \theta$ to determine their strategies. Then, by the logic just presented, it is best for bidder i to also select $b^i = \theta^i$. Therefore, the bidding function $b^i = \beta(\theta^i) = \theta^i$, for all $i \in \{1, \ldots, N\}$, determines a (symmetric) Bayes-Nash equilibrium of the second-price auction. Note that this mechanism leads to full information revelation in the sense that all agents directly reveal their private information to the seller. The price the seller pays for this information (relative to knowing everything for free) is equal to the difference between the highest and the second-highest bid. □

---

4. The problem of selling an item to a single potential buyer is considered in chapter 5.
5. In the case where several bidders submit the same winning bid, the item is allocated randomly among them, at equal odds.

The problem of equilibrium multiplicity, discussed in example 4.1, is compounded in dynamic games where information is incomplete. Subgame perfection as refinement has very little bite, because (proper) subgames can start only at points where all players have perfect information about the status of the game. When player A possesses private information, then player B moving after A does not know which type of player A he is dealing with and therefore tries to infer information from the history of the game about A's type. The multiplicity arises because what happens on the equilibrium path may depend strongly on what happens off the equilibrium path, that is, what the players expect would happen if they deviated from their equilibrium actions. The notion of equilibrium path becomes clear when reconsidering the ultimatum game in example 4.1. In that game, Bert's strategy is a complete contingent plan, which specifies his action for every possible offer that Ann could make. *On* the equilibrium path she makes only one offer, and Bert will respond to precisely that offer. All of his other possible responses to Ann's other possible offers are *off* the equilibrium path. Thus, in terms of Nash equilibrium, Bert's anticipated *off-equilibrium behavior* may change Ann's *in-equilibrium behavior*. Indeed, when Bert threatens to use a (not subgame-perfect) threshold strategy, Ann's best response is to offer him exactly that threshold, because Ann then fears that Bert would reject lower offers. These off-equilibrium-path beliefs usually drive the multiplicity of equilibria in dynamic games of incomplete information, as can be seen in the following example.

Example 4.3 (Job Market Signaling) (Spence 1973) Consider a job applicant whose productivity in terms of his profit-generation ability is either $\theta_L = 1$ or $\theta_H = 2$. While the information about his productivity type $\theta \in \{\theta_L, \theta_H\}$ is private, the worker does have the option to acquire a publicly observable education level e at the cost $C(e, \theta) = e/\theta$, which has no influence on his productivity. With complete information, a firm considering to hire the applicant would offer a wage $w(\theta) = \theta$, compensating the worker exactly for his productivity.[6] If the two types choose the same education level, so that the firm cannot distinguish between them, then the firm offers a wage of $\bar{\theta} = (\theta_L + \theta_H)/2 = 3/2$, corresponding to

---

6. This implicitly assumes that enough firms are competing for workers, so that none is able to offer wages below a worker's expected productivity and still expect to hire workers.

the worker's *expected* productivity. The low-type worker would be willing to acquire an education level of up to $\bar{e}_L = 1/2$ to attain this pooling equilibrium, whereas the high-type worker would be willing to get an education of up to $\bar{e}_H = 1$ in order to end up at a separating equilibrium with different education levels. Note that in a separating equilibrium, the low-type worker would never acquire any education. The high type has to acquire at least $\underline{e}_H = 1/2$ in order to discourage the low type from trying to pool by matching the high type's education. What actually happens in equilibrium depends decisively on the firm's interpretation of out-of-equilibrium actions. For example, it would be legitimate for the firm to interpret any worker with an education level off the equilibrium path as a low type. With these somewhat extreme out-of-equilibrium beliefs, it is possible to construct any pooling equilibrium where both worker types acquire education level $e^* \in [0, \bar{e}_L]$ as well as any separating equilibrium where the workers acquire the education levels $e_L^* = 0$ and $e_H^* \in [\underline{e}_H, \bar{e}_H]$, respectively. □

Section 4.2 provides an introduction to games in discrete time with complete and incomplete information. Based on this body of classical game theory, continuous-time differential games are discussed in section 4.3. In a differential game all players' strategies are functions of continuous time and their reactions may be instantaneous, unless there is an imperfect information structure which, for example, could include a delay in player i's noticing player j's actions. To deal with excessive equilibrium multiplicity in dynamic games where strategies can depend on the full history of past actions and events,[7] when dealing with differential games one often requires equilibria to satisfy a Markov property in the sense that current actions can depend only on current states or, in other words, that all the relevant history of the game is included in the current state. A further simplification is to assume that all players condition their strategies on time rather than on the state, which leads to open-loop strategies. By contrast, a closed-loop strategy can be conditioned on all the available information, which usually includes the state variable. The refinement of subgame perfection in the context of differential games leads to the concept of a Markov-perfect equilibrium. The discussion also includes some examples of non-Markovian equilibria using, for example, trigger strategies, where a certain event (such as

---

7. See, e.g., the folk theorems in proposition 4.5 and remark 4.4 about equilibria in infinitely repeated games.

a player's deviation from the equilibrium path) would cause a regime shift in the players' behavior.

4.2 Fundamental Concepts[8]

4.2.1 Static Games of Complete Information

In the absence of any time dependence, a game Γ in normal form is fully specified by a *player set* $\mathcal{N}$, an *action set* $\mathcal{A}^i$ for each player $i \in \mathcal{N}$, and a *payoff function* $U^i : \mathcal{A} \to \mathbb{R}$ for each player $i \in \mathcal{N}$, where $\mathcal{A} = \prod_{i \in \mathcal{N}} \mathcal{A}^i$ is the set of (pure) *strategy profiles* $a = (a^i)_{i \in \mathcal{N}}$. Thus,

$$\Gamma = (\mathcal{N}, \mathcal{A}, \{U^i(\cdot)\}_{i \in \mathcal{N}}) \tag{4.1}$$

fully describes a static game of complete information. The following two basic assumptions govern the players' interaction in the game Γ:

- *Rationality* Each player $i \in \mathcal{N}$ chooses his action a^i in the normal-form game Γ so as to maximize his payoff $U^i(a^i, a^{-i})$ given the other players' strategy profile $a^{-i} = (a^j)_{j \in \mathcal{N} \setminus \{i\}}$.
- *Common knowledge* Each player $i \in \mathcal{N}$ knows the rules of the game Γ (all of its elements) and knows that the other players know the rules of the game and that they know that he knows the rules of the game and that he knows that they know that he knows, and so on.[9]

The following simple but very important example illustrates the notation and describes how one might reach a prediction about the outcome of a game, both in terms of payoffs and in terms of the eventually implemented strategy profile.

Example 4.4 (Prisoner's Dilemma) Let $\mathcal{N} = \{1, 2\}$ be a set of two prisoners that are under suspicion of having committed a crime together. During their separate but simultaneous interrogations each prisoner can choose either to cooperate (C), that is, to deny all charges, or to defect (D), that is, to admit all charges, incriminating the other prisoner. Each prisoner $i \in \mathcal{N}$ has an action set of the form $\mathcal{A}^i = \{C, D\}$, so $\mathcal{A} = \{C, D\} \times \{C, D\} = \{(C, C), (C, D), (D, C), (D, D)\}$ is the set of all strategy profiles. The prisoners' payoffs are specified as follows:

---
8. Readers already familiar with the basics of game theory can skip directly to section 4.3 without loss in continuity.
9. When this generally infinite belief hierarchy is interrupted at a finite level, then the game will be of bounded rationality, which is beyond the scope of this book. Geanakoplos (1992) summarized the interesting consequences of the common-knowledge assumption in economics.

	Prisoner 2	
	C	D
Prisoner 1 C	(1, 1)	(−1, 2)
D	(2, −1)	(0, 0)

In this payoff matrix, the entry $(-1, 2)$ for the strategy profile (C, D) means that $U^1(C, D) = -1$ and $U^2(C, D) = 2$. The other entries have analogous interpretations. It is easy to see that when fixing, say, prisoner 2's action, prisoner 1 is better off choosing D instead of C, because $u^1(D, a^2) > u^1(C, a^2)$, for all $a^2 \in \mathcal{A}^2$. By symmetry prisoner 2 is also always best off to choose D, so the only reasonable prediction about the outcome of this game is that both prisoners will choose D, leading to the payoff vector $(0, 0)$. This famous game is usually referred to as *prisoner's dilemma* because both players, caught in their strategic interdependence, end up with a payoff vector worse than their socially optimal payoff vector of $(1, 1)$ when they both cooperate. The key reason for the socially suboptimal result of this game is that when one prisoner decides to cooperate, the other prisoner invariably prefers to defect. □

As shown in example 4.4, it is useful to decompose a strategy profile $a \in \mathcal{A}$ into player i's action a^i and all other players' strategy profile $a^{-i} = (a^j)_{j \in \mathcal{N} \setminus \{i\}}$, so it is customary in game theory to write

$$a = (a^i, a^{-i}) = (a^j)_{j \in \mathcal{N}}$$

for any $i \in \mathcal{N}$. Using this notation, one can introduce John Nash's notion of an equilibrium as the leading prediction about the outcome of the game Γ. A strategy profile $a^* = (a^{i*})_{i \in \mathcal{N}}$ is a *Nash equilibrium* (NE) if

$$\forall i \in \mathcal{N}: \quad U^i(a^{i*}, a^{-i*}) \geq U^i(a^i, a^{-i*}), \quad \forall a^i \in \mathcal{A}^i. \tag{4.2}$$

This means that at a Nash-equilibrium strategy profile a^* each player i maximizes his own payoff given that all other players implement the strategy profile a^{-i*}. In other words, for any player i there is no strategy $a^i \in \mathcal{A}^i$ such that he strictly prefers strategy profile (a^i, a^{-i*}) to the Nash-equilibrium strategy profile $a^* = (a^{i*}, a^{-i*})$. Another common way to express the meaning of a Nash equilibrium is to recognize that (4.2) is equivalent to the following simple statement:

No player wants to unilaterally deviate from a Nash-equilibrium strategy profile.

Game Theory

The following classic example demonstrates that Nash equilibria do not have to be unique.

Example 4.5 (Battle of the Sexes) Consider two players, Ann and Bert, each of whom can choose between two activities, "go dancing" (D) or "go to the movies" (M). Their payoffs are as follows:

		Bert	
		D	M
Ann	D	(2, 1)	(0, 0)
	M	(0, 0)	(1, 2)

It is straightforward to verify that any strategy profile in which both players choose the same action is a Nash equilibrium of this *coordination game*. If players are allowed to randomize over their actions, an additional Nash equilibrium in *mixed* strategies can be identified. For each player $i \in \mathcal{N} = \{\text{Ann}, \text{Bert}\}$, let

$$\Delta(\mathcal{A}^i) = \{(\pi, 1-\pi) : \pi = \text{Prob}(\text{Player } i \text{ plays } D) \in [0,1]\}$$

denote an augmented strategy space such that $p^i = (p_D^i, p_M^i) \in \Delta(\mathcal{A}^i)$ represents the probability distribution with which player i chooses the different actions in the augmented game

$$\Gamma_\Delta = \left(\mathcal{N}, \prod_{i \in \mathcal{N}} \Delta(\mathcal{A}^i), \left\{ \bar{U}^i : \prod_{j \in \mathcal{N}} \Delta(\mathcal{A}^j) \to \mathbb{R} \right\}_{i \in \mathcal{N}} \right),$$

where player i's expected payoff,

$$\bar{U}^i(p^i, p^{-i}) = p_D^i p_D^{-i} U^i(D, D) + p_M^i p_M^{-i} U^i(M, M)$$
$$= 3 p_D^i p_D^{-i} + (1 - p_D^i - p_D^{-i}) U^i(M, M),$$

takes into account the payoffs shown in the matrix and the fact that $p_M^i = 1 - p_D^i$. Given the other player's strategy p^{-i}, player i's *best-response correspondence* is

$$\text{BR}^i(p^{-i}) = \arg\max_{p^i \in \Delta(\mathcal{A}^i)} \bar{U}^i(p^i, p^{-i}) = \begin{cases} \{0\} & \text{if } 3 p_D^{-i} - U^i(M, M) < 0, \\ [0,1] & \text{if } 3 p_D^{-i} - U^i(M, M) = 0, \\ \{1\} & \text{otherwise.} \end{cases}$$

In the context of Ann and Bert's coordination game this means that player i's best response is to do what the other player is sufficiently likely to do, unless the other player makes player i *indifferent* over all possible probability distributions in $\Delta(\mathcal{A}^i)$. By definition, a Nash equilibrium $p^* = (p^{i*}, p^{-i*})$ of Γ_Δ is such that

$$p^{i*} \in \mathrm{BR}^i(p^{-i*}), \quad \forall i \in \mathcal{N}.$$

Continuing the previous argument, if both players make each other indifferent, by choosing $p_D^{\mathrm{Ann}} = U^{\mathrm{Bert}}(M,M)/3 = 2/3$ and $p_D^{\mathrm{Bert}} = U^{\mathrm{Ann}}(M,M)/3 = 1/3$, one obtains the Nash equilibrium $p^* = (p^{\mathrm{Ann}*}, p^{\mathrm{Bert}*})$ of Γ_Δ, with

$$p^{\mathrm{Ann}*} = (2/3, 1/3) \quad \text{and} \quad p^{\mathrm{Bert}*} = (1/3, 2/3).$$

Both players' corresponding Nash-equilibrium payoffs, $\bar{U}^{\mathrm{Ann}}(p^*) = \bar{U}^{\mathrm{Bert}}(p^*) = 2/3$, are less than the equilibrium payoffs under either of the two pure-strategy Nash equilibria of Γ. Note also that the latter equilibria reappear as Nash equilibria $((1,0),(1,0))$ and $((0,1),(0,1))$ of the augmented game Γ_Δ, which therefore has three (mixed-strategy) Nash equilibria in total. □

Based on the insights obtained in example 4.5, it is useful to extend the definition of a Nash equilibrium to allow for players' randomizations over actions. For this, let $\Delta(\mathcal{A}^i)$ be the space of all probability measures P^i defined on the standard probability space $(\mathcal{A}^i, \mathcal{F}^i, \mathcal{P})$, where $\mathcal{F}^i$ is an appropriate σ-algebra over $\mathcal{A}^i$,[10] such that

$$\int_{\mathcal{A}^i} dP^i(a^i) = 1.$$

A *mixed-strategy Nash equilibrium* of Γ is a (pure-strategy) Nash equilibrium of the augmented game

$$\Gamma_\Delta = \left(\mathcal{N}, \prod_{i \in \mathcal{N}} \Delta(\mathcal{A}^i), \left\{ \bar{U}^i : \prod_{j \in \mathcal{N}} \Delta(\mathcal{A}^j) \to \mathbb{R} \right\}_{i \in \mathcal{N}} \right),$$

where for any $P \in \prod_{i \in \mathcal{N}} \Delta(\mathcal{A}^i)$ player i's expected payoff is

---

10. It is implicitly assumed that player i's action set $\mathcal{A}^i$ is closed under the operation of union, intersection, and difference (i.e., it forms a ring). For details on measure theory and on how to slightly relax this assumption (to semirings), see, e.g., Kirillov and Gvishiani (1982).

Game Theory

$$\bar{U}^i(p) = \int_{\mathcal{A}} U^i(a)\, dP(a).$$

It is clear that the distinction between mixed-strategy and pure-strategy Nash equilibria depends only on the viewpoint, since any mixed-strategy Nash equilibrium is defined in terms of a pure-strategy equilibrium of a suitably augmented game.

In actual games, it is sometimes difficult to interpret the meaning of a mixed-strategy equilibrium that places positive probability mass on more than one action, especially if the game is only played once. Yet, the main reason for introducing the possibility of randomization is to *convexify* the players' action spaces in order to guarantee the existence of a Nash equilibrium, at least in mixed strategies. The following example shows that the existence of a Nash equilibrium in pure strategies cannot be taken for granted.

Example 4.6 (Matching Pennies) Two agents, 1 and 2, play a game where each player chooses simultaneously one side of a penny, either "heads" (H) or "tails" (T). Player 1 wins both pennies if the players have chosen matching sides; otherwise player 2 wins the pennies. The net payoffs in this zero-sum game are as follows:

		Player 2	
		H	T
Player 1	H	(1, −1)	(−1, 1)
	T	(−1, 1)	(1, −1)

It is easy to see that there does exist a pure-strategy equilibrium of this matching-pennies game. Indeed, player 1 would always like to imitate player 2's strategy, whereas player 2 would then want to deviate and choose a different side, so unilateral deviations cannot be excluded from any strategy profile in $\mathcal{A} = \{(H,T)\} \times \{(H,T)\}$. The only mixed-strategy equilibrium is such that both players randomize so as to choose H and T with equal probabilities. □

Proposition 4.1 (Existence of a Nash Equilibrium) Let

$$\Gamma = \left(\mathcal{N}, \mathcal{A} = \prod_{i \in \mathcal{N}} \mathcal{A}^i, \{U^i(\cdot)\}_{i \in \mathcal{N}} \right)$$

be a normal-form game, where the player set $\mathcal{N} \neq \emptyset$ is finite and each action set $\mathcal{A}^i \neq \emptyset$ is finite-dimensional, convex, and compact for $i \in \mathcal{N}$. If in addition each player i's payoff function $U^i(a^i, a^{-i})$ is continuous in $a = (a^i, a^{-i})$ and quasi-concave in a^i, then Γ has a (pure-strategy) Nash equilibrium.

Proof Let BR : $\mathcal{A} \rightrightarrows \mathcal{A}$ be the set-valued best-response correspondence for Γ, defined by

$$\mathrm{BR}(a) = (\mathrm{BR}^i(a^{-i}))_{i \in \mathcal{N}}, \qquad \forall a \in \mathcal{A}.$$

A Nash equilibrium $a^* \in \mathcal{A}$ is by definition a fixed point of BR($\cdot$), that is, it is such that

$$a^* \in \mathrm{BR}(a^*).$$

Since all players' payoff functions are by assumption continuous and their (nonempty) action sets compact, by the Weierstrass theorem (proposition A.10 in appendix A) the image BR(a) is nonempty for any strategy profile $a \in \mathcal{A}$. The Berge maximum theorem (proposition A.15) further implies that BR(a) is compact-valued and upper semicontinuous. Last, $U^i(a^i, a^{-i})$ is quasi-concave in a^i, so for any player $i \in \mathcal{N}$, it is

$$\hat{a}^i, \check{a}^i \in \mathrm{BR}^i(a^{-i}) \Rightarrow \theta \hat{a}^i + (1-\theta)\check{a}^i \in \mathrm{BR}^i(a^{-i}), \qquad \forall \theta \in (0,1),$$

whence for any $a \in \mathcal{A}$:

$$\hat{a}, \check{a} \in \mathrm{BR}(a) \Rightarrow \theta \hat{a} + (1-\theta)\check{a} \in \mathrm{BR}(a), \qquad \forall \theta \in (0,1).$$

Thus, BR($\cdot$) is an upper semicontinuous mapping with convex and compact values in $2^\mathcal{A}$, where the set of strategy profiles is convex and compact. By the Kakutani fixed-point theorem (proposition A.17), there exists a point $a^* \in \mathcal{A}$ such that $a^* \in \mathrm{BR}(a^*)$. ∎

Corollary 4.1 (Existence of a Mixed-Strategy Nash Equilibrium) Any normal-form game $\Gamma = (\mathcal{N}, \mathcal{A} = \prod_{i \in \mathcal{N}} \mathcal{A}^i, \{U^i : \mathcal{A} \to \mathbb{R}\}_{i \in \mathcal{N}})$ with a finite player set $\mathcal{N}$ and a finite set of strategy profiles $\mathcal{A}$ has a mixed-strategy Nash equilibrium.

Proof The existence of a mixed-strategy Nash equilibrium of the normal-form game Γ is by definition equivalent to the existence of a (pure-strategy) Nash equilibrium of the augmented game Γ_Δ, which satisfies the assumptions of proposition 4.1. ∎

Example 4.7 (Cournot Oligopoly) Consider N firms, each of which sells a differentiated product on a common market. Given the other firms' strategy profile q^{-i}, each firm $i \in \mathcal{N} = \{1, \ldots, M\}$ simultaneously chooses its output quantity $q^i \geq 0$ so as to maximize its profits

$$\Pi^i(q^i, q^{-i}) = P^i(q^i, q^{-i})q_i - C^i(q^i),$$

where the inverse demand curve $P^i(q^i, q^{-i})$ (continuous, decreasing in its first argument) describes the nonnegative price firm i can obtain for its product as a function of all firms' output decisions, and $C^i(q^i)$ is a continuously differentiable, increasing, convex cost function such that $C^i_{q^i}(0) = 0$ and the Inada conditions $C^i_{q^i}(0) = 0$ and $C^i(\infty) = \infty$ are satisfied.[11] If firm i's revenue $R^i(q^i, q^{-i}) = P^i(q_i, q^{-i})q^i$ is quasi-concave in q^i for all $q^i \geq 0$ and bounded for all (q^i, q^{-i}), then firm i's profit function $\Pi^i(q^i, q^{-i})$ is quasi-concave and thus by proposition 4.1 a Nash equilibrium of this Cournot oligopoly game does exist. The special case where $P^i(q^i, q^{-i}) = 1 - Q$ with $Q = q^1 + \cdots + q^N$ and $C^i(q^i) = cq^i$ with $c \in (0, 1)$ a constant marginal cost parameter yields that $q^* = (q^{i*})_{i \in \mathcal{N}}$, with $q^{i*} \equiv (1-c)/(N+1)$, is the unique Nash equilibrium of this game.[12] □

Remark 4.1 (Wilson's Oddness Result) Wilson (1971) showed that almost any finite game (as in corollary 4.1) has an odd number of mixed-strategy Nash equilibria, in the sense that if a given game happens to have an even number of mixed-strategy Nash equilibria, then a small random perturbation of the players' payoffs will produce with probability 1 a new game with an odd number of mixed-strategy Nash equilibria. For instance, after the two pure-strategy Nash equilibria in the battle-of-the-sexes game discussed in example 4.5 have been determined, the oddness theorem suggests that unless the game is singular there exists another Nash equilibrium, in mixed strategies. □

4.2.2 Static Games of Incomplete Information

In many real-world games the publicly available information about players' preferences may be limited. For example, a bidder in an auction

---

11. *Inada conditions*, such as strict monotonicity and derivatives of either zero or infinity toward the interval boundaries, ensure that solutions to optimization problems are attained at the interior of their domains. From the Inada conditions and the properties of the revenue function, it is possible to conclude that firm i can restrict its attention to the convex compact action set $\mathcal{A}^i = [0, \bar{q}^i]$ for some appropriate positive constant $\bar{q}^i, i \in \mathcal{N}$.
12. The Banach fixed-point theorem (proposition A.3) can sometimes be used to guarantee the existence of a unique Nash equilibrium (see example A.4).

may not know the other bidders' payoff functions or budget constraints. In fact, he might not even be sure how many other bidders there are in the first place. In such games of incomplete information, one continues to maintain the assumption of rationality and common knowledge as formulated in section 4.2.1. Yet, it is necessary relax the degree to which information about the elements of the game (the players, their action sets, and their payoff functions) is available. For this, it is convenient to encapsulate all the not commonly available information about a player i as a point θ^i in a *type space* Θ^i, which is referred to as the player's *type*. In general, the type space could be infinite-dimensional, but in most practically relevant situations the type space can be taken as a subset of a finite-dimensional Euclidean space. Before the game starts, *ex ante*, all players assume that the types are jointly distributed, with the cumulative distribution function (cdf) $F(\theta) = \text{Prob}(\tilde{\theta} \leq \theta)$,[13] where

$$\theta = (\theta^i)_{i \in \mathcal{N}} \in \Theta = \prod_{i \in \mathcal{N}} \Theta^i.$$

Assuming that each player θ^i observes his own type, his beliefs about the other players' types are given by the conditional distribution $F^i(\theta^{-i}) = F(\theta^{-i}|\theta^i)$, which is obtained using Bayesian updating. Because of the players' use of information to perform Bayesian updates, a game with incomplete information is also commonly referred to as a *Bayesian game*. A Bayesian game in normal form is a collection

$$\Gamma_B = (\mathcal{N}, \mathcal{A}, \Theta, \{U^i : \mathcal{A} \times \Theta \to \mathbb{R}\}_{i \in \mathcal{N}}, F : \Theta \to [0,1]). \tag{4.3}$$

Note that for simplicity the influence of the types is limited to the players' payoff functions, even though it could in principle also figure in the explicit description of the player set and the action sets. A game with more general dependences can be rewritten in the current form, for instance, by including an additional player (referred to as Nature) whose type determines all the remaining components of the game and whose

---
13. In principle, it is possible to assume that each player i has a different joint distribution $F^i(\theta)$ of types in mind (given that this difference of opinion is publicly known; otherwise prior beliefs over prior beliefs are needed). However, since the subsequent arguments remain essentially unaffected, this added complexity is dropped. Aumann (1976) showed that in the absence of any strategic considerations, individuals sharing statistical information are in fact not able to disagree about their prior beliefs. Hence, assuming that all agents have the same beliefs about the joint type distribution amounts to requiring that all of them have initial access to the same public information and that any additional information that is obtained privately by agent i is part of his type θ^i.

action set is a singleton. The concept of a Nash equilibrium generalizes to Bayesian games as follows. A *Bayes-Nash equilibrium* of Γ_B is a strategy profile $\alpha^* = (\alpha^{i*})_{i \in \mathcal{N}}$ with $\alpha^{i*} : \Theta^i \to \mathcal{A}^i$ such that

$$\alpha^{i*}(\theta^i) \in \arg\max_{a^i \in \mathcal{A}^i} \bar{U}^i(a^i, \alpha^{-i*}, \theta^i), \qquad \forall \theta^i \in \Theta^i, \qquad \forall i \in \mathcal{N},$$

where player i's expected utility conditional on his own type is given by

$$\bar{U}^i(a^i, \alpha^{-i*}, \theta^i) = E[U^i(a^i, \alpha^{-i*}(\tilde{\theta}^{-i}), \theta^i, \tilde{\theta}^{-i}) | \theta^i]$$

$$= \int_{\Theta^{-i}} U^i(a^i, \alpha^{-i*}(\theta^{-i}), \theta^i, \theta^{-i}) \, dF(\theta^{-i}|\theta^i),$$

where $\alpha^{-i*}(\theta^{-i}) = (\alpha^j(\theta^j))_{j \in \mathcal{N} \setminus \{i\}}$ and $F(\theta^{-i}|\theta^i)$ denotes a conditional distribution function.

Example 4.8 (First-Price Auction) Consider $N \geq 2$ bidders participating in a first-price auction for a certain good. Each bidder $i \in \mathcal{N} = \{1, \ldots, N\}$ has a private valuation $\theta^i \in \Theta^i = [0, 1]$, and it is assumed that the good cannot be resold. Each bidder believes that the type vector $\theta = (\theta^1, \ldots, \theta^N) \in [0, 1]^N$ is distributed according to the cdf $F(\theta) = G^N(\theta^i)$, where $G : [0, 1] \to [0, 1]$ is a continuous, increasing cdf; that is, the types can be viewed as an independently and identically distributed sample of the distribution G. Each bidder i simultaneously chooses a bid $a^i \in \mathcal{A}^i = [0, 1]$. Given the other bidders' strategy profile $\alpha^{-i}(\theta^{-i})$, player i's best-response correspondence is

$$\mathrm{BR}^i(\theta^i) = \arg\max_{a^i \in [0,1]} \{(\theta^i - a^i) \mathrm{Prob}(\max_{j \in \mathcal{N} \setminus \{i\}} \{\alpha^j(\tilde{\theta}^j)\} \leq a^i | \theta^i)\},$$

for all $\theta^i \in [0, 1]$. Provided that each bidder $j \neq i$ follows the same symmetric bidding strategy $\alpha^j(\theta^j) \equiv \beta(\theta^j)$, which is increasing in his type θ^j, it is possible to invert β and obtain that

$$\mathrm{BR}^i(\theta^i) = \arg\max_{a^i \in [0,1]} \{(\theta^i - a^i) G^{N-1}(\beta^{-1}(a^i))\}, \quad \forall \theta^i \in [0, 1].$$

A necessary optimality condition for bidder i's bid a^{i*} to lie in its best response $\mathrm{BR}^i(\theta^i)$ is that

$$0 = \frac{d}{da^i}\bigg|_{a^i = a^{i*}} \{(\theta^i - a^i) G^{N-1}(\beta^{-1}(a^i))\}$$

$$= -G^{N-1}(\beta^{-1}(a^{i*})) + (\theta^i - a^{i*}) \cdot \frac{(N-1) G^{N-2}(\beta^{-1}(a^{i*}))}{\dot{\beta}(\beta^{-1}(a^{i*}))/g(\beta^{-1}(a^{i*}))}, \quad (4.4)$$

using the inverse function theorem (proposition A.8), and where g is the positive probability density corresponding to the distribution G. In a symmetric Nash equilibrium, bidder i's bid a^{i*} will be equal to $\beta(\theta^i)$, so that $\beta^{-1}(a^{i*}) = \beta^{-1}(\beta(\theta^i)) = \theta^i$, and relation (4.4) becomes

$$0 = -G^{N-1}(\theta^i) + (N-1)\frac{(\theta^i - \beta(\theta^i))G^{N-2}(\theta^i)}{\dot{\beta}(\theta^i)/g(\theta^i)}, \qquad \forall \theta^i \in [0,1],$$

or equivalently, a linear ordinary differential equation (ODE) of the form

$$\dot{\beta}(\theta^i) + (N-1)\frac{g(\theta^i)}{G(\theta^i)}\beta(\theta^i) = (N-1)\frac{\theta^i g(\theta^i)}{G(\theta^i)}, \qquad \forall \theta^i \in [0,1].$$

With the insight that the only amount a bidder with zero valuation can bid in equilibrium is zero, which implies the initial condition $\beta(0) = 0$, the Cauchy formula yields the equilibrium bidding function[14]

$$\beta(\theta^i) = \theta^i - \frac{\int_0^{\theta^i} G^{N-1}(\vartheta)\,d\vartheta}{G^{N-1}(\theta^i)}, \qquad \forall \theta^i \in [0,1], \qquad (4.5)$$

which fully describes the symmetric Bayes-Nash equilibrium, with $\alpha^{i*}(\theta^i) \equiv \beta(\theta^i)$. Note also that $\beta(\theta^i)$ is an increasing function, justifying *ex post* the initial monotonicity assumption that led to this solution. □

Example 4.9 (Cournot Oligopoly with Cost Uncertainty) Consider $N \geq 2$ identical firms selling a homogeneous good in a common market. Each firm $i \in \mathcal{N} = \{1, \ldots, N\}$ decides about its output $q^i \geq 0$, which costs $C(q^i, \theta^i) = \theta^i q^i$ to produce, where the marginal cost parameter $\theta^i \in [0,1]$ belongs to the firm's private information. The vector $\theta = (\theta^1, \ldots, \theta^N)$ follows *ex ante* the symmetric joint distribution $F(\theta)$. Thus, each firm i, by observing its own cost information, is generally able to infer some information about the other firms' costs as well. The market price $P(Q) = 1 - Q$ depends only on the firms' aggregate output $Q = q^1 + \cdots + q^N$. Looking for a symmetric Bayes-Nash equilibrium,

---

14. Specifically, the Cauchy formula (2.9) gives

$$\beta(\theta^i) = \left(\int_0^{\theta^i} (N-1)\frac{\vartheta g(\vartheta)}{G(\vartheta)} \exp\left[\int_0^{\vartheta} (N-1)\frac{g(s)}{G(s)}ds\right]d\vartheta\right) \exp\left[-\int_0^{\theta^i} (N-1)\frac{g(\vartheta)}{G(\vartheta)}d\vartheta\right]$$

$$= \left(\int_0^{\theta^i} (N-1)\vartheta g(\vartheta) G^{N-2}(\vartheta)d\vartheta\right)\left(\frac{1}{G^{N-1}(\theta^i)}\right) = \frac{\int_0^{\theta^i} \vartheta\left(\frac{d}{d\vartheta}G^{N-1}(\vartheta)\right)d\vartheta}{G^{N-1}(\theta^i)},$$

for all $\theta^i \in [0,1]$, which after an integration by parts simplifies to the expression in (4.5).

given that any firm $j \neq i$ follows the strategy $q^{j*} = \alpha^0(\theta^j)$, firm i's profit-maximization problem,

$$\alpha^0(\theta^i) \in \arg\max_{q^i \in [0,1]} \int_{[0,1]^{N-1}} \left(1 - q^i - \sum_{j \in \mathcal{N} \setminus \{i\}} \alpha^0(\theta^j) - \theta^i \right) q^i dF(\theta^{-i} | \theta^i)$$

determines the missing function $\alpha^0 : [0,1] \to \mathbb{R}_+$. Indeed, the first-order necessary optimality condition yields that

$$0 = 1 - 2\alpha^0(\theta^i) - (N-1)\bar{\alpha}^0(\theta^i) - \theta^i, \qquad \forall \theta^i \in [0,1],$$

as long as $\alpha^0(\theta^i) \in (0,1)$, where

$$\bar{\alpha}^0(\theta^i) = \int_{[0,1]^{N-1}} \alpha^0(\theta^j) \, dF(\theta^{-i} | \theta^i), \qquad \forall \theta^i \in [0,1], \qquad \forall j \in \mathcal{N} \setminus \{i\}.$$

Combining the last two relations, and using the abbreviation

$$\bar{\theta} = E[\tilde{\theta}^i] = \int_{[0,1]^N} \theta^i dF(\theta),$$

for all $i \in \mathcal{N}$, one obtains

$$\bar{\alpha}^0(\theta^i) \equiv (1 - \bar{\theta}) \left(\frac{1 - E[\tilde{\theta}^j | \theta^i]}{1 - \bar{\theta}} - \frac{N-1}{N+1} \right).$$

This yields the interior solution

$$\alpha^0(\theta^i) = \frac{1 - \bar{\theta}}{2} \left(\frac{1 - \theta^i}{1 - \bar{\theta}} - \frac{N-1}{2} \left(\frac{1 - E[\tilde{\theta}^j | \theta^i]}{1 - \bar{\theta}} - \frac{N-1}{N+1} \right) \right), \qquad \forall \theta^i \in [0,1],$$

provided that $\alpha^0(\theta^i) \in (0,1)$ for almost all $\theta^i \in [0,1]$, that is, at an interior solution.[15] The Bayes-Nash equilibrium α^* with $\alpha^{i*} = \alpha^0$ specializes to the symmetric Nash equilibrium in the corresponding game with complete information, when $c = \bar{\theta} \equiv E[\tilde{\theta}^j | \theta^i] \equiv \theta^i \in (0,1)$, which implies that $\alpha^0(\theta^i) \equiv (1-c)/(N+1)$, as in example 4.7. □

Proposition 4.2 (Existence of a Bayes-Nash Equilibrium) A Bayesian game Γ_B of the form (4.3) with a finite player set $\mathcal{N} = \{1, \ldots, N\}$ (with $N \geq 2$), compact action sets $\mathcal{A}^1, \ldots, \mathcal{A}^N$, compact finite-dimensional

---

15. To guarantee that the equilibrium strategy α^0 takes only values at the interior of the action set with probability 1, it is necessary that $E[\tilde{\theta}^j | \theta^i]$ be bounded from above by $\bar{\theta} + 2(1 - \bar{\theta})/(N+1)$.

type spaces $\Theta^1, \ldots, \Theta^N$, and utility functions $U^i(\cdot, \theta)$, which are continuous for all $\theta \in \Theta$, has a Bayes-Nash equilibrium (in behavioral (mixed) strategies).

Proof See Balder (1988) for a proof; he defines a *behavioral (mixed) strategy* for player i as a transition-probability measure between the spaces Θ^i and $\mathcal{A}^i$. ∎

4.2.3 Dynamic Games of Complete Information

A dynamic game evolves over several time periods and can be described by an extensive form, which, depending on past actions and observations as well as on current and expected future payoffs, for each time period specifies if and how a player can choose an action. It is clear that because of the generality of this situation the study of a given dynamic game in extensive form can be very complicated. Since the purpose here is to motivate and provide background for the theory of differential games, attention is restricted to games with a fairly simple dynamic structure.

Extensive-Form Games An extensive-form description of a dynamic game specifies what players know in each stage of the game, when it is their turn to play, and what actions they can take. Before introducing the general elements of this description it is useful to consider a simple example.

Example 4.10 (Entry Game) Consider two firms, 1 and 2. Firm 1 is a potential entrant and at time $t = 0$ decides about entering a market (i.e., select the action e) or not (i.e., select the action $\bar{e}$). Firm 2 is an incumbent monopolist who can observe firm 1's action, and at time $t = 1$ chooses to either start a price war and thus to fight the entrant (i.e., select the action f) or to accommodate firm 1 and not fight (i.e., select the action $\bar{f}$). Assume that firm 1 obtains a zero payoff if it does not enter and a payoff of either -1 or 1 if it enters the market, depending on whether firm 2 decides to fight or not. Firm 2, on the other hand, obtains a payoff of 2 if firm 1 does not enter and otherwise a payoff of -1 when fighting or a payoff of 1 when not fighting. The game tree in figure 4.2 depicts the sequence of events as well as the firms' payoffs. At each node a firm makes a decision, until a *terminal node* with payoffs is reached. A firm's strategy consists of a *complete contingent plan* for each of its decision nodes, no matter if that node is reached in equilibrium or not. For example, when firm 1 decides not to enter the market, then in reality

Game Theory

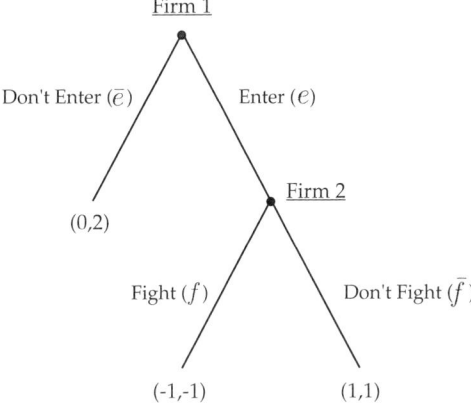

Figure 4.2
Game tree for the entry game.

there is nothing to decide for firm 2. Yet, in order for firm 1 to be able to come to the decision of not entering, it needs to form an expectation about what firm 2 would do if firm 1 decided to enter the market. The normal-form representation of the game in a payoff matrix is as follows:

		Firm 2	
		f	$\bar{f}$
Firm 1	e	$(-1,-1)$	$(1,1)$
	$\bar{e}$	$(0,2)$	$(0,2)$

It is easy to verify that there are two pure-strategy Nash equilibria, $(e,\bar{f})$ and $(\bar{e},f)$.[16] While the first of these two Nash equilibria seems plausible, there clearly is a problem with the equilibrium where firm 1 decides not to enter the market based on firm 2's threat of fighting. Indeed, given the intertemporal structure of the game, once firm 1 actually enters the market, it is better for firm 2 not to fight (leading to a payoff of 1 instead of −1). Thus, firm 2's threat of fighting is not credible and should be eliminated. □

---

16. This game has no additional Nash equilibria, even in mixed strategies. Thus, with an even number of equilibria it is degenerate in view of Wilson's oddness result (see remark 4.1). The degeneracy is created by firm 2's indifference about its action when firm 1 does not enter the market.

This example shows that the concept of a Nash equilibrium is generally too weak for dynamic games because it may lead to *noncredible threats* or, more generally, to *time inconsistencies*. A noncredible threat arises because of a lack in commitment ability of the player that issues the threat. A well-known example with a surprising consequence of this lack in commitment ability is the following.

Example 4.11 (Intertemporal Pricing and the Coase Conjecture) A seller with market power is able to sell her goods at prices above marginal cost. While this seems to guarantee the seller a strictly positive net payoff, the following informal argument by Coase (1972) shows that this does not have to be the case, when the seller can sell her goods at any point $t \geq 0$ in *continuous* time.[17] The intuition is that the monopolist at time $t = 0$ is competing with a copy of its own product that is sold at time $t = \Delta > 0$. Of course, any consumer, when looking at the product today versus the product tomorrow will agree that the product at time $t = \Delta$ is not quite as good as the product now at time $t = 0$.[18] However, this difference in product-quality difference vanishes when Δ tends to zero. Thus, as $\Delta \to 0^+$ arbitrarily many copies of virtually the same product will be available in any fixed time interval, so the resulting perfect competition must drive the monopolist's price down to marginal cost. Hence, in a perfect world with continuous money and time increments, a seller's ability to adjust her prices over time is more of a curse than a blessing, because of the seller's lack of commitment power. This Coase problem can be ameliorated by renting the product instead of selling it or by making binding promises about future production (e.g., by issuing a limited edition). Perishable products also increase the monopolist's commitment power (e.g., when selling fresh milk) as well as adjustment costs for price changes (e.g., due to the necessity of printing a new product catalogue). The ability to commit to a price path from the present into the future is a valuable asset for a seller; the question of commitment as a result of the consumers' option of intertemporal arbitrage (i.e., they can choose between buying now or later) is important for any dynamic pricing strategy, at least when the available information is fairly complete. Note that the Coase problem is not significant in

---
17. The Coase conjecture was proved by Stokey (1981), Bulow (1982), and Gül et al. (1986) in varying degrees of generality.
18. The reason for this may be both time preference and quality preference. By waiting, the consumer incurs, on the one hand, an opportunity cost of not being able to use the product and, on the other hand, a cost due to a decay in the product's quality (e.g., due to perishability or technological obsolescence).

Game Theory 169

situations when consumers are nonstrategic (i.e., not willing or able to wait). □

The previous two examples underline the importance of time-consistency issues. The lack of players' ability to commit to their intertemporal strategic plans weakens the plausibility of Nash equilibria that rely on this commitment. The additional requirement of *subgame perfection* eliminates time inconsistencies. A *subgame* of a dynamic game is a game that arises after a certain time $t \geq 0$ has passed in the original dynamic game, which up to that instant could have taken any arbitrary feasible path. The subgame therefore starts at a certain node in the extensive-form description of the game. A *subgame-perfect Nash equilibrium* is a strategy profile that when restricted to any subgame induces a Nash equilibrium of the subgame.

Example 4.12 (Entry Game, Continued) The only "proper" subgame, namely, a subgame that is not the game itself, is the subgame that starts at firm 2's decision node (see figure 4.3). The strategy profile $(\bar{e}, f)$ of the original game does not induce a Nash equilibrium of this subgame, since f does not maximize firm 2's payoff, given that firm 1 decided to enter the market. Hence, the Nash equilibrium $(\bar{e}, f)$, which contains the noncredible threat, is not subgame-perfect. □

It is now possible to formally describe a (finite) *extensive-form game*,

$$\Gamma_E = (\mathcal{N}, \mathcal{K}, \mathcal{Z}, \mathcal{A}, \mathcal{H}, \mathsf{a}(\,\cdot\,), \mathsf{h}(\,\cdot\,), \nu(\,\cdot\,), \pi(\,\cdot\,), \sigma(\,\cdot\,), \{U^i : \mathcal{Z} \to \mathbb{R}\}_{i \in \mathcal{N}}),$$

where $\mathcal{N} = \{1, \ldots, N\}$ is the set of $N \geq 1$ players; $\mathcal{K} = \{\kappa_1, \ldots, \kappa_K\}$ is a set of $K \geq 1$ nodes; $\mathcal{Z} \subset \mathcal{K}$ is a set of *terminal nodes*; $\mathcal{A}$ is a set of all players' actions; $\mathcal{H} \subset 2^{\mathcal{K}}$ is a set of *information sets* (usually a partition of the set of *nonterminal nodes* $\mathcal{K} \setminus \mathcal{Z}$); $\mathsf{a} : \mathcal{K} \to \mathcal{A}$ is a function that assigns to each node k an action $\mathsf{a}(k)$ that leads to it from its predecessor (with $\mathsf{a}(\kappa_1) = \emptyset$ for the initial node k_0); $\mathsf{h} : \mathcal{K} \setminus \mathcal{Z} \to \mathcal{H}$ is a function that assigns each nonterminal node to an information set; $\nu : \mathcal{H} \to \mathcal{N}$ is a *player function* that assigns to each information set a player whose turn it is to take an action; $\pi : \mathcal{K} \to \mathcal{K}$ is a *predecessor function* that specifies the (unique) predecessor of each node (with $\emptyset$ being the predecessor of the initial node $\kappa_1 \in \mathcal{K}$); $\sigma : \mathcal{K} \rightrightarrows \mathcal{K}$ is a (set-valued) *successor function* that specifies the set of successor nodes for each node (with $\sigma(\mathcal{Z}) = \{\emptyset\}$); and the *payoff functions* U^i specify for each terminal node $z \in \mathcal{Z}$ and each player $i \in \mathcal{N}$ a payoff $U^i(z)$. To understand the meaning of all the elements of the extensive-form description of a dynamic game, consider again the simple game in example 4.10.

Example 4.13 (Entry Game, Continued) An extensive-form representation Γ_E of the entry game has the elements $\mathcal{N} = \{1,2\}$, $\mathcal{K} = \{\kappa_1, \ldots, \kappa_5\}$, $\mathcal{Z} = \{\kappa_3, \kappa_4, \kappa_5\}$, $\mathcal{H} = \{\{\kappa_1\}, \{\kappa_2\}\}$, $h(\kappa_i) = \{\kappa_i\}$, and $v(\{\kappa_i\}) = i$ for $i \in \{1,2\}$. The functions a, π, σ are specified in table 4.1, and the nodes and players' payoffs are depicted in figure 4.3. □

As becomes clear from example 4.10, even for simple dynamic games a full-scale specification of all the elements of its extensive-form representation is rather cumbersome. In most practical applications it is therefore simply omitted, and one relies on an informal description of the game together with a basic game tree such as the one shown in figure 4.2 for example 4.10. This game is termed a dynamic game with

Table 4.1
Extensive-Form Representation of the Entry Game

Node	κ_1	κ_2	κ_3	κ_4	κ_5
$a(\kappa_i)$	∅	e	$\bar{e}$	f	$\bar{f}$
$\pi(\kappa_i)$	∅	κ_1	κ_1	κ_2	κ_2
$\sigma(\kappa_i)$	$\{\kappa_2, \kappa_3\}$	$\{\kappa_4, \kappa_5\}$	∅	∅	∅

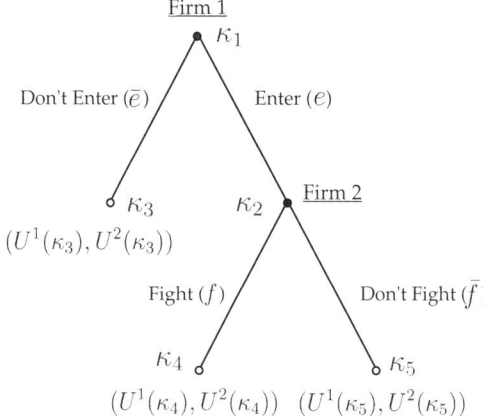

Figure 4.3
Game tree with node specification for the entry game.

Game Theory

perfect information, since all the players' information sets contained at most one node. If information sets contain sometimes more than one node, the game is called a dynamic game with *imperfect information*. Thus, while information in both types of games is complete, we can distinguish between cases where information is perfect and others where it is not. The role of information sets and imperfect information becomes clear when representing a simple static game of complete information in extensive form.

Example 4.14 (Battle of the Sexes in Extensive Form) Ann and Bert's game in example 4.5 can be represented in extensive form as in figure 4.4. The information set containing two of Bert's decision nodes means that at the time of taking a decision, Bert does not know what action Ann has decided to take. This dynamic game is therefore equivalent to the original simultaneous-move game. Note also that it is possible to switch Ann and Bert in the diagram and obtain an equivalent representation of the game; extensive-form representations are in general not unique. Finally, note that when information is perfect in the sense that all information sets are singletons, and Bert knows what Ann has decided to do (as in figure 4.4b), then Ann has a definite advantage in moving first because she is able to commit to going dancing, which then prompts Bert to go dancing as well, in the only subgame-perfect Nash equilibrium of

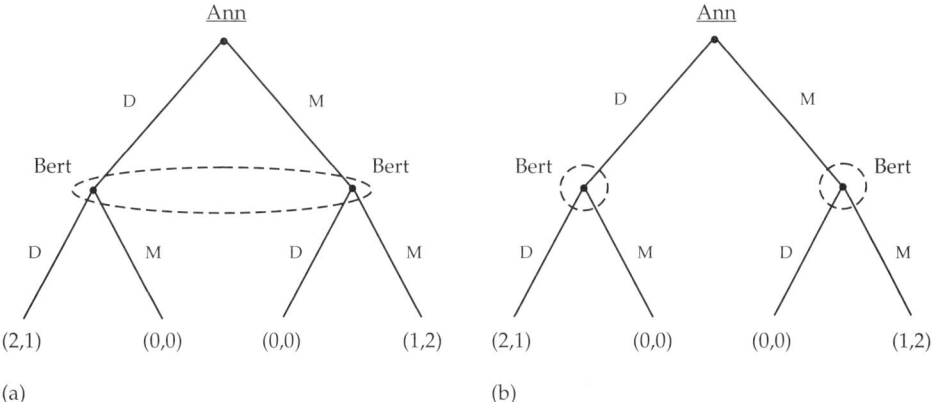

Figure 4.4
Battle-of-the-sexes game in extensive form: (*a*) without commitment (one information set), and (*b*) with commitment (two information sets).

this perfect-information, sequential-move version of the battle-of-the-sexes game. □

Proposition 4.3 (Existence of a Subgame-Perfect Nash Equilibrium)
(Zermelo 1913) (1) Every finite extensive-form game (of perfect information) has a pure-strategy Nash equilibrium that can be derived using backward induction. (2) If no player has the same payoffs at two terminal nodes, then there is a unique subgame-perfect Nash equilibrium (which can be derived through backward induction).

Proof (Outline) Consider the dynamic game Γ_E.

(1) Any subgame of Γ_E can be represented in normal form, which by proposition 4.1 has a Nash equilibrium. Thus, by backward induction it is possible to find a Nash equilibrium of Γ_E.

(2) If players are not indifferent between terminal nodes, then at each node there exists a strictly dominant choice of action; randomization between different actions is therefore never optimal. Hence, any subgame-perfect Nash equilibrium must be in pure strategies. Because of the strict dominance at each decision node, the subgame-perfect Nash equilibrium must also be unique. ∎

In order to verify that a given strategy profile (which for each player $i \in \mathcal{N}$ contains a mapping α^i from all his information sets to feasible actions) of the dynamic game Γ_E constitutes a subgame-perfect Nash equilibrium, the following result is of great practical significance.

Proposition 4.4 (One-Shot Deviation Principle) In an extensive-form game Γ_E a strategy profile $\alpha^* = (\alpha^{i*})_{i \in \mathcal{N}}$ is a subgame-perfect Nash equilibrium if and only if it satisfies the one-shot deviation condition: no player can gain by deviating from α^{*i} at one single information set while conforming to it at all other information sets.

Proof (Outline) ⇒: The necessity of the one-shot deviation condition follows directly from the definition of a subgame-perfect Nash equilibrium. ⇐: If the one-shot deviation condition is satisfied for α^* when α^* is not a subgame-perfect Nash equilibrium, then there exists a decision node at which some player i has a better response than the one prescribed by α^{i*}. But then choosing that response and conforming to α^{i*} thereafter would violate the one-shot deviation condition. Hence, the strategy profile α^* must be a subgame-perfect Nash equilibrium. ∎

Example 4.15 (Centipede Game) Each of two players, 1 and 2, has a starting capital of one dollar. The players take turns (starting with player 1), choosing either "continue" (C) or "stop" (S). Upon a player's selecting C, one dollar is transferred by an independent third party from that player's capital to the other player's capital and one additional dollar is added to the other player's capital. The game stops when one player chooses S or if both players' payoffs are at least $100. Figure 4.5 shows the corresponding game tree. Now consider the strategy profile α, which is such that both players always choose C, and use the one-shot deviation principle to check if it constitutes a subgame-perfect Nash equilibrium. Since player 2 finds deviating in the last period, that is, playing $(C, C, \ldots, C, S)$ instead of $(C, C, \ldots, C)$, profitable, the one-shot deviation condition in proposition 4.4 is not satisfied and α cannot be a subgame-perfect Nash equilibrium. Via backward induction, it is easy to determine that the unique subgame-perfect Nash equilibrium is of the form $[(S, S, \ldots, S); (S, S, \ldots, S)]$, namely, both agents choose S at each of their respective decision nodes. □

Remark 4.2 (Infinite-Horizon One-Shot Deviation Principle) The one-shot deviation principle in proposition 4.4 extends to infinite-horizon games, provided that they are continuous at infinity, that is, if two strategy profiles agree in the first T periods, then the absolute difference in each player's payoff converges to zero as T goes to infinity. This condition is satisfied if the players' stage-game payoffs are uniformly bounded, as long as the *discount factor* δ^i applied to player i's payoffs between periods lies in $(0, 1)$, for all i. □

Repeated Games The simplest way to extend the notion of a complete-information normal-form game (see section 4.2.1) to $T > 1$ discrete time periods is to repeat a one-period stage game Γ of the form (4.1) for T times. The resulting repeated game Γ^T is referred to as a *supergame*. At

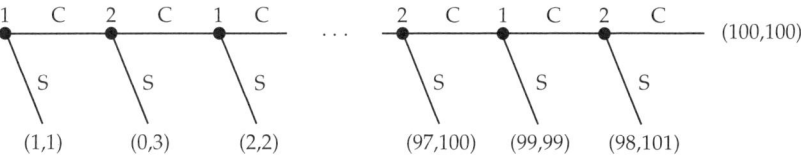

Figure 4.5
Centipede game.

each time $t \in \{0, \ldots, T-1\}$ any player $i \in \mathcal{N}$ selects an action $a_t^i \in \mathcal{A}^i$. Player i can condition his action on the information $s_t^i \in \mathcal{S}_t^i$ available to him, where $\mathcal{S}_t^i$ is player i's (measurable) *observation* (or *sample*) *space* at time t. A player's observation s_t^i may include the players' past action profiles (up to time $t-1$, for $t \geq 1$) and the realizations of certain publicly observable random events. The set-valued function that maps the available information to a probability distribution over player's i's observation space $\mathcal{S}_t^i$ is called player i's *information structure*.

For simplicity, it is assumed in the discussion of repeated games that each player's information is complete. That is, at time $t \in \{0, \ldots, T-1\}$, player i's observation space $\mathcal{S}_t^i$ corresponds to the set H_t of possible time-t histories h_t, where

$$\mathsf{H}_0 = \emptyset, \qquad \mathsf{H}_t = \mathsf{H}_{t-1} \times \mathcal{A}, \qquad \forall t \in \{1, \ldots, T-1\},$$

and

$$\mathsf{h}_t = (a_0, \ldots, a_{t-1}) \in \mathsf{H}_t \subset \mathsf{H}.$$

Thus, a player's strategy α^i is a mapping from the set H of all possible *histories* of the supergame, with

$$\mathsf{H} = \bigcup_{t=0}^{T-1} \mathsf{H}_t,$$

to time-t actions in $\mathcal{A}^i$.[19] Let $\delta \in (0,1)$ be a per-period *discount factor*, common to all players. For a given strategy profile $\alpha = (\alpha^i)_{i \in \mathcal{N}}$, we set $u(0) = (u^i(0))_{i \in \mathcal{N}} = \alpha^i(\emptyset)$ and $u(t) = (u^i(t))_{i \in \mathcal{N}} = \alpha(\{\mathsf{h}_t, u(t-1)\})$ for $t \in \{1, \ldots, T-1\}$; there player i's *average payoff* is given by

$$J_{\mathrm{Avg}}^i(u^i|u^{-i}) = \frac{1-\delta}{1-\delta^{T+1}} \sum_{t=0}^{T} \delta^t U^i(u(t)) \to (1-\delta) \sum_{t=0}^{\infty} \delta^t U^i(u(t)), \quad \text{as } T \to \infty.$$

The intuition is that when player i obtains a constant payoff $U^i(u(t)) = c$ in each period $t \geq 0$, then his average payoff will be equal to c as well. In this way one can directly compare a player's average payoffs to his stage-game payoffs, irrespective of the discounting between periods. A *Nash equilibrium* of the supergame Γ^T is a strategy profile $\alpha^* = (\alpha^{i*})_{i \in \mathcal{N}}$ such that

---
19. A generalization of this definition that includes *mixed*-strategy profiles is straightforward and therefore omitted.

$$\alpha^{i*}(h_t) \in \arg\max_{a^i \in \mathcal{A}^i}\{U^i(a^i, \alpha^{-i*}(h_t)) + V^i(t, (h_t, (a^i, \alpha^{-i*}(h_t))))\},$$

where

$$V^i(t, h_t) = \sum_{s=t+1}^{T-1} \delta^{t-s} U^i(a_s^*)$$

with

$$a_s^* = \alpha^*(h_s^*), \qquad h_{s+1}^* = (h_s^*, \alpha^*(h_s^*)), \qquad h_t^* = h_t, \quad s \in \{t+1, \ldots, T-1\}.$$

Remark 4.3 (Augmented History) In some situations it is useful to include publicly observable realizations of random variables (e.g., a coin toss, or the number of currently observable sunspots) in the players' observations because they may allow the players to coordinate their actions and thus effectively enlarge the set of attainable payoff profiles. Given time-t realizations of a random process $\tilde{\omega}_t$, let

$$\hat{h}_t = (a_0, \ldots, a_{t-1}; \omega_0, \ldots, \omega_{t-1}) \in \hat{H}_t \subset \hat{H}$$

be an *augmented* time-t history, and

$$\hat{H} = \bigcup_{t=0}^{T-1} \hat{H}_t$$

be the set of all such augmented histories, where

$$\hat{H}_t = H_t \times \Omega^t, \quad \forall t \in \{1, \ldots, T-1\}.$$

Analogous to the earlier definition, an *augmented strategy profile* is of the form $\hat{\alpha} = (\hat{\alpha}^i)_{i \in \mathcal{N}}$, where

$$\hat{\alpha}^i : \hat{H} \to \mathcal{A}^i$$

for all $i \in \mathcal{N}$. □

Example 4.16 (Finitely Repeated Prisoner's Dilemma) Consider a T-fold repeated prisoner's dilemma game with a stage-game payoff matrix as in example 4.4. To obtain a subgame-perfect Nash equilibrium of the supergame, by proposition 4.3 one can use backward induction starting at $t = T$. Clearly, in the last period both prisoners choose D, which then fixes the terminal payoffs at time $t = T - 1$, so that again both prisoners will choose D. Continuing this argument (e.g., by induction) yields that for any finite time horizon T, the unique subgame-perfect Nash equilibrium of the supergame is for both players to play D (i.e., to defect) in all periods. □

For infinitely repeated games, backward induction cannot be used to obtain a subgame-perfect Nash equilibrium. Yet, by threatening a lower future payoff it may be possible to induce other players to deviate from a "myopic" stage-game Nash equilibrium. Depending on the threats used, different outcomes can be attained (see, e.g., proposition 4.5). Note that the game does not even have to be really infinite: a positive probability of continuation in each period is enough to yield an equivalent analysis. For instance, if the continuation probability p is constant across periods, then one may be able to consider $\hat{\delta} = \delta p$ instead of δ as the per-period discount factor over an infinite time horizon; this is often referred to as *stochastic discounting*.

Example 4.17 (Infinitely Repeated Prisoner's Dilemma) Consider an infinitely repeated prisoner's dilemma, obtained by letting T in example 4.16 go to infinity. The one-shot deviation principle (proposition 4.4 and remark 4.2) can be used to show that one subgame-perfect Nash equilibrium of the supergame is that both players choose D in every period. If the players can condition their strategies on histories, then other subgame-perfect Nash equilibria are also possible. For example, as long as $\delta > 1/2$, the following grim-trigger strategy profile (for all $i \in \mathcal{N}$) constitutes a subgame-perfect Nash equilibrium:

- Player i chooses C in the first period.
- Player i continues to choose C, as long as no player has deviated to D in any earlier period.
- If the opponent chooses D, then player i plays D always (i.e., for the rest of the game).

If both players conform to the grim-trigger strategy, then each player's average payoff is 1. To show that the preceding grim-trigger strategy profile constitutes a subgame-perfect Nash equilibrium, consider a one-shot deviation in period t, which yields a payoff of

$$(1-\delta)(1+\delta+\cdots+\delta^{t-1}+2\delta^t+0+0+\cdots) = 1-\delta^t(2\delta-1) < 1,$$

as long as $\delta > 1/2$. Now one must check that in the subgame in which both players play D neither has an incentive to deviate. But it has already been shown that the stationary strategy profile in which all players always play D is a subgame-perfect Nash equilibrium. Indeed,

Game Theory

all individually rational payoff vectors can be implemented using a grim-trigger strategy (see proposition 4.5 and remark 4.4).[20] □

The set of individually rational payoffs is

$$\mathcal{V} = \{(v^1, \ldots, v^N) \in \mathbb{R}^N : \exists a \in \mathcal{A} \text{ s.t. } U^i(a) \geq v^i \geq \underline{v}^i, \ \forall i \in \mathcal{N}\},$$

where

$$\underline{v}^i = \min_{a^{-i} \in \mathcal{A}^{-i}} \{\max_{a_i \in \mathcal{A}^i} U^i(a^i, a^{-i})\}$$

is player i's *minmax payoff*. Payoff vectors v such that each player i's payoff v^i is strictly greater than his minmax payoff $\underline{v}^i$ are *strictly* individually rational. Figure 4.6 illustrates the set $\mathcal{V}$ of all individually rational payoff vectors in the context of example 4.17. Furthermore, the set

$$\mathcal{R} = \{(v^1, \ldots, v^N) : \exists \text{ NE } e^* \text{ of } \Gamma \text{ and } \exists a \in \mathcal{A} \text{ s.t. } U^i(a) \geq v^i \geq U^i(e^*)\} \subseteq \mathcal{V}$$

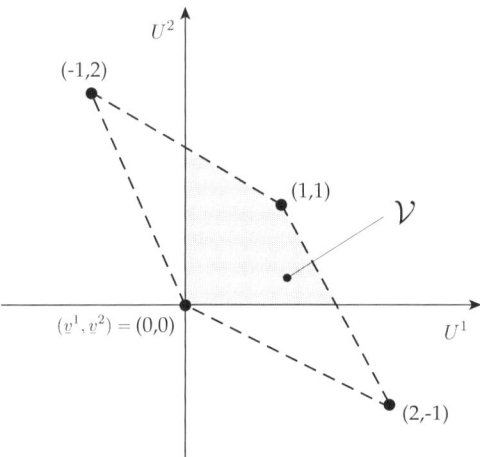

Figure 4.6
Convex hull of the stage-game payoff vectors, and set of individually rational payoffs.

---

20. Axelrod (1984) reports the results of an experiment where subjects (including prominent economists) were asked to specify a strategy for a repeated two-player prisoner's dilemma game so as to maximize the average performance against all other submitted strategies. The winning entry, by the mathematician Anatol Rapoport, was tit-for-tat, i.e., a strategy that prescribes cooperation in the first period and from then on copies the opponent's previous-period action. While the strategy never wins, it tends to perform very well on average. This was confirmed in a repeat experiment where the same entry won again. For more details, see Dixit and Nalebuff (1991, ch. 4).

is the set of Nash-reversion payoffs. The following Nash-reversion folk theorem[21] provides a simple implementation of equilibrium payoff vectors in $\mathcal{R}$.

Proposition 4.5 (Nash-Reversion Folk Theorem) (Friedman 1971)
For any Nash-reversion payoff $\pi \in \mathcal{R}$ there is a constant $\underline{\delta} \in (0,1)$ such that for any common discount factor $\delta \in (\underline{\delta}, 1)$, there exists a subgame-perfect Nash equilibrium of the supergame $\Gamma^{\infty}(\delta)$ with payoffs equal to π.

Proof (Outline) Consider a (possibly correlated) stage-game strategy profile a such that $v = (U^1(a), \ldots, U^N(a)) \in \mathcal{R}$. The following strategy profile induces a subgame-perfect Nash equilibrium in the supergame:

- Start playing a_i and continue doing so as long as a was played in the previous period.
- If in the previous period at least one player deviated, then each player plays a dominated Nash-equilibrium strategy profile e^* for the rest of the game.

This strategy profile indeed constitutes a subgame-perfect Nash equilibrium, since

$$\max_{\hat{a} \in A} U^i(\hat{a}) + \frac{\delta U^i(e^*)}{1-\delta} \leq \frac{U^i(a)}{1-\delta}$$

as long as $\delta \in (0,1)$ is large enough. The rest follows using the one-shot deviation principle. ∎

Remark 4.4 (Subgame-Perfect Folk Theorem) (Fudenberg and Maskin 1986) The set $\mathcal{R}$ of Nash-reversion payoffs is a subset of the set $\mathcal{V}$ of all individually rational payoffs. The following folk theorem states that all (strictly) individually rational payoffs can be implemented using an appropriate subgame-perfect strategy profile, provided that the players are patient enough and provided that the set of individually rational payoffs is of full dimension. This is remarkable because the implementable payoff vectors may be strictly smaller than the smallest Nash-equilibrium payoff vector.

If $dim(\mathcal{V}) = N$, then for any $v = (v^1, \ldots, v^N) \in \mathcal{V}$ with $v^i > \underline{v}^i$ there is a constant $\underline{\delta}(v) \in (0,1)$ such that for any $\delta \in (\underline{\delta}(v), 1)$, there exists a subgame-perfect

---

21. The term *folk theorem* stems from the fact that its content was known (i.e., it was part of "folk wisdom") before a proof appeared in the literature.

Nash equilibrium (in mixed strategies) of $\Gamma^\infty(\delta)$ with an (average) expected payoff of v.

The proof of this result is constructive and requires trigger strategies with *finite-length* punishment phases (including punishment for the failure to punish), followed by an indefinite reward phase (with extra payoffs for those who punished). Abreu et al. (1994) have shown that the dimensionality condition can be relaxed to $\dim(\mathcal{V}) \geq N - 1$. □

Remark 4.5 (Equilibrium Multiplicity and Institutional Design)
Kreps (1990, 95–128) noted that one of the major problems plaguing game theory is the generic lack of predictive power due to equilibrium multiplicity. The preceding folk theorems underline this deficiency. On the other hand, the explicit equilibrium constructions in the proof of each result can provide valuable clues as to how to design self-enforcing institutions, in the sense that no explicit contracts are needed for the players' repeated interaction, only a common expectation that a certain equilibrium will be played. The latter view of institutions as a self-enforcing common set of expectations corresponds to a modern definition of institutions by Aoki (2001). As a natural refinement for equilibria one can, for example, focus on equilibria that produce Pareto-optimal outcomes, which (by definition) cannot be improved upon for any player without making another player worse off. This refinement is often referred to as Pareto perfection. □

Example 4.18 (Repeated Cournot Duopoly) Consider an infinitely repeated Cournot duopoly, where two firms, 1 and 2, produce homogeneous widgets in respective quantities q^1 and q^2. Firm i's production cost is $C(q^i) = cq^i$ (with constant marginal cost, $c > 0$). The inverse market demand is specified as $P(Q) = a - Q$, where $a > c$ and $Q = q^1 + q^2$.

(1) *Duopoly.* The unique Nash equilibrium of the stage game is given by $q_c^1 = q_c^2 = (a-c)/3$, yielding profits of $\pi_c^1 = \pi_c^2 = (a-c)^2/9$ for the firms.

(2) *Monopoly.* If the two firms merge, they can improve stage-game profits by producing half of the monopoly quantity each, that is, they choose $q_m^1 = q_m^2 = (a-c)/4$ so as to obtain $\pi_m^1 = \pi_m^2 = (a-c)^2/8 > \pi_i^c$. Note that the monopoly outcome is Pareto-dominant (from the firms' point of view); however, without a contract, each firm could improve its profit unilaterally by deviating; in other words, it is not a Nash equilibrium of the stage game: the best response to monopoly quantity would be $BR^i(q_m^{-i}) = ((a-c) - q_m^{-i})/2 = 3(a-c)/8 > q_c^i > q_m^i$, leading to

deviation profits of $\bar{\pi}^i = 9(a-c)^2/64 > \pi_m^i$. Furthermore, a collusion is possible in this game if both firms are patient enough, that is, if the firms' common discount factor δ is close enough to 1. Consider the following Nash-reversion strategy for firm i:

- Produce q_m^i in the first period and continue doing so as long as the observed outcome in the previous period is (q_m^1, q_m^2).
- If the outcome in the previous period is different from (q_m^1, q_m^2), then always choose q_c^i thereafter.

Using the one-shot deviation principle (see proposition 4.4 and remark 4.2), it is straightforward to verify that this strategy profile constitutes a subgame-perfect Nash equilibrium of the infinite-horizon supergame. Indeed, the payoff difference from a deviation,

$$\Delta^i = \left(\bar{\pi}^i + \frac{\delta \pi_c^i}{1-\delta}\right) - \frac{\delta \pi_m^i}{1-\delta}$$

$$= \left(\frac{9(a-c)^2}{64} + \frac{\delta(a-c)^2}{9(1-\delta)}\right) - \frac{(a-c)^2}{8(1-\delta)} < 0 \Leftrightarrow \delta > \frac{9}{17}$$

is negative as long as δ is close enough to 1, since $\pi_m^i > \pi_c^i$. □

4.2.4 Dynamic Games of Incomplete Information

Preliminaries Recall that a game is of perfect information if each information set contains a single node; otherwise it is of imperfect information. A game is of complete information if all players know all relevant information about each other; otherwise it is of incomplete information. It turns out that games of imperfect information are sufficient to represent all games provided one introduces an additional player, referred to as Nature, who selects the player types following a mixed strategy with probability weights that implement the players' beliefs.

Proposition 4.6 (Equivalence of Incomplete and Imperfect Information) (Harsanyi 1967) Any game of incomplete information can be rewritten as a game of imperfect information.

Proof (Outline) Given an arbitrary game of incomplete information, one can introduce an additional player, called Nature (N_0). Player N_0 is the first to move, and her actions follow all other players' beliefs: in fact, N_0 randomizes over the player types in Θ. Any move by Nature corresponds to a particular type realization; however, players cannot observe that move, and thus their respective information sets contain

Game Theory

all possible nodes that N_0's choice could lead to. Clearly this is a game of imperfect information, equivalent to the given game of incomplete information. ∎

In dynamic games with incomplete information, the concept of *Bayesian perfection* strengthens the Bayes-Nash equilibrium (see section 4.2.2) by requiring that players have beliefs about the probability that each particular decision node has been reached in equilibrium. A belief is thereby a probability distribution over the set of nodes in a given information set. A strategy profile (together with a belief system) constitutes a *perfect Bayesian equilibrium* (PBE) of a game of incomplete information if the following four requirements are satisfied:

• At each information set, the player with the move must have a belief (a probability distribution) about which node in his information set has been reached.

• Given their beliefs, all players' strategies (complete contingent plans) must be sequentially rational, that is, the actions taken at all information sets by players with the move must be optimal.

• On any equilibrium path (information sets reached with positive probability in a given equilibrium), beliefs must be determined by Bayes' rule and the players' equilibrium strategies.

• Off any equilibrium path, beliefs are determined by Bayes' rule and the players' equilibrium strategies, where possible.

Signaling Games When decision-relevant information is held privately by individual agents, an uninformed decision maker (the principal) may be able to elicit credible revelation of this private information by designing an appropriate incentive-compatible screening mechanism (see chapter 5). For example, a sales manager may be able to elicit truthful revelation of a potential buyer's willingness to pay by proposing to him a menu of purchase contracts, indexed by different price-quality (or price-quantity) tuples. However, in cases when the decision maker is unable (or unwilling) to create such a mechanism, it may be in (at least some of) the agents' best interest to take the initiative and send messages to the principal in an attempt to credibly convey their private information. When doing so, the parties engage in *signaling*.[22]

---

22. The term *signaling* in this context was coined by Michael Spence (1973). For this discovery he was awarded the 2001 Nobel Memorial Prize in Economics, together with George Akerlof and Joseph Stiglitz.

Despite the risk of somewhat confusing the issue, it can be noted that in many practical problems private information is held by all parties, dissolving the fine distinction between signaling and screening. For instance, information about a product's true quality might be held by a salesman, whereas each potential buyer best knows her own willingness to pay for the product as a function of its quality. To maximize profits the salesman could attempt to design an optimal screening mechanism (see chapter 5), but then he might still be unable to sell some of his products if he cannot credibly communicate quality information (see example 4.19). To avoid such market failure through adverse selection, the seller may attempt to actively signal the product's true quality to consumers, for instance, by offering a (limited) product warranty as part of each sales contract.

This section focuses on the signaling issue and neglects the screening issue in the trade situation with bilateral private information. Thus, attention is limited to situations where the only private information is held by the party that engages in signaling.

Consider the following canonical two-stage signaling game. In the first stage a sender S sends a message $s \in \mathcal{S}$ to a receiver R. The sender's private information can be summarized by S's type $\theta \in \Theta$. R's prior beliefs about the distribution of types in the type space are common knowledge and given by a probability distribution μ. In the second stage, player R, after receiving the message, possibly updates her beliefs about S's type (resulting in posterior beliefs $p(\theta|s)$ contingent on the observed $s \in \mathcal{S}$) and takes an action $a \in \mathcal{A}$. At this point, player S obtains a utility payoff described by the function $u : \mathcal{A} \times \mathcal{S} \times \Theta \to \mathbb{R}$, and player R obtains a utility payoff given by $v : \mathcal{A} \times \mathcal{S} \times \Theta \to \mathbb{R}$. For simplicity it is assumed in this section that the action space $\mathcal{A}$, the signal space $\mathcal{S}$, and the type space Θ are all finite.

Example 4.19 (Adverse Selection: Market for Lemons) Akerlof (1970) described a market for used cars, in which sellers S offer either one of two possible car types, lemons (L for low quality) or peaches (H for high quality). A car's true quality (or type) $\theta \in \{\theta_L, \theta_H\} = \Theta$ is observed only by the seller, and any buyer thinks that with probability $\mu \in (0,1)$ the car will be a desirable peach. The buyers R have valuations $v(\theta_H) > v(\theta_L)$ for the two goods, and the sellers have valuations $u(\theta_H) > u(\theta_L)$ such that $v(\theta) > u(\theta)$ for $\theta \in \Theta$ (i.e., there exist gains from trade).[23] In the

---

23. The buyer's action set $\mathcal{A}$ is given by {"buy," "don't buy"}; gains from trade can be realized only when a car is sold.

absence of signaling, that is, if there is no credible product information, both qualities are traded in a common market, and the expected value of a good for a potential buyer becomes $\bar{v} = (1-\mu)v(\theta_L) + \mu v(\theta_H)$. If the sellers offering the high-quality goods have a private valuation $u(\theta_H) > \bar{v}$ for these goods, there will be no more high-quality goods traded, and the lemons take over the market completely. This yields a socially inefficient outcome as a result of the buyers' adverse selection in response to the sellers' private quality information. Even though transactions between agents may be mutually beneficial, only limited trade may occur as a result of private information that the sellers possess about the quality of the good. If a seller can send a message $s \in S$ (e.g., the length of warranty offered with the car) before the other party decides about buying the car, then this may induce a *separating equilibrium*, in which warranty is offered only with peaches, not lemons. A *pooling equilibrium* occurs if no warranty is offered at all (if expensive for sellers), or if warranty is offered with both types of cars (if cheap for sellers). To demonstrate the potential effects of warranty, assume that a seller of a used car of quality $\theta \in \{\theta_L, \theta_H\}$ is able to offer warranty to the buyer. Assume that providing such warranty services incurs an expected cost of $c(\theta)$ with $c(\theta_L) > c(\theta_H) \geq 0$, that is, providing warranty services for low-quality cars is more expensive in expectation than for high-quality cars. Warranty would entitle the buyer to a technical overhaul or a full refund of the purchase price (within a reasonable time interval) if the car were found by the buyer to be of lower quality than θ_H.[24] Further assume that there are many identical car dealers engaging in price competition; then in a separating equilibrium (in which only sellers of high-quality cars offer warranty) the price charged for a high-quality car is $p_H = u(\theta_H) + c(\theta_H)$, and for a low-quality car it is $p_L = u(\theta_L)$. Note that the seller of a low-quality car, by deviating, could obtain a payoff of $p_H - c(\theta_L)$ instead of zero. Hence, a separating equilibrium in which only high-quality cars are offered with warranty exists if and only if

$$p_H - c(\theta_L) = u(\theta_H) - [c(\theta_L) - c(\theta_H)] \leq 0 \qquad (4.6)$$

---

24. Note the implicit assumption that the seller has no reason to believe that the buyer would engage in any negligent activity with the car before requesting a full refund. It is also implicitly assumed that both parties agree on what constitutes a car of quality θ_H, and that quality is observable to the owner within the warranty period.

and

$$v(\theta_H) \geq u(\theta_H) + c(\theta_H). \tag{4.7}$$

Under the last two conditions market failure through adverse selection can be prevented. Nevertheless, the unproductive investment $c(\theta_H)$ in the warranty services for high-quality cars is wasted and accounts for the inefficiency generated by the information asymmetry in this signaling game. If $c(\theta_H) = 0$, then the separating equilibrium is efficient. If only (4.6) fails to hold, then low-quality cars can be bought without risk at the higher price p_H, and there thus exists a pooling equilibrium in which *all* cars are offered with warranty at price p_H. If (4.7) fails to hold, then high-quality cars cannot be offered with warranty; depending on (4.6), low-quality cars might still be sold with a warranty. □

Definition 4.1 A *(behavioral) strategy* for player S of type $\theta \in \Theta$ is a function $\sigma : \mathcal{S} \times \Theta \to [0,1]$, which assigns probability $\sigma(s,\theta)$ to sending message $s \in \mathcal{S}$, where

$$\sum_{s \in \mathcal{S}} \sigma(s,\theta) = 1,$$

for all $\theta \in \Theta$. Similarly, a (behavioral) strategy for player R is a function $\alpha : \mathcal{A} \times \mathcal{S} \to [0,1]$, which assigns probability $\alpha(a,s)$ to action $a \in \mathcal{A}$ given that message $s \in \mathcal{S}$ has been received and which satisfies

$$\sum_{a \in \mathcal{A}} \alpha(a,s) = 1,$$

for all $s \in \mathcal{S}$.

Definition 4.2 (Bayes-Nash Equilibrium) The strategy profile (σ, α) constitutes a *Bayes-Nash equilibrium* of the canonical signaling game if

$$\sigma(s,\theta) > 0 \Rightarrow s \in \arg\max_{\hat{s} \in \mathcal{S}} \left\{ \sum_{a \in \mathcal{A}} \alpha(a,\hat{s}) u(a,\hat{s},\theta) \right\}, \tag{4.8}$$

and for each $s \in \mathcal{S}$ for which $\nu(s) = \sum_{\theta \in \Theta} \sigma(s,\theta) \mu(\theta) > 0$,

$$\alpha(a,s) > 0 \Rightarrow a \in \arg\max_{\hat{a} \in \mathcal{A}} \left\{ \sum_{\theta \in \Theta} p(\theta|s) v(\hat{a},s,\theta) \right\}, \tag{4.9}$$

where $p(\theta|s)$ is obtained via Bayesian updating,

$$p(\theta|s) = \frac{\sigma(s,\theta)\mu(\theta)}{\nu(s)}, \qquad (4.10)$$

whenever possible, that is, for all $s \in \mathcal{S}$ for which $\nu(s) > 0$.

Definition 4.3 A Bayes-Nash equilibrium (σ, α) of the signaling game is called *separating equilibrium* if each type sends a different message. It is called *pooling equilibrium* if (σ, α) is such that there is a single signal $s_0 \in \mathcal{S}$ sent by all types, namely, $\sigma(s_0, \theta) = 1$ for all $\theta \in \Theta$. Otherwise it is called *partially separating*.

Since in a separating equilibrium each type sends different signals, for any $\theta \in \Theta$ there exists a collection of pairwise disjoint sets $\mathcal{S}_\theta \subset \mathcal{S}$ such that $\cup_{\theta \in \Theta} \mathcal{S}_\theta = \mathcal{S}$, such that $\sum_{s \in \mathcal{S}_\theta} \sigma(s, \theta) = 1$.

In the definition of the Bayes-Nash equilibrum, the posterior *belief system p* is not a part of the equilibrium. In addition, off the equilibrium path, that is, for messages $s \in \mathcal{S}$, for which $\nu(s) = 0$, the receiver's posterior beliefs $p(\cdot|s)$ are not pinned down by Bayesian updating (4.10). Since the freedom in choosing these out-of-equilibrium beliefs may result in a multitude of Bayes-Nash equilibria, it is useful to make the belief system p, which specifies posterior beliefs for *any* $s \in \mathcal{S}$, a part of the equilibrium concept.

Definition 4.4 (Perfect Bayesian Equilibrium) The tuple (σ, α, p) is a *perfect Bayesian equilibrium* of the canonical signaling game if the strategy profile (σ, α) and the belief system p satisfy conditions (4.8)–(4.10).

Example 4.20 Consider a signaling game with type space $\Theta = \{\theta_1, \theta_2\}$, message space $\mathcal{S} = \{s_1, s_2\}$, and action set $\mathcal{A} = \{a_1, a_2\}$ (figure 4.7). Player R's prior beliefs about the type distribution are given by $\mu_k = \text{Prob}(\theta_k)$ for $k \in \{1, 2\}$ with $\mu_k \in [0, 1]$ and $\mu_1 + \mu_2 = 1$. If a sender of type θ_k plays strategy $\{\sigma(s_i, \theta_k)\}_{i,k}$, then upon observing $s \in \mathcal{S}$, player R's posterior beliefs are given by

$$p(\theta_k|s_i) = \frac{\sigma(s_i, \theta_k)\mu_k}{\sigma(s_i, \theta_1)\mu_1 + \sigma(s_i, \theta_2)\mu_2},$$

for all $i, k \in \{1, 2\}$, provided that the denominator $\nu(s_i) = \sigma(s_i, \theta_1)\mu_1 + \sigma(s_i, \theta_2)\mu_2$ is positive. If $\nu(s_i) = 0$, then no restriction is imposed on player R's posterior beliefs. First examine *pooling equilibria*, in which both sender types send s_i in equilibrium, $\sigma(s_i, \theta_k) = 1$ and $\sigma(s_{-i}, \theta_k) = 0$. In that

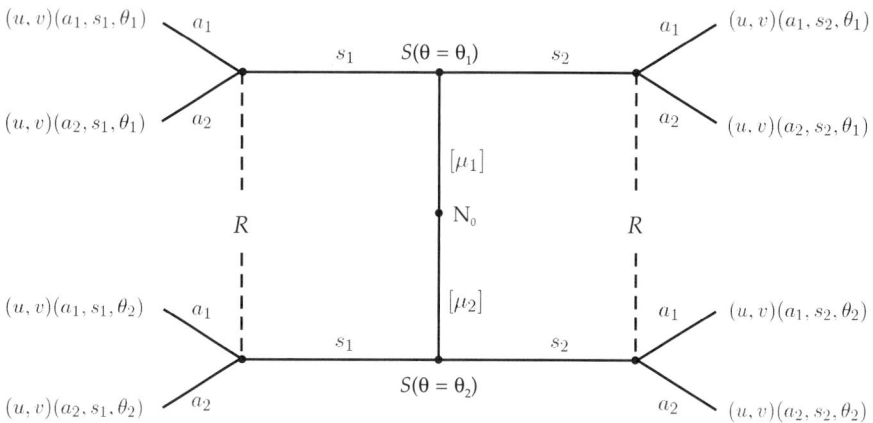

Figure 4.7
Signaling game with $|\mathcal{A}| = |\mathcal{S}| = |\Theta| = 2$ (see example 4.20).

case player R's posterior beliefs are only partly determined, since on the equilibrium path $p(\theta_k|s_i) = \mu_k$, whereas off the equilibrium path $p(\theta_k|s_{-i})$ cannot be pinned down by Bayesian updating. The reason for the latter is that given the sender's strategy, receiving a message s_{-i} corresponds to a zero-probability event. If $q_k = p(\theta_k|\theta_{-i})$ denote the sender's off-equilibrium-path beliefs, then naturally $q_k \in [0,1]$ and $q_1 + q_2 = 1$. The receiver's equilibrium strategy α off the equilibrium path is such that $\alpha(a_j, s_{-i}) > 0$ implies

$$q_1 v(a_j, s_{-i}, \theta_1) + q_2 v(a_j, s_{-i}, \theta_2) \geq q_1 v(a_{-j}, s_{-i}, \theta_1) + q_2 v(a_{-j}, s_{-i}, \theta_2),$$

for any $j \in \{1, 2\}$. In other words, an action can only be included in player R's mixed-strategy profile if it maximizes her expected payoff conditional on having observed s_{-i}. Similarly, on the equilibrium path player R's strategy α is such that $\alpha(a_j, s_i) > 0$ implies

$$\mu_1 v(a_j, s_i, \theta_1) + \mu_2 v(a_j, s_i, \theta_2) \geq \mu_1 v(a_{-j}, s_i, \theta_1) + \mu_2 v(a_{-j}, s_i, \theta_2).$$

A strict preference, either on or off the equilibrium path, for an action a_j over a_{-j} yields a pure-strategy equilibrium on the part of the receiver, which means that R puts all of the probability mass on a_j. It is important to note that off the equilibrium path the optimal action generally depends on R's posterior beliefs $q = (q_1, q_2)$, whence the usefulness of including posterior beliefs in the description of the equilibrium. While the Bayes-Nash equilibrium does not impose sequential rationality on the receiver (her actions do not have to be consistent with *any* off-equilibrium beliefs), the perfect Bayesian equilibrium requires that her action be utility-maximizing conditional on having formed a belief. To determine all pooling equilibria, one then needs to verify that (4.8) holds, that is,

$$s_i(q) \in \arg\max_{s \in \{s_1, s_2\}} \{\alpha(a_1, s; q)u(a_1, s, \theta_k) + \alpha(a_2, s; q)u(a_2, s, \theta_k)\},$$

for all $k \in \{1, 2\}$.

Now consider *separating equilibria*, in which different types send different messages. Without loss of generality assume that sender type θ_k sends s_k, that is, $\sigma(s_k, \theta_k) = 1$ and $\sigma(s_{-k}, \theta_k) = 0$. Then the receiver can perfectly infer the sender's type from the observed message, so her posterior beliefs are given by $p(\theta_k|s_k) = 1$ and $p(\theta_k|s_{-k}) = 0$, at least as long as she believes that both types can occur ($\mu_k \in (0,1)$).[25] Player R's equilibrium strategy α is now such that

$$\alpha(a_j, s_k) > 0 \Rightarrow a_j \in \arg\max_{a \in \{a_1, a_2\}} v(a, s_k, \theta_k),$$

for all $j \in \{1, 2\}$. As before one needs to verify that (4.8) holds (noting that $\sigma(s_k, \theta_k) = 1$), namely,

$$s_k \in \arg\max_{s \in \{s_1, s_2\}} \{\alpha(a_1, s)u(a_1, s, \theta_k) + \alpha(a_2, s)u(a_2, s, \theta_k)\},$$

for $k \in \{1, 2\}$, to determine all separating equilibria. There may exist partially separating equilibria, in which $\sigma(s_i, \theta_k) \in (0,1)$ for all $i, k \in \{1, 2\}$. □

To deal with the potential multiplicity of signaling equilibria, consider an equilibrium refinement. Let

$$\text{BR}(\hat{\Theta}, s) = \bigcup_{p: p(\hat{\Theta}|s) = 1} \arg\max_{a \in \mathcal{A}} \sum_{\theta \in \hat{\Theta}} p(\theta|s) v(a, s, \theta)$$

be the receiver's set of pure-strategy best responses conditional on observing the message s which the receiver believes comes from a sender with a type in $\hat{\Theta} \subseteq \Theta$, implying that her posterior beliefs p are such

---

25. If the receiver believes that sender type θ_k never occurs ($\mu_k = 0$), then her posterior beliefs upon observing s_k cannot be determined by Bayesian updating, in which case, similar to the off-equilibrium-path reasoning in pooling equilibria, the receiver can have arbitrary posterior beliefs $p(\theta_l|s_k)$ (for $l \in \{1,2\}$) as to which type sent an "impossible" message.

that

$$p(\hat{\Theta}|s) = \sum_{\theta \in \hat{\Theta}} p(\theta|s) = 1.$$

The set $\mathrm{BR}(\hat{\Theta}, s) \subseteq \mathcal{A}$ contains all the receiver's actions that might be optimal conditional on having observed s and the receiver's having formed a belief which presumes the sender's type θ to lie in the set $\hat{\Theta}$. Given a perfect Bayesian equilibrium (σ, α, p), let

$$u^*(\theta) = \sum_{(a,s) \in \mathcal{A} \times \mathcal{S}} \alpha(a,s)\sigma(s,\theta)u(a,s,\theta)$$

be sender type θ's expected equilibrium payoffs. Hence, any sender of type θ in the set

$$\bar{\Theta}_s = \{\theta \in \Theta : u^*(\theta) > \max_{a \in \mathrm{BR}(\Theta,s)} u(a,s,\theta)\}$$

would *never* send message s, since it would result in a payoff strictly less than his expected equilibrium payoff $u^*(\theta)$. The set $\Theta \setminus \bar{\Theta}_s$ therefore contains all the types that could reasonably be expected to send the message s in equilibrium. Hence, any equilibrium in which a sender of a type in $\Theta \setminus \bar{\Theta}_s$ obtains, by sending the message s, an expected equilibrium payoff strictly below the worst payoff he could rationally expect from the receiver (conditionally on her observation of s) seems counterintuitive. Based on this reasoning, Cho and Kreps (1987) introduced an intuitive criterion that can be used to eliminate signaling equilibria in which the sender could increase his equilibrium payoff by deviating (taking into account that the receiver cannot rationally play a strategy that is never a best response).

4.3 Differential Games

Let (t_0, x_0) be given initial data, consisting of an initial time $t_0 \in \mathbb{R}$ and an initial state $x_0 \in \mathbb{R}^n$, and let $T > t_0$ be a (possibly infinite) time horizon. A *differential game (of complete information)* in normal form is given by

$$\Gamma(t_0, x_0) = (\mathcal{N}, \{\mathcal{U}^i(\cdot)\}_{i \in \mathcal{N}}, \{J^i(u^i|\mu^{-i}(\cdot))\}_{i \in \mathcal{N}}),$$

where $\mathcal{N} = \{1,\ldots,N\}$ is a finite set of $N \geq 2$ players, and where each player $i \in \mathcal{N}$ chooses his control (or action) $u^i(t) = \mu^i(t, x(t)) \in \mathbb{R}^{m_i}$ for all $t \in [0, T)$ so as to maximize his objective functional

$$J^i(u^i|\mu^{-i}) = \int_{t_0}^{T} h^i(t, x(t), u^i(t), \mu^{-i}(t, x(t))) \, dt,$$

subject to the control constraint

$$u^i(t) \in \mathcal{U}^i(t, x(t)), \quad \forall t \in [t_0, T),$$

and subject to the system equation in integral form (as in (2.17))

$$x(t) = x_0 + \int_{t_0}^{t} f(s, x(s), u^i(s), \mu^{-i}(s, x(s))) \, ds, \quad \forall t \in [t_0, T),$$

given the other players' strategy profile $\mu^{-i}(\cdot)$. The reason for considering the right-open interval $[t_0, T)$ instead of the closed interval $[0, T]$ is to suggest with this notation that all the developments here also apply to the infinite-horizon case where $T = \infty$. Let $m = m_1 + \cdots + m_N$. The functions $f : \mathbb{R}^{1+n+m} \to \mathbb{R}^n$ and $h^i : \mathbb{R}^{1+n+m} \to \mathbb{R}$ are assumed to be continuously differentiable, and the upper semicontinuous set-valued mapping $\mathcal{U}^i : \mathbb{R}^{1+n} \rightrightarrows \mathbb{R}^{m_i}$, with $(t, x) \mapsto \mathcal{U}^i(t, x)$,[26] is assumed to have nonempty, convex, and compact images, for all $i \in \mathcal{N}$.

4.3.1 Markovian Equilibria

A strategy-profile $\mu^*(\cdot) = (\mu^{i*}(\cdot))_{i \in \mathcal{N}}$ is a (Markovian) Nash equilibrium of the differential game $\Gamma(t_0, x_0)$ if there exists a state trajectory $x^*(t), t \in [0, T)$, such that for each $i \in \mathcal{N}$ the control $u^{i*}(t) = \mu^i(t, x^*(t)), t \in [t_0, T)$, solves player i's optimal control problem

$$J(u^i|\mu^{-i*}) = \int_{t_0}^{T} h^i(t, x^*(t), u^i(t), \mu^{-i*}(t, x^*(t))) \, dt \longrightarrow \max_{u^i(\cdot)}, \quad (4.11)$$

$$\dot{x}^{*(t)} = f(t, x^*(t), u^i(t), \mu^{-i*}(t, x^*(t))), \quad x^*(t_0) = x_0, \quad (4.12)$$

$$u^i(t) \in \mathcal{U}^i(t, x^*(t)), \quad \forall t, \quad (4.13)$$

$$t \in [t_0, T], \quad (4.14)$$

---

26. The control-constraint set $\mathcal{U}^i(t, x)$ can usually be represented in the form $\mathcal{U}^i(t, x) = \{u^i = (u^{1i}, u^{2i}) \in \mathbf{L}_\infty : R^i(t, x, u^1) \geq 0, \ u^{2i} \in \mathcal{U}^{2i}(t)\}$, analogous to the relations (3.37)–(3.38) of the general optimal control problem discussed in section 3.4.

in the sense that $(x^*(t), u^{i*}(t))$, $t \in [t_0, T)$, is an optimal state-control trajectory of (4.11)–(4.14). The Nash-equilibrium strategy profile μ^* is called an *open-loop* Nash equilibrium if it is independent of the state; otherwise it is called a *closed-loop* Nash equilibrium.

Remark 4.6 (Open-Loop vs. Closed-Loop Nash Equilibria) In an open-loop Nash equilibrium any player $i \in \mathcal{N}$ commits to a control trajectory over the entire time horizon, that is, $u^{i*}(t) = \mu^{i*}(t)$, for a.a. $t \in [t_0, T)$. In a closed-loop Nash equilibrium player i takes into account the fact that all other players use the state when determining their equilibrium strategy profile $u^{-i*}(t) = \mu^{-i*}(t, x^*(t))$. Because any open-loop Nash equilibrium can be viewed as a closed-loop Nash equilibrium with trivial state dependence, it is evident that the class of closed-loop Nash equilibria is richer than the class of open-loop Nash equilibria. The key difference between the two concepts is that a functional dependence on the state of the *other* players' equilibrium strategy profile prompts player i to take into account his actions' effect on the state because of the anticipated reactions to variations of the state (in addition to the direct payoff externalities from the other players' actions). Figure 4.8 provides some additional intuition of open-loop versus closed-loop strategy profiles in the differential game $\Gamma(t_0, x_0)$. □

Remark 4.7 (Markovian Strategies) The term *Markovian* in the definition of closed-loop (and open-loop) Nash equilibria refers to the fact that the strategies do not depend on the history other than through the current state.[27] In general, one can expect non-Markovian, that is, history-dependent Nash equilibria (see section 4.2.3). The fact that a strategy profile is Markovian means that all relevant memory in the system is carried by the current state; it does *not* mean that there is no memory in the system. Thus, the system equation determines the way history influences the players' closed-loop strategies. □

Time Consistency and Subgame Perfection
As in section 4.2.1, the differential game $\Gamma(t, x)$ is a *subgame* of the differential game $\Gamma(t_0, x_0)$ if $t \geq t_0$ and there exists an admissible and feasible strategy profile such that the system can be controlled from (t_0, x_0) to the event (t, x) (see remark 3.3). A Markovian Nash equilibrium μ^*

---

27. More generally, a stochastic process has the *Markov property*, named after the Russian mathematician Andrey Markov, if the conditional distribution of future states depends solely on the current state.

Game Theory

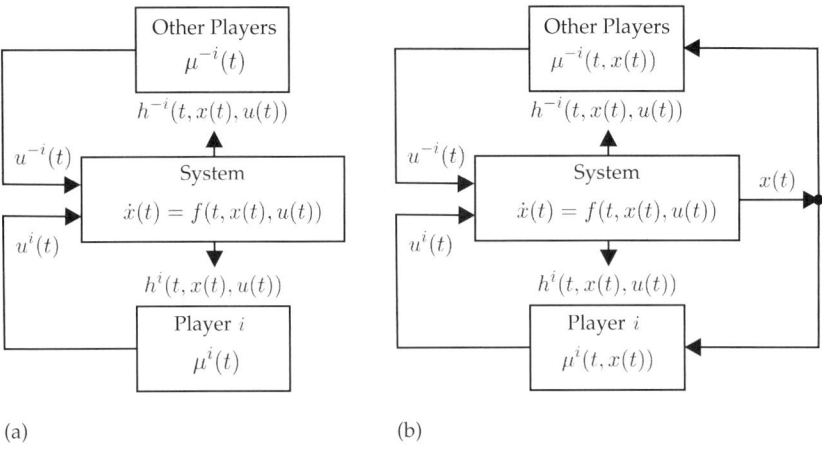

Figure 4.8
Differential game with (a) open-loop and (b) closed-loop strategies.

of $\Gamma(t_0, x_0)$ with associated (by proposition 2.3 unique) state trajectory

$$x^*(t) = x_0 + \int_{t_0}^{t} f(s, x^*(s), \mu^*(s, x^*(s))) \, ds, \qquad \forall t \in [t_0, T), \tag{4.15}$$

is called *time-consistent* if the restriction $\mu^*|_{[t,T)\times\mathbb{R}^n}$ is a Markovian Nash equilibrium of $\Gamma(t, x^*(t))$, for all $t \in [t_0, T)$. The Markovian Nash equilibrium μ^* of $\Gamma(t_0, x_0)$ is called *subgame-perfect* if the restriction $\mu^*|_{[t,T)\times\mathbb{R}^n}$ is a Markovian Nash equilibrium of any subgame $\Gamma(t, x)$ of $\Gamma(t_0, x_0)$. A subgame-perfect Markovian Nash equilibrium is also referred to as a *Markov-perfect* Nash equilibrium. It is evident that Markov perfection implies time consistency. The converse is not true in general, because Markov perfection requires an equilibrium to not rely on any noncredible threats *off* the equilibrium path as well as *on* the equilibrium path, whereas time consistency imposes a condition only *on* the equilibrium path, not considering the situation after an albeit unexpected deviation.[28]

Proposition 4.7 (Time Consistency) Any Markovian Nash equilibrium μ^* of the differential game $\Gamma(t_0, x_0)$ is time-consistent.

Proof The proof proceeds via contradiction. If μ^* is not time-consistent, then there exists $\hat{t} \in (t_0, T)$ such that $\mu^*|_{[\hat{t},T)\times\mathbb{R}^n}$ is not a Markovian Nash

---

28. For the notion of noncredible threats, see example 4.10.

equilibrium of $\Gamma(\hat{t}, x^*(\hat{t}))$, where $x^*(t)$, $t \in [\hat{t}, T)$, is as in (4.14). Hence, for some player $i \in \mathcal{N}$ an alternative strategy $\hat{\mu}^i(t, \hat{x}^*(t))$ yields a higher payoff than $\mu^{i*}(t, x^*(t))$, where $\hat{x}^*$ is the state trajectory under the strategy profile $(\hat{\mu}^i, \mu^{-i*})$. But this implies that μ^* cannot be a Nash equilibrium, since player i could unilaterally improve his payoff by switching to $\hat{\mu}^i$ for all times $t \geq \hat{t}$. ∎

The following example, which builds on several examples in section 35, illustrates the difference between open-loop and closed-loop Markovian equilibria as well as the difference between time consistency and subgame perfection.

Example 4.21 (Joint Exploitation of an Exhaustible Resource) In the same setting as in exercise 3.2, consider $N \geq 2$ identical agents who, starting at time $t = 0$, exploit a nonrenewable resource of initial quantity $x_0 > 0$, so that at given time $t = T$ in the future none of the resource is left. Each agent $i \in \mathcal{N} = \{1, \ldots, N\}$ chooses a consumption rate $c^i(t) \in [0, \bar{c}/N]$, where $\bar{c} > (r/\rho) x_0 / (1 - e^{(r/\rho)T})$ is a maximum allowable extraction rate. The evolution of the resource stock $x(t)$ solves the initial value problem (IVP)

$$\dot{x}(t) = -\sum_{j=1}^{N} c^j(t), \qquad x(0) = x_0,$$

for all $t \in [0, T]$, provided that the feasibility constraint

$$c^i(t) \in [0, \mathbf{1}_{\{x(t) \geq 0\}} \bar{c}]$$

is satisfied a.e. on $[0, T]$, for all $i \in \mathcal{N}$. Assume that each agent experiences the (instantaneous) utility $U(y) = y^{1-\rho}$ when consuming the resource at a nonnegative rate y, consistent with a constant relative risk aversion of $\rho > 1$ (see example 3.5). Agent i determines his consumption $c^i(\cdot)$ so as to maximize the discounted utility

$$J(c^i) = \int_0^T e^{-rt} U(c^i(t)) \, dt,$$

where $r > 0$ is the discount rate. First determine the unique open-loop Nash-equilibrium strategy profile c^*. For this, introduce agent i's current-value Hamiltonian

$$\hat{H}^i(t, x, c, v^i) = U(c^i) - v^i \sum_{j=1}^{N} c^j,$$

where v^i is his (current-value) adjoint variable and $c = (c^1, \ldots, c^N)$ is a strategy profile. Let $(x^*(t), c^*(t))$, $t \in [0, T]$, be an open-loop Nash-equilibrium state-control trajectory. As in example 3.4, using the Pontryagin maximum principle (PMP),

$$c^{i*}(t) = c_0 e^{-(r/\rho)t} = \left(\frac{x_0}{N}\right) \frac{(r/\rho) e^{-(r/\rho)t}}{1 - e^{-(r/\rho)T}}, \quad \forall t \in [0, T].$$

The constant c_0 (the same for all agents) is determined by the endpoint constraint $x^*(T) = 0$, so that

$$x_0 = \sum_{j=1}^{N} \int_0^T c^{j*}(t)\, dt = \frac{N c_0}{(r/\rho)} (1 - e^{-(r/\rho)T}),$$

resulting in the open-loop Nash-equilibrium trajectory

$$x^*(t) = x_0 \frac{e^{-(r/\rho)t} - e^{-(r/\rho)T}}{1 - e^{-(r/\rho)T}}, \quad \forall t \in [0, T].$$

Thus, the open-loop Nash equilibrium leads to a socially optimal exploitation of the nonrenewable resource, just as in example 3.5. This equilibrium is also time-consistent, that is, if after time t the system is started in the state $x^*(t)$, then the corresponding open-loop Nash equilibrium on $[t, T]$ is the same as the restriction of the open-loop Nash equilibrium on $[0, T]$ to the interval $[t, T]$. On the other hand, the open-loop Nash equilibrium is not subgame-perfect, because after an unexpected deviation, for instance, from $x^*(t)$ to some other state $\xi \in (0, x_0)$, agents do not condition their subsequent consumption choices on the new state ξ but only on the elapsed time t.

Now consider a closed-loop Nash equilibrium with an affine feedback law of the form

$$\mu^i(x) = \alpha^i + \beta^i x, \quad \forall x \in [0, x_0],$$

where α^i, β^i are constants. Taking into account the symmetry, set $\alpha \equiv \alpha^i$ and $\beta \equiv \beta^i$, so agent i's current-value Hamiltonian becomes

$$\hat{H}^i(t, x, (c^i, \mu^{-i}), v^i) = U(c^i) - v^i(c^i + (N-1)(\alpha + \beta x)).$$

The corresponding adjoint equation is then $\dot{v}^i = (r + (N-1)\beta) v^i$, so

$$v^i(t) = v_0^i e^{\hat{r}t},$$

where $\hat{r} = r + (N-1)\beta$. Thus, agent i's closed-loop Nash-equilibrium consumption becomes

$$\hat{c}^{i*}(t) = \left(\frac{x_0}{N}\right) \frac{(\hat{r}/\rho)e^{-(\hat{r}/\rho)t}}{1 - e^{-(\hat{r}/\rho)T}}, \qquad \forall t \in [0,T].$$

On the other hand, using the given feedback law, the closed-loop state trajectory $\hat{x}^*(t)$, $t \in [0,T]$, solves the IVP

$$\dot{x} = -N(\alpha + \beta x), \quad x(0) = x_0,$$

so

$$\hat{x}^*(t) = x_0 - \frac{\alpha}{\beta}(1 - e^{-N\beta t}), \quad \forall t \in [0,T].$$

The closed-loop state trajectory $\hat{x}^*(t)$, $t \in [0,T]$, under the strategy profile $\hat{c}^* = (\hat{c}^1, \ldots, \hat{c}^N)$ is the same as the open-loop state trajectory after replacing r by $\hat{r}$. Combining this with the previous expression for $\hat{x}^*$ yields

$$\alpha = \left(\frac{x_0}{N}\right) \frac{(\hat{r}/\rho)}{1 - e^{-(\hat{r}/\rho)T}} \quad \text{and} \quad \beta = \frac{(\hat{r}/\rho)}{N},$$

where

$$\frac{\hat{r}}{\rho} = \frac{r}{\rho - \left(1 - \frac{1}{N}\right)}.$$

The closed-loop Nash equilibrium is Markov-perfect, for its feedback law is conditioned directly on the state. Also note that in the closed-loop Nash equilibrium the agents tend to exploit the resource faster than when all players are committed up to time $t = T$. The lack of commitment in the closed-loop Nash equilibrium leads to an overexploitation of the resource compared to the welfare-maximizing solution that is implemented by an open-loop equilibrium carrying with it the ability to fully commit to a control trajectory at the beginning of the time interval. Figure 4.9 contrasts the open-loop and closed-loop equilibrium trajectories. □

Example 4.22 (Linear-Quadratic Differential Game) Consider $N \geq 2$ players with quadratic objective functionals whose payoffs depend on the evolution of a linear system. Given a time horizon $T > 0$, each player $i \in \mathcal{N} = \{1, \ldots, N\}$ chooses a control $u^i(t)$, $t \in [0,T)$, in a convex,

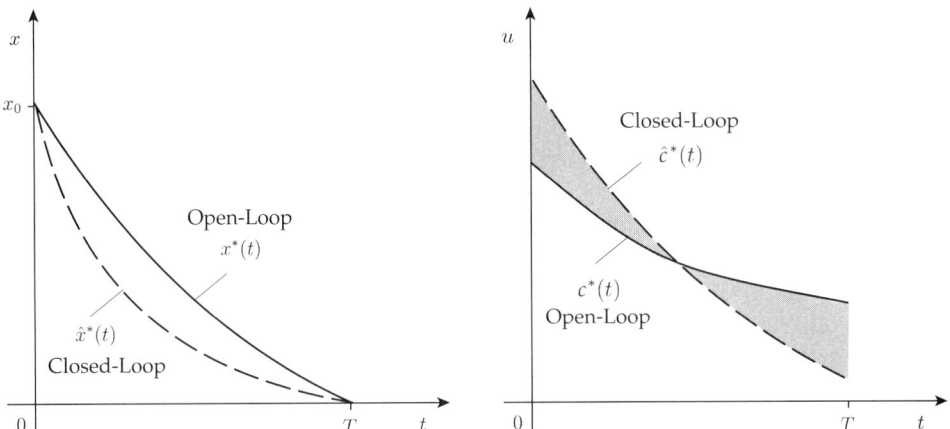

Figure 4.9
State and control trajectories in an open-loop and a closed-loop Markovian Nash equilibrium of a joint exploitation game (see example 4.21).

compact subset $\mathcal{U}^i$ of $\mathbb{R}^{m_i}$. It is assumed to be large enough to allow for an effectively unconstrained optimization. The evolution of the state variable $x(t) \in \mathbb{R}^n$ is governed by the linear system equation

$$\dot{x}(t) = A(t)x(t) + \sum_{i=1}^{N} B^i(t)u^i(t),$$

where the continuous bounded matrix functions A, B^i are such that $A(t) \in \mathbb{R}^{n \times n}$ and $B^i(t) \in \mathbb{R}^{n \times m_i}$, for all $t \in [0, T)$. Given the other players' strategy profile $\mu^{-i}(t, x)$, player i solves the optimal control problem

$$J(u^i|\mu^{-i}) = \int_0^T h^i(t, x(t), u^i(t), \mu^{-i}(t, x(t))) \, dt - e^{-rT} x'(T) K^i x(T) \underset{u^i(\cdot)}{\longrightarrow} \max,$$

$$\dot{x}(t) = A(t)x(t) + \sum_{i=1}^{N} B^i(t)u^i(t), \qquad x(0) = x_0,$$

$$u^i(t) \in \mathcal{U}^i, \qquad \forall\, t,$$

$$t \in [0, T],$$

where

$$h^i(t, x, u) = -e^{-rt} \left[x' R^i(t) x + \sum_{j=1}^{N} (u^j)' S^{ij}(t) u^j \right],$$

$r \geq 0$ is a common discount rate, and $R(t)$, $S^{ij}(t)$ are continuous bounded matrix functions with values in $\mathbb{R}^{n \times n}$ and $\mathbb{R}^{m_j \times m_j}$, respectively. The terminal cost matrix $K^i \in \mathbb{R}^{n \times n}$ is symmetric positive definite. The setup is analogous to the linear-quadratic regulator problem in example 3.3. Using the PMP or the Hamilton-Jacobi-Bellman (HJB) equation it is possible to explicitly determine open-loop and closed-loop equilibria of this linear-quadratic differential game (see exercise 4.1). □

4.3.2 Non-Markovian Equilibria

The discussion of repeated games in section 4.2.3 showed how important the players' available information structure can be for the construction of Nash equilibria, of which, according to the various folk theorems (e.g., proposition 4.5), there can in principle be many. Now consider the concept of information structure, which for each player i defines what is known at a given time t about the state of the system and the history of the players' actions. For any player $i \in \mathcal{N}$, let

$$\mathcal{I}^i : \{(t, u(\,\cdot\,)) : u(\,\cdot\,) \in \mathbf{L}_\infty([t_0, T], \mathbb{R}^m), \, t \in [t_0, T]\} \to \mathcal{S}^i$$

be a mapping from the set of available data to his observation space $\mathcal{S}^i$, called player i's *information structure* (IS).[29] The observation space $\mathcal{S} = \mathcal{S}^1 \times \cdots \times \mathcal{S}^N$ is a subset of a finite-dimensional Euclidean space. The (combined) information structure $\mathcal{I} = \mathcal{I}^1 \times \cdots \times \mathcal{I}^N$ is called *causal* (or *nonanticipatory*) if

$$\mathcal{I}(t, u(\,\cdot\,)) = \mathcal{I}(t, u(\,\cdot\,)\big|_{[0,t)}), \quad \forall t \in [t_0, T].$$

A causal information structure does not use any future information. The information structure $\mathcal{I}$ is called *regular* if for any admissible $\hat{u}(\,\cdot\,), u(\,\cdot\,) \in \mathbf{L}_\infty([t_0, T], \mathbb{R}^m)$:

$$\int_{t_0}^T \|\hat{u}(s) - u(s)\| ds = 0 \Rightarrow \mathcal{I}(t, \hat{u}(\,\cdot\,)) = \mathcal{I}(t, u(\,\cdot\,)), \quad \forall t \in [t_0, T].$$

A regular information structure produces therefore the same observations for any two strategy profiles which are identical except on a zero-measure set of time instances. Last, a (*pure*) *strategy profile* of a differential game $\Gamma_\mathcal{I}(t_0, x_0)$ with information structure $\mathcal{I}$ is a mapping $\mu = (\mu^1, \ldots, \mu^N) : [t_0, T) \times \mathcal{S} \to \mathbb{R}^m$. A definition of a Nash equilibrium in

---

29. In contrast to the repeated games discussed in section 4.2.3, assume here (for simplicity) that the information structure maps to a stationary observation space, which then corresponds essentially to the space of all histories in the earlier discussion.

this context is completely analogous to the earlier definitions of Nash equilibrium, and is therefore omitted.

Example 4.23 (Information Structures) Let $i \in \{1, \ldots, N\}$, $N \geq 2$.

- *Markovian IS* $\mathcal{I}^i(t, u(\,\cdot\,)) \equiv x(t)$.
- *Delayed-state IS* Given a delay $\delta \in (0, T - t_0)$, consider

$$\mathcal{I}^i(t, u(\,\cdot\,)) \equiv \begin{cases} x_0 & \text{if } t \in [t_0, t_0 + \delta], \\ x(t - \delta) & \text{if } t \in [t_0 + \delta, T). \end{cases}$$

This IS depends only on the past, that is, it is causal; it is also regular. However, when the delay δ is negative, then the IS $\mathcal{I}^i(t, u(\,\cdot\,))$, $t \in [t_0, T)$, becomes noncausal and continues to be regular.

- *Delayed-control IS* Let $\delta \in (t_0, T - t_0)$ be a given delay, as in the last example. The IS $\mathcal{I}^i(t, u(\,\cdot\,)) = u^j(\max\{t_0, t - \delta\})$ for some $j \in \{1, \ldots, N\} \setminus \{i\}$ is causal but not regular. If the delay is negative, the IS is neither causal nor regular.
- *Sampled-observation IS* Given the time instances $t_1, \ldots, t_\kappa \in (t_0, T)$ with $t_{k-1} < t_k$ for all $k \in \{2, \ldots, \kappa\}$, consider $\mathcal{I}^i(t, u(\,\cdot\,)) \equiv \{x(t_k) : t_k \leq t\}$. This IS is causal and regular.

The next example provides an information structure that commonly arises in hierarchical play with commitment. □

Example 4.24 (Stackelberg Leader-Follower Games) Consider a game with two players, 1 and 2, where player 1 in the role of the leader first commits to a control path $u^1(t)$, $t \in [t_0, T)$. Player 2, the follower, observes this strategic preannouncement and chooses his payoff-maximizing response, $u^2(t)$, $t \in [t_0, T)$. The corresponding (anticipatory but regular) information structure $\mathcal{I} = (\mathcal{I}^1, \mathcal{I}^2)$ is such that $\mathcal{I}^1(t, u(\,\cdot\,)) = \{\emptyset, u^1(\,\cdot\,)\}$ and $\mathcal{I}^2(t, u(\,\cdot\,)) \in \{u^1(\,\cdot\,), u(\,\cdot\,)\}$. In other words, the follower knows the leader's entire control trajectory, while the leader chooses his strategy without knowing the follower's strategy.[30] This generally creates a time-consistency problem, similar to the discussion of this phenomenon in section 4.2.3. □

Example 4.25 (Cournot-Stackelberg Duopoly) Building on the hierarchical information structure in the last example, consider two identical firms, 1 and 2, with firm 1 as leader and firm 2 as follower. Each firm $i \in \{1, 2\}$ chooses a production output $u^i \geq 0$ at the cost $C(u^i) =$

---

30. Of course, the leader can anticipate the follower's response, but he is unable to change his own strategy intermittently; that is, the leader has no recourse.

$(u^i)^2/2$ on the infinite time interval $[0, \infty)$, so as to maximize its objective functional

$$J^i(u^i|u^{-i}) = \int_0^\infty e^{-rt}[p(t)u^i(t) - C(u^i)]\,dt,$$

where $r > 0$ is a common discount factor. The price process $p(t)$ is determined as solution of the IVP

$$\dot{p} = 1 - p - u^1(t) - u^2(t), \quad p(0) = p_0,$$

for a given initial value $p_0 > 0$. To determine an open-loop Nash equilibrium of this hierarchical game, one first solves the follower's optimal control problem given the leader's control trajectory $u^1(t)$, $t \geq 0$. The corresponding current-value Hamiltonian is

$$\hat{H}^1(p, u^1, u^2, v^2) = pu^2 - \frac{(u^2)^2}{2} + v^2(1 - p - u^1 - u^2),$$

where v^2 is the current-value adjoint variable. Using the PMP, the maximality condition yields $u^2 = p - v^2$, and the adjoint equation becomes

$$\dot{v}^2 = (2 + r)v^2 - p(t), \quad \forall t \in [0, \infty).$$

The leader can now solve its own optimal control problem, taking into account the anticipated actions by the follower. Firm 1's corresponding current-value Hamiltonian is

$$\hat{H}^2(p, u^1, u^2, v^1, v^2) = pu^1 - \frac{(u^1)^2}{2} + v_1^1(1 - p - u^1 - u^2)$$
$$+ v_2^1((2 + r)v^2 - p),$$

where $v^1 = (v_1^1, v_2^1)$ is the leader's adjoint variable. It is important to note that the leader takes into account the evolution of the follower's adjoint variable and thus works with an augmented state variable $x = (p, v^2)$ instead of just with p which the follower uses. The PMP yields the maximality condition $u^1 = p - v_1^1$ and the adjoint equations

$$\dot{v}_1^1 = (3 + r)v_1^1 + v_2^1 - p(t),$$
$$\dot{v}_2^1 = -v_1^1 - 2v_2^1,$$

for all $t \in [0, \infty)$. Because the Hamiltonian system of adjoint equations and state equation (for the price) amounts to a system of linear ODEs, it is possible to find an explicit solution; for computational details in a similar setting, see exercise 4.2. The turnpike price,

$$\bar{p}^* = \frac{(5+2r)(2+r)}{21+23r+6r^2},$$

is obtained by computing the equilibrium $(\bar{p}^*, \bar{v}^1, \bar{v}^2)$ of the Hamiltonian system, which also yields the corresponding long-run equilibrium production levels,

$$\bar{u}^{1*} = \bar{p}^* - \bar{v}^1_1 = \frac{3+3r+r^2}{9+10r+3r^2} \quad \text{and} \quad \bar{u}^{2*} = \bar{p}^* - \bar{v}^2 = \frac{2+3r+r^2}{9+10r+3r^2}.$$

It is interesting to compare these last results to the turnpike price $\hat{p}^* = (2+r)/(4+3r)$ and the long-run production levels $\hat{u}^{1*} = \hat{u}^{2*} = (1+r)/(4+3r)$ in an open-loop Nash equilibrium without hierarchical play:

$$\bar{p}^* < \hat{p}^* \quad \text{and} \quad \bar{u}^{2*} < \hat{u}^{i*} < \bar{u}^{1*},$$

that is, the leader produces more than any of the firms in the simultaneous-move equilibrium, which does not afford the leader the possibility to anticipate the follower's reaction completely. In the long run, the resulting total output is higher and thus the market price lower than in the non-hierarchical open-loop Nash equilibrium. □

Trigger-Strategy Equilibria The relevance and intuition of trigger-strategy equilibria was first discussed in section 4.2.1. The general intuition carries over to differential games, yet it is necessary to be careful about defining what exactly a deviation means. The reason is that, for example, deviations from a given target strategy profile $\hat{u}(t)$, $t \in [t_0, T)$, at single time instances are not payoff-relevant and should therefore not trigger any response.

Example 4.26 (Prisoner's Dilemma as Differential Game) In what follows, a static two-agent prisoner's dilemma game is generalized to a suitable infinite-horizon differential game Γ, and subgame-perfect trigger-strategy equilibria of Γ are derived with cooperation (on the equilibrium path), assuming that both players have a common positive discount rate r. In this it is assumed that each player experiences a detection lag, and that he can condition his time-t action on the entire history of play up to time $t \geq 0$. Describe the set of (average) payoff vectors that can be implemented using such trigger-strategy equilibria, depending on r. Following is the payoff matrix for a standard single-period prisoner's dilemma game, where $u^i \in [0,1]$ is player i's chosen probability of cooperating (i.e., playing C), for $i \in \{1,2\}$ instead of defecting (i.e., playing D):

		Player 2	
		(u_2)	$(1-u_2)$
		C	D
Player 1	(u_1) C	$(1,1)$	$(-1,2)$
	$(1-u_1)$ D	$(2,-1)$	$(0,0)$

Consider a dynamic version of the prisoner's dilemma stage game over an infinite time horizon,[31] in which both players care about their respective average payoffs. This game can be written as a differential game by realizing that, given the strategy $u^j : \mathbb{R} \to [0,1]$ of player $j \in \{1,2\}$ (an essentially bounded, measurable function), player $i \in \{1,2\} \setminus \{j\}$ solves the (somewhat degenerate) optimal control problem

$$J^i(u^i|u^j) \longrightarrow \max_{u^i(\cdot)}$$

s.t. $\dot{x}_i = 1 - u^j, \quad x(0) = 0,$

$u^i(t) \in [0,1], \quad \forall t \geq 0,$

where

$$J^i(u^i|u^j) = r \int_0^\infty e^{-rt}(u^i(t)u^j(t) + 2u^j(t)(1-u^i(t)) - u^i(t)(1-u^j(t))) \, dt.$$

The state $x_i(t)$, $t \geq 0$, measures the duration over which player j has, on average, not cooperated before time t. Though it is not directly relevant for player i's objective functional, keeping track of this state may allow player i to condition his strategy on player j's cumulative behavior as part of a Markov-perfect equilibrium.

Note first that if both players restrict attention to stationary strategy profiles $u(t) \equiv u_0 = (u_0^i, u_0^j) \in [0,1]^2$, the payoffs $J^i(u_0^i|u_0^j)$ will be identical to those shown in the preceding payoff matrix. Thus, the differential

---

31. Equivalently, consider a finite-horizon game in which a *random* length T of the time horizon is exponentially distributed, so that its hazard rate is constant.

Game Theory

game Γ is indeed a generalization of the static game, when players are free to choose nonstationary strategy profiles. Now consider different types of equilibria of Γ.

Open-Loop Equilibria The current-value Hamiltonian for player i's optimal control problem is

$$\hat{H}^i(t, x, u, v^i) = u^i u^j + 2u^j(1-u^i) - u^i(1-u^j) + v^i_i(1-u^i) + v^i_j(1-u^j),$$

where $x = (x_1, x_2)$ is the state of the system. The PMP yields the following necessary optimality conditions for any optimal state-control trajectory $(x^*(t), u^{i*}(t))$, $t \geq 0$:

- Adjoint equation

$$\dot{v}^i(t) = rv^i(t), \qquad \forall t \geq 0.$$

- Transversality

$$e^{-rt}v^i(t) \to 0 \quad \text{as} \quad t \to \infty.$$

- Maximality

$$u^{i*}(t) \in \begin{cases} \{0\} & \text{if } v^i_j(t) < -1 \\ [0,1] & \text{if } v^i_j(t) = -1 \\ \{1\} & \text{if } v^i_j(t) > -1 \end{cases} = \arg \max_{u^i \in [0,1]} \hat{H}^i(t, x^*(t), u^i, u^{j*}(t), v^i(t)),$$

for all $t \geq 0$.

The adjoint equation together with the transversality condition implies that $v^i(t) \equiv 0$, so the maximality condition entails that $u^{i*}(t) \equiv 0$ for $i \in \{1, 2\}$. Thus, the unique open-loop Nash equilibrium of the differential game Γ is for both players to always defect.

Closed-Loop Equilibria Now examine closed-loop equilibria of the form $\mu^{i*}(t, x) = 1 - [\text{sgn}(x_i)]_+$. The intuition behind this strategy is that any significant (i.e., not on a zero-measure time interval) deviation from cooperation by player j is noticed by player i, and subsequently punished by player i's playing $u^i = 0$ (i.e., defect) for all future times. Indeed, if player j plays μ^{j*}, then it is best for player i to play μ^{i*}, since that yields a payoff of 1, while any deviation from this strategy yields a payoff of zero. Note also that this closed-loop equilibrium is Markovian and regular. It implements perpetual cooperation on the equilibrium path.

Trigger-Strategy Equilibria Assuming that each player experiences a common positive detection lag δ, and that he can condition his time-t

action on the information $\mathcal{I}(t, u(\,\cdot\,)) = \{u(s) : 0 \leq s \leq [t - \delta]_+\}$ up to time $t \geq 0$, subgame-perfect trigger-strategy equilibria with cooperation on the equilibrium path can be sustained by the threat of minmax punishment (see example 4.17) off the equilibrium path. For this, let any admissible reference function $\hat{u} = (\hat{u}^1, \hat{u}^2) : \mathbb{R}_+ \to [0, 1]^2$ be given, and consider the strategy profile $\mu^* = (\mu^{i*}, \mu^{j*})$ with

$$u^i(t) \equiv \mu^{i*}(t, \mathcal{I}(t, u(\,\cdot\,))|\hat{u}(\,\cdot\,)) = \begin{cases} 1 & \text{if } \int_0^{[t-\delta]_+} \|u(s) - \hat{u}(s)\| ds = 0, \\ 0 & \text{otherwise.} \end{cases}$$

Note that because the system does not explicitly depend on time, if it is optimal for player i to deviate on an interval $[\tau, \tau + \delta]$ (i.e., playing a deviating strategy $\check{u}^i$) for some $\tau \geq 0$, then it is also optimal to deviate on the interval $[0, \delta]$. Player i's corresponding deviation payoff $J^i_{\text{dev}}(\check{u}|u^j)$ is bounded, as

$$0 \leq J^i_{\text{dev}}(\check{u}^i|u^j) \leq 2r \int_0^\delta e^{-rs} ds = 2\left(1 - e^{-r\delta}\right) \leq J^i(\hat{u}^i|\hat{u}^j) \leq 2,$$

provided that

$$0 < \delta \leq \frac{1}{r} \min \left\{ \ln\left(\frac{2}{2 - J^i(\hat{u}^i|\hat{u}^j)}\right), \ln\left(\frac{2}{2 - J^j(\hat{u}^j|\hat{u}^i)}\right) \right\} \equiv \bar{\delta}_{\hat{u}}.$$

Note that $u(t) = \hat{u}(t)$, $t \geq 0$, on the equilibrium path, so that it is possible to implement any individually rational average payoff vector $v = (v^i, v^j) \in \mathcal{V}$, where

$$\mathcal{V} = \{v \in \mathbb{R}^2_{++} : v \in \text{co}(\{0, (-1, 2), (2, -1), (1, 1)\})\}$$

is the intersection of the convex hull of the static payoff vectors and the positive quadrant of $\mathbb{R}^2$. In other words, as long as the reference strategy profile $\hat{u} = (\hat{u}^i, \hat{u}^j)$ is individually rational, so that

$$(J^i(\hat{u}^i|\hat{u}^j), J^j(\hat{u}^j|\hat{u}^i)) \in \mathcal{V},$$

it can be implemented as a trigger-strategy equilibrium. In addition, one can obviously implement $\hat{u}(t) \equiv 0$, which corresponds to the standard prisoner's dilemma outcome (and the open-loop equilibrium). □

4.4 Notes

The theory of games dates back to von Neumann (1928) and von Neumann and Morgenstern (1944). Good introductory textbooks on noncooperative game theory are by Fudenberg and Tirole (1991),

Gibbons (1992), and for differential games, Başar and Olsder (1995), and Dockner et al. (2000). Mailath and Samuelson (2006) give an introduction to repeated games, including issues concerning asymmetric information.

Nash (1950) introduced the modern notion of equilibrium that is widely used in noncooperative game theory. Radner and Rosenthal (1982) provided sufficient conditions for the existence of pure-strategy Bayes-Nash equilibria. Milgrom and Weber (1985) relaxed those conditions by allowing for distributional strategies. Their findings were further generalized by Balder (1988), whose main result essentially corresponds to proposition 4.2. Time-consistency problems in leader-follower games such as economic planning were highlighted by Kydland and Prescott (1977); for an overview of such issues in the context of nonrenewable resources, see Karp and Newbery (1993). Signaling games were first discussed by Spence (1973) and appear in many economic contexts, for instance, advertising (Kihlstrom and Riordan 1984).

4.5 Exercises

4.1 (Linear-Quadratic Differential Game) Consider the differential game in example 4.22.

a. Determine the (unique) open-loop Nash equilibrium.

b. Determine a closed-loop Nash equilibrium.

c. Explain the difference between the two equilibria in parts a and b.

4.2 (Cournot Oligopoly) In a market where all $N \geq 2$ firms are offering homogeneous products, the evolution of the "sticky" market price $p(t)$ as a function of time $t \geq 0$ is described by the IVP

$$\dot{p}(t) = f(p(t), u^1(t), \ldots, u^N(t)), \qquad p(0) = p_0,$$

where the initial price $p_0 > 0$ and the continuously differentiable excess demand function $f : \mathbb{R}^{1+N} \to \mathbb{R}$ are given, with

$$f(p, u^1, \ldots, u^N) = \alpha \left(a - p - \sum_{i=1}^{N} u^i \right).$$

The constant $\alpha > 0$ is a given adjustment rate, and $a > 0$ represents the known market potential. Each firm $i \in \{1, \ldots, N\}$ produces the output $u^i(t) \in \mathcal{U} = [0, \bar{u}]$, given the large capacity limit $\bar{u} > 0$, resulting in the production cost

$$C(u^i) = cu^i + \frac{(u^i)^2}{2},$$

where $c \in [0, a)$ is a known constant. Given the other firms' strategy profile $u^{-i}(t) = \mu^{-i}(t, p(t))$, $t \geq 0$, each firm i maximizes its infinite-horizon discounted profit,

$$J^i(u^i) = \int_0^\infty e^{-rt}(p(t)u^i(t) - C(u^i(t)))\, dt,$$

where $r > 0$ is a common discount rate.

a. Formulate the differential game $\Gamma(p_0)$ in normal form.

b. Determine the unique Nash equilibrium u_0 (together with an appropriate initial price p_0) for a static version of this Cournot game, where each firm i can choose only a constant production quantity $u_0^i \in \mathcal{U}$ and where the price p_0 is adjusted only once, at time $t = 0$, and remains constant from then on.

c. Find a symmetric open-loop Nash equilibrium $u^*(t)$, $t \geq 0$, of $\Gamma(p_0)$, and compute the corresponding equilibrium turnpike $(\bar{p}^*, \bar{u}^*)$, namely, the long-run equilibrium state-control tuple. Compare it to your solution in exercise 4.2.a and explain the intuition behind your findings.

d. Find a symmetric Markov-perfect (closed-loop) Nash equilibrium $\mu^*(t, p)$, $(t, p) \in \mathbb{R}_+^2$, of $\Gamma(p_0)$, and compare it to your results in part c.

e. How do your answers in parts b–d change as the market becomes competitive, as $N \to \infty$?

4.3 (Duopoly Pricing Game) Two firms in a common market are competing on price. At time $t \geq 0$, firm $i \in \{1, 2\}$ has a user base $x_i(t) \in [0, 1]$. Given the firms' pricing strategy profile $p(t) = (p^i(t), p^j(t))$,[32] $t \geq 0$, this user base evolves according to the ODE

$$\dot{x}_i = x_i(1 - x_i - x_j)[\alpha(x_i - x_j) - (p^i(t) - p^j(t))], \qquad x_i(0) = x_{i0},$$

where $j \in \{1, 2\} \setminus \{i\}$, and the initial user base $x_{i0} \in (0, 1 - x_{j0})$ is given. Intuitively, firm i's installed base increases if its price is smaller than $\alpha(x_i - x_j) + p^j$, where the constant $\alpha \geq 0$ determines the importance of the difference in installed bases as brand premium. For simplicity,

---

32. It is without loss of generality to assume that $p(t) \in [0, P]^2$, where $P > 0$ can be interpreted as a (sufficiently large) maximum willingness to pay.

assume that the firms are selling information goods at zero marginal cost. Hence, firm i's profit is

$$J^i(p^i|p^j) = \int_0^\infty e^{-rt} p^i(t) [\dot{x}_i(t)]_+ \, dt,$$

where $r > 0$ is a given common discount rate.

a. Formulate the differential game $\Gamma(x_0)$ in normal form.

b. Show that any admissible state trajectory $x(t)$ of the game $\Gamma(x_0)$ moves along the curve $C(x_0) = \{(x_1, x_2) \in [0,1]^2 : x_1 x_2 = x_{10} x_{20}\}$.

c. Show that an open-loop state-control trajectory $(x^*(t), p^*(t))$ in the game $\Gamma(x_0)$ is characterized as follows.
- If $x_{10} = x_{20}$, then $p^*(t) \equiv 0$ and $x^*(t) \equiv x_0$.
- If $x_{i0}^* > x_{j0}^*$, then $p^{j*}(t) = 0$, and $(x_i^*(t), -x_j^*(t))$ increases along the curve $C(x_0)$, converging to the stationary point

$$\bar{x} = (\bar{x}_i, \bar{x}_j) = \left(\frac{1 - \sqrt{1 - 4x_{i0}x_{j0}}}{2}, \frac{1 + \sqrt{1 - 4x_{i0}x_{j0}}}{2} \right).$$

d. Plot a typical Nash-equilibrium state-control trajectory.

e. Provide an intuitive interpretation of the Nash-equilibrium strategy profile of $\Gamma(x_0)$ determined earlier, and discuss the corresponding managerial conclusions. What features of the game are not realistic?

4.4 (Industrial Pollution) Consider a duopoly in which at time $t \geq 0$ each firm $i \in \{1,2\}$ produces a homogeneous output of $q^i(t) \in [0, y_i]$, where $y_i \geq 0$ is its capacity limit. Given a total output of $Q = q^1 + q^2$, the market price is given by $p(Q) = [1 - Q]_+$. The aggregate output Q causes the emission of a stock pollutant $x(t)$, which decays naturally at the rate $\beta > 0$. Given the initial stock $x(0) = x_0 > 0$, the evolution of the pollutant is described by the initial value problem

$$\dot{x}(t) = Q(t) - \beta x(t), \qquad x(0) = x_0,$$

for all $t \geq 0$. The presence of the pollutant exerts an externality on both firms, as firm i's total cost

$$C^i(x, q^i) = c^i q^i + \gamma^i \frac{x^2}{2}$$

depends not only on its own production but also on the accumulated stock of pollution. The constant $c^i \in (0,1)$ is a known marginal production cost, and $\gamma^i \in \{0,1\}$ indicates if firm i cares about pollution ($\gamma^i = 1$) or not ($\gamma^i = 0$). At time t, firm i's total profit is

$$\pi^i(x,q) = p(q^1 + q^2) \cdot q^i - C^i(x, q^i),$$

where $q = (q^1, q^2)$. At each time $t \geq 0$, firm i chooses the capacity-expansion rate $u^i(t) \in [0, \bar{u}]$ to invest in expanding its capacity $y(t)$, which evolves according to

$$\dot{y}_i(t) = u^i(t) - \delta y_i(t), \quad y(0) = y_0,$$

where $\delta > 0$ is a depreciation rate, $y_{i0} > 0$ is firm i's initial capacity level, and $\bar{u} > 0$ is a (large) upper bound on the capacity-expansion rate. The cost of expanding capacity at the rate u^i is $K(u^i) = \kappa(u^i)^2/2$, where $\kappa > 0$ is the marginal expansion cost. Assume that both firms maximize their total discounted payoffs, using the common discount rate $r > 0$.[33] (*Hint:* Always check first if $q^i = y_i$ in equilibrium.)

Part 1: Symmetric Environmental Impact

Both firms care about pollution, $(\gamma^1, \gamma^2) = (1,1)$.

a. Derive the solution to a static version of the duopoly game in which all variables and their initial values are constants.

b. Formulate the differentiable game $\Gamma(x_0, y_0)$ and derive an open-loop Nash equilibrium.

c. Determine a closed-loop Nash equilibrium of $\Gamma(x_0, y_0)$ in affine strategies.

d. Compare the results you obtained in parts a–c.

Part 2: Asymmetric Environmental Impact

Only firm 1 cares about pollution, $(\gamma^1, \gamma^2) = (1,0)$.

e. Derive static, open-loop, and closed-loop equilibria of $\Gamma(x_0, y_0)$, as in part 1, and discuss how an asymmetric environmental impact changes the outcome of the game compared to the situation with a symmetric environmental impact.

f. Discuss possible public-policy implications of your findings in parts 1 and 2.

---

33. The cost $K(u^i)$ is sometimes referred to as internal adjustment cost.

5 Mechanism Design

See first that the design is wise and just:
that ascertained, pursue it resolutely;
do not for one repulse forego the purpose
that you resolved to effect.
—William Shakespeare

This chapter reviews the basics of static mechanism design in settings where a principal faces a single agent of uncertain type. The aim of the resulting screening contract is for the principal to obtain the agent's type information in order to avert adverse selection (see example 4.19), maximizing her payoffs. Nonlinear pricing is discussed as an application of optimal control theory.

5.1 Motivation

A decision maker may face a situation in which payoff-relevant information is held privately by another economic agent. For instance, suppose the decision maker is a sales manager. In a discussion with a potential buyer, she is thinking about the right price to announce. Naturally, the client's private value for a product is a piece of *hidden information* that the manager would love to know before announcing her price. It would prevent her from announcing a price that is too high, in which case there would be no trade, or a price that is too low, in which case the buyer is left with surplus that the seller would rather pocket herself. In particular, the decision maker would like to charge a person with a higher value more for the product than a person with a lower value, provided she could at least cover her actual marginal cost of furnishing the item.

Thus, the key question is, What incentive could an economic agent have to reveal a piece of hidden information if an advantage could be obtained by announcing something untruthful? More specifically, could the decision maker devise a *mechanism* that an agent might find attractive enough to participate in (instead of ignoring the decision maker and doing something else) and that at the same time induces revelation of private information, such as the agent's willingness to pay? In order for an agent to voluntarily disclose private information to a decision maker (the *principal*), he would have to be offered a nonnegative *information rent*, which he would be unable to obtain without revealing his private information. An appropriate *screening* mechanism should make it advantageous for any agent type (compared to his status quo or an outside option) to disclose his private information even when this information is likely to be used against him. The *revelation principle* (see proposition 5.1) guarantees that without loss of generality the principal can limit her search of appropriate mechanisms to those in which any agent would find it optimal to announce his private information truthfully (*direct* mechanisms). An agent's piece of private information is commonly referred to as his *type*. The sales manager, before announcing a price for the product, thus wishes to know the type of the potential buyer, allowing her to infer his willingness to pay. As an example, if there are two or more agent types competing for the item, then a truth-revealing mechanism can be implemented using a second-price auction (see example 4.2).[1] The next section provides a solution for the sales manager's mechanism design problem when the buyer's private information is binary (i.e., when there are only two possible types).

5.2 A Model with Two Types

Assume that a sales manager (seller) faces a buyer of type θ_L or θ_H, whereby $\theta_L < \theta_H$. The buyer's type $\theta \in \{\theta_L, \theta_H\} = \Theta$ is related to his willingness to pay in the following way: if the seller (the principal) announces a price (or transfer) t for a product of quality $x \in \mathbb{R}$,[2] the (type-dependent) buyer's utility is equal to zero if he does not buy (thus exercising his outside option), and it is

$$U(x,\theta) - t \geq 0 \tag{5.1}$$

if he does buy. The function $U : \mathcal{X} \times \Theta \to \mathbb{R}$ (with $\mathcal{X} = [\underline{x}, \bar{x}] \subset \mathbb{R}$ and $-\infty < \underline{x} < \bar{x} < \infty$), assumed to be strictly increasing in (x, θ) and

---

1. The basic ideas of mechanism design discussed in this chapter remain valid for the design of mechanisms with multiple agents, such as auctions.
2. Instead of quality one can, equivalently, also think of quantity as instrument for the screening mechanism (see example 5.2).

Mechanism Design

concave in x, represents the buyer's preferences.[3] In order to design an appropriate screening mechanism that distinguishes between the two types, the seller needs a contracting device (i.e., an *instrument*), such as a product characteristic that she is able to vary. The sales contract could then specify the product characteristic, say, the product's quality x, for which the buyer would need to pay a price $t = \tau(x)$.[4] The idea for the design of a screening mechanism is that the seller proposes a *menu of contracts* containing variations of the instrument, from which the buyer is expected to select his most desirable one. Since there are only two possible types, the seller needs at most two different contracts, indexed by the quality on offer, $x \in \{x_L, x_H\}$. The buyer, regardless of type, cannot be forced to sign the sales contract: his participation is voluntary. Thus, inequality (5.1) needs to be satisfied for any participating type $\theta \in \{\theta_L, \theta_H\}$. Furthermore, at the price t_H a buyer of type θ_H should prefer quality x_H,

$$U(x_H, \theta_H) - t_H \geq U(x_L, \theta_H) - t_L, \tag{5.2}$$

and at price t_L a buyer of type L should prefer x_L, so

$$U(x_L, \theta_L) - t_L \geq U(x_H, \theta_L) - t_H. \tag{5.3}$$

Assume that the unit cost for a product of quality x is $c(x)$, where $c : \mathbb{R} \to \mathbb{R}$ is a strictly increasing continuous function. The contract-design problem is to choose $\{(t_L, x_L), (t_H, x_H)\}$ (with $t_L = \tau(x_L)$ and $t_H = \tau(x_H)$) so as to maximize the manager's expected profit,

$$\bar{\Pi}(t_L, x_L, t_H, x_H) = (1-p)\left[t_L - c(x_L)\right] + p\left[t_H - c(x_H)\right], \tag{5.4}$$

where $p = \text{Prob}(\tilde{\theta} = \theta_H) = 1 - \text{Prob}(\tilde{\theta} = \theta_L) \in (0, 1)$ denotes the seller's prior belief about the probability of being confronted with type θ_H as opposed to θ_L.[5] The optimization problem,

$$\max_{\{(t_L, x_L),(t_H, x_H)\}} \bar{\Pi}(t_L, x_L, t_H, x_H), \tag{5.5}$$

is subject to the *individual-rationality* (or *participation*) constraint (5.1) as well as the *incentive-compatibility* constraints (5.2) and (5.3). The general

---

3. It is assumed here that the buyer's preferences are quasilinear in money. Refer to footnote 1 in chapter 4 for the definition of the utility function as representation of preferences.
4. It is important to note that the instrument needs to be *contractable*, i.e., *observable* by the buyer and *verifiable* by a third party, so that a sales contract specifying a payment $\tau(x)$ for a quality x can be enforced by a benevolent court of law.
5. Since the buyer's type is unknown to the seller, she treats $\tilde{\theta}$ as a *random variable* with realizations in the *type space* Θ.

solution to this *mechanism design problem* may be complicated and depends on the form of U. Its solution is simplified when U has increasing differences in (x, θ). In other words, assume that $U(x, \theta_H) - U(x, \theta_L)$ is increasing in x, or equivalently, that

$$\hat{x} \geq x \Rightarrow U(\hat{x}, \theta_H) - U(x, \theta_H) \geq U(\hat{x}, \theta_L) - U(x, \theta_L), \tag{5.6}$$

for all $\hat{x}, x \in \mathcal{X}$. Condition (5.6) implies that the marginal gain from additional quality is greater for type θ_H (the high type) than for type θ_L (the low type). To further simplify the principal's constrained optimization problem, one can show that the low type's participation constraint is binding. Indeed, if this were not the case, then $U(x_L, \theta_L) - t_L > 0$ and thus,

$$U(x_H, \theta_H) - t_H \geq U(x_L, \theta_H) - t_L \geq U(x_L, \theta_L) - t_L > 0,$$

which would allow the principal to increase prices for both the high and the low type because neither type's participation constraint is binding. As an additional consequence of this proof the individual-rationality constraint for the high type can be neglected, but is binding for the low type. This makes the principal's problem substantially easier. Another simplification is achieved by noting that the high type's incentive-compatibility constraint (5.2) must be active. If this were not true, then

$$U(x_H, \theta_H) - t_H > U(x_L, \theta_H) - t_L \geq U(x_L, \theta_L) - t_L = 0,$$

whence it would be possible to increase t_H without breaking (5.1) for the high type: a contradiction. Moreover, it is then possible to neglect (5.3), since incentive compatibility for the low type is implied by the fact that (5.2) is binding and the sorting condition (5.6) holds, $t_H - t_L = U(x_H, \theta_H) - U(x_L, \theta_H) \geq U(x_H, \theta_L) - U(x_L, \theta_L)$. The last inequality with $\theta_H > \theta_L$ also implies that $x_H > x_L$. To summarize, one can therefore drop the high type's participation constraint (5.1) and the low type's incentive-compatibility constraint (5.3) from the principal's program, which is now constrained by the high type's incentive-compatibility constraint,

$$t_H - t_L = U(x_H, \theta_H) - U(x_L, \theta_H), \tag{5.7}$$

and the low type's participation constraint,

$$U(x_L, \theta_L) = t_L. \tag{5.8}$$

Mechanism Design

Equations (5.7) and (5.8) allow substituting t_L and t_H into the manager's expected profit (5.4). With this, the contract-design problem (5.5), subject to (5.1)–(5.3), can be reformulated as an unconstrained optimization problem,

$$\max_{x_L, x_H \in \mathcal{X}} \{(1-p)[U(x_L, \theta_L) - c(x_L)]$$

$$+ p[U(x_H, \theta_H) - c(x_H) - (U(x_L, \theta_H) - U(x_L, \theta_L))]\}.$$

The problem therefore decomposes into the two independent maximization problems,

$$x_H^* \in \arg\max_{x_H \in \mathcal{X}} \{U(x_H, \theta_H) - c(x_H)\} \tag{5.9}$$

and

$$x_L^* \in \arg\max_{x_L \in \mathcal{X}} \left\{ U(x_L, \theta_L) - c(x_L) - \frac{p}{1-p}(U(x_L, \theta_H) - U(x_L, \theta_L)) \right\}. \tag{5.10}$$

From (5.9)–(5.10) the principal can determine t_H^* and t_L^* using (5.7)–(5.8). In order to confirm that indeed $x_H^* > x_L^*$, as initially assumed, first consider the *first-best* solution to the mechanism design problem, $\{(t_L^{FB}, x_L^{FB}), (t_H^{FB}, x_H^{FB})\}$, that is, the solution under full information. Indeed, if the principal knows the type of the buyer, then

$$x_j^{FB} \in \arg\max_{x_j \in \mathcal{X}} \{U(x_j, \theta_j) - c(x_j)\}, \quad j \in \{L, H\}, \tag{5.11}$$

and

$$t_j^{FB} = U(x_j, \theta_j), \quad j \in \{L, H\}. \tag{5.12}$$

Comparing (5.9) and (5.11) yields that $x_H^* = x_H^{FB}$. In other words, even in the presence of hidden information *the high type will be provided with the first-best quality level*. As a result of the supermodularity assumption (5.6) on U, the first-best solution $x^{FB}(\theta)$ is increasing in θ. Hence, $\theta_L < \theta_H$ implies that $x_L^{FB} = x^{FB}(\theta_L) < x^{FB}(\theta_H) = x_H^{FB}$. In addition, supermodularity of U implies that for the low type the *second-best* solution x_L^* in (5.10) cannot exceed the first-best solution x_L^{FB} in (5.11), since $U(x_L, \theta_H) - U(x_L, \theta_L)$, a nonnegative function, increasing in x_L, is subtracted from the first-best maximand in order to obtain the second-best solution (which therefore cannot be larger than the first-best solution). Hence, it is

$$x_H^* = x_H^{FB} > x_L^{FB} \geq x_L^*. \tag{5.13}$$

Thus, in a hidden-information environment the low type is furnished with an inefficient quality level compared to the first-best. Moreover, the low type is left with zero surplus (since $t_L^* = U(x_L^*, \theta_L)$), whereas the high type enjoys a positive information rent (from (5.7) and (5.12)),

$$t_H^* = t_H^{FB} - \underbrace{\left(U(x_L^*, \theta_H) - U(x_L^*, \theta_L)\right)}_{\text{Information Rent}}. \tag{5.14}$$

The mere possibility that a low type exists thus exerts a *positive externality* on the high type, whereas the net surplus of the low type remains unchanged (and equal to zero) when moving from the principal's first-best to her second-best solution (figure 5.1).

If the principal's prior belief is such that she thinks the high type is very likely (i.e., p is close enough to 1), then she adopts a *shutdown solution*, in which she effectively stops supplying the good to the low type. Let $\underline{x} = \min \mathcal{X}$ be the lowest quality level (provided at cost $c(\underline{x})$).

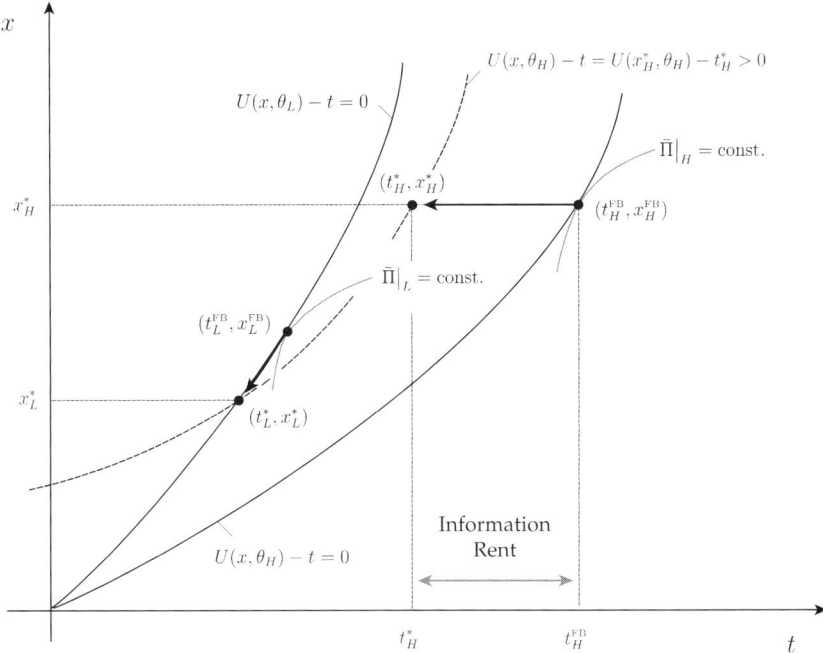

Figure 5.1
First-best and second-best solution of the model with two types.

Then (5.10) implies that for

$$p \geq \frac{U(\underline{x}, \theta_L) - c(\underline{x})}{U(\underline{x}, \theta_H) - c(\underline{x})} \equiv p_0 \qquad (5.15)$$

the principal shuts down (charges zero price for no product or a costless minimum-quality product), that is, she starts selling exclusively to the high type. In that case, the high type's information rent collapses to zero, as he is unable to derive a positive externality from the now valueless (for the principal) low type. The following example clarifies some of the general notions introduced in this section for a widely used parametrization of the two-type model.

Example 5.1 Assume that the consumer's private value for the product is proportional to both his type θ and the product's quality (or quantity) x,[6] so that $U(x, \theta) = \theta x$, whereby $\theta \in \{\theta_L, \theta_H\}$ with $\theta_H > \theta_L > 0$ and $x \in [0, \bar{x}]$ with some (large enough) maximum achievable quality level $\bar{x}$. The cost of a product of quality x is assumed to be quadratic, $c(x) = \gamma x^2 / 2$ for some positive constant $\gamma \geq \theta_H / \bar{x}$ (so that $\bar{x} \geq \theta_H / \gamma$). Note first that $U(x, \theta)$ exhibits increasing differences in (x, θ), since $U(x, \theta_H) - U(x, \theta_L) = x(\theta_H - \theta_L)$ is increasing in x so that condition (5.6) is indeed satisfied. Assuming a principal's prior $p \in (0, 1)$ of the same form as before, the general results obtained earlier can be used to get

$$x_H^* = x_H^{FB} = \theta_H / \gamma$$

and

$$x_L^* = \frac{1}{\gamma} \left[\theta_L - \frac{p}{1-p} (\theta_H - \theta_L) \right]_+ < x^{FB}$$

whereby $x_j^{FB} = \theta_j / \gamma$ for $j \in \{L, H\}$. Let

$$p_0 = \frac{\theta_L}{\theta_H}$$

denote the threshold probability for the high type as in (5.15): for $p \geq p_0$, it is $x_L^* = 0$. In other words, if the high type is more likely than p_0, then the principal offers a single product of efficient high quality while

---

6. In this formulation, the type parameter θ can be interpreted as the marginal utility of quality (or quantity), $\theta = U_x(x, \theta)$. The higher the type for a given quality level, the higher the marginal utility for extra quality. The underlying heuristic is that "power users" are often able to capitalize more on quality improvements (or larger quantities, e.g., more bandwidth) than less sophisticated, occasional users.

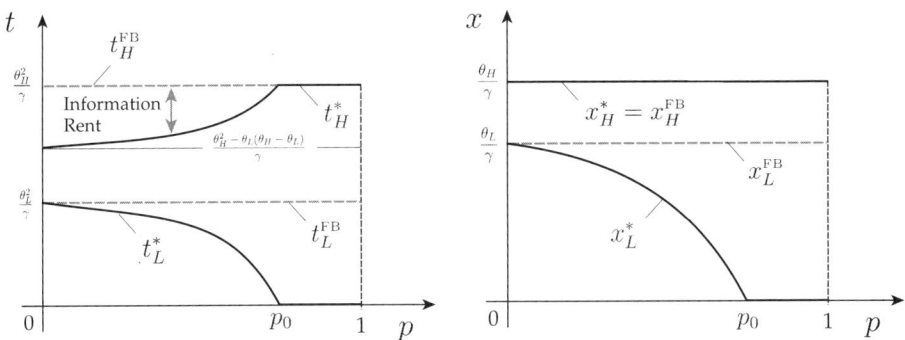

Figure 5.2
Comparison of first-best and second-best solutions in terms of expected profit ($\bar{\Pi}^{FB}$ vs. $\bar{\Pi}^*$) and expected welfare ($\bar{W}^{FB}$ vs. $\bar{W}^*$) (see example 5.1).

the low type is excluded from the market. The corresponding second-best prices are given by $p_L^* = U(x_L^*, \theta_L) = x_L^* \theta_L$ and $p_H^* = p_L^* + (U(x_H^*, \theta_H) - U(x_L^*, \theta_H)) = p_L^* + x_H^*(\theta_H - \theta_L)/(1-p)$ (Figure 5.2), whence the principal's expected profit under this optimal screening mechanism becomes

$$\bar{\Pi}^* = \bar{\Pi}(t_L^*, x_L^*, t_H^*, x_H^*) = \begin{cases} \dfrac{\theta_L^2 + p\theta_H^2 - 2p\theta_L\theta_H}{2\gamma(1-p)} & \text{if } p \leq p_0 \\ \dfrac{p\theta_H^2}{2\gamma} & \text{otherwise.} \end{cases}$$

By contrast, the first-best profit is

$$\bar{\Pi}^{FB} = \bar{\Pi}(t_L^{FB}, x_L^{FB}, t_H^{FB}, x_H^{FB}) = \frac{1}{2\gamma}(\theta_L^2 + p(\theta_H^2 - \theta_L^2)).$$

In the absence of type uncertainty, for $p \in \{0,1\}$, it is $\bar{\Pi}^* = \bar{\Pi}^{FB}$. With uncertainty, for $p \in (0,1)$, it is $\bar{\Pi}^* < \bar{\Pi}^{FB}$. Figure 5.3 shows how $\bar{\Pi}^*$ and $\bar{\Pi}^{FB}$ differ as a function of p. Now consider the social welfare (i.e., the sum of the buyer's and seller's surplus in expectation) as a function of p. In the absence of hidden type information, the seller is able to appropriate all the surplus in an efficient manner, and thus first-best expected welfare, $\bar{W}^{FB}$, equals first-best expected profit, $\bar{\Pi}^{FB}$. The second-best expected welfare, $\bar{W}^* = \bar{\Pi}^* + p(U(x_H^*, \theta_H) - t_H^*)$, is not necessarily monotonic in p: as p increases the seller is able to appropriate more information rent from the high type, while at the same time losing revenue from the low type, for which she continues decreasing quality (as a function of p) until the shutdown point p_0 is reached, at which all low types are excluded from

Mechanism Design

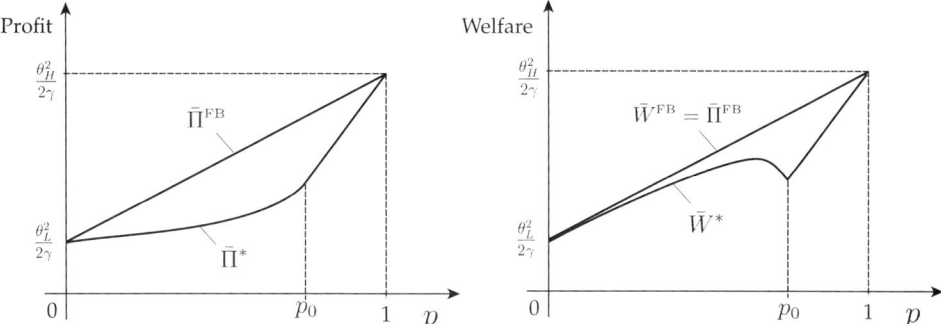

Figure 5.3
Comparison of first-best and second-best solutions in terms of expected profit ($\bar{\Pi}^{FB}$ vs. $\bar{\Pi}^*$) and expected welfare ($\bar{W}^{FB}$ vs. $\bar{W}^*$) (see example 5.1).

the market. From then on, high types are charged the efficient price, and second-best welfare linearly approaches the first-best for $p \to 1^-$. □

5.3 The Screening Problem

Now consider the screening problem in a more abstract mechanism design setting. A principal faces one agent of unknown type $\theta \in \Theta = [0, 1]$ and can offer him a contract (t, x), where $x \in \mathbb{R}$ is a consumption input for the agent provided by the principal and $t \in \mathbb{R}$ denotes a monetary transfer from the agent to the principal. Assume that for all $\theta \in \Theta$ an agent's preference order over allocations (t, x) can be obtained by evaluating $U(t, x, \theta)$, where $U : \mathbb{R}^2 \times \Theta \to \mathbb{R}$ is a sufficiently smooth utility function that is increasing in x, θ, concave in x, and decreasing in t.

Assumption 5.1 $U_t < 0 < U_x, U_\theta;\ U_{xx} \leq 0$.

The principal typically controls the agent's choice of x; however, as part of the mechanism the principal can *commit* to a certain set of rules for making the allocation decision (see footnote 9). Her prior beliefs about the distribution of agents on Θ are described in terms of a cumulative distribution function $F : \Theta \to [0, 1]$. Before the principal provides the agent's consumption input, the agent decides about his participation in the mechanism, and he can send a *message* $m \in \mathfrak{M}$ to the principal, whereby the *message space* $\mathfrak{M}$ is a (measurable) set specified by the principal. For instance, if the principal is a sales manager, as before, the set $\mathfrak{M}$ might contain the different products on offer.[7] For simplicity assume that the message space contains a null message of the type "I would not like to participate." Allowing the agent not to participate means that the principal needs to consider the agent's individual-rationality constraint when designing her mechanism.

Definition 5.1 A *mechanism* $\mathcal{M} = (\mathfrak{M}, \mathfrak{a})$ consists of a (compact) *message space* $\mathfrak{M} \neq \emptyset$ and an *allocation function* $\mathfrak{a} : \mathfrak{M} \to \mathbb{R}^2$ that assigns an allocation $\mathfrak{a}(\mathfrak{m}) = (t, x)(\mathfrak{m}) \in \mathbb{R}^2$ to any message $\mathfrak{m} \in \mathfrak{M}$.

Now consider a (dynamic) game, in which the principal proposes a mechanism $\mathcal{M}$ to an agent. The game usually has three periods. In the first period, the principal commits to $\mathcal{M} = (\mathfrak{M}, \mathfrak{a})$ and the agent decides whether to participate in the mechanism. In case of nonparticipation,[8] both the agent and the principal obtain zero payoffs from their outside options, and the game ends. Otherwise, in the second period, the agent selects a message $m \in \mathfrak{M}$ so as to maximize his utility (which results in his incentive-compatibility constraint). In the third period, the allocation $\mathfrak{a}(m)$ specified by the mechanism is implemented, and the game ends. The relevant equilibrium concept for this dynamic game under incomplete information is either the Bayes-Nash equilibrium or the perfect Bayesian equilibrium (see section 4.2.3).

Let the principal's preferences over allocations (t, x) for an agent of type θ be represented by a (sufficiently smooth) utility function $V : \mathbb{R}^2 \times \Theta \to \mathbb{R}$. The problem of finding a mechanism $\mathcal{M}$ that (in expectation) maximizes the principal's utility can be greatly simplified using the revelation principle, which is essentially due to Gibbard (1973), Green and Laffont (1979), and Myerson (1979). It is presented here in a simplified one-agent version.

Proposition 5.1 (Revelation Principle) If for a given mechanism $\mathcal{M} = (\mathfrak{M}, \mathfrak{a})$ an agent of type $\theta \in \Theta$ finds it optimal to send a message $\mathfrak{m}^*(\theta)$, then there exists a *direct revelation mechanism* $\mathcal{M}^d = (\Theta, \mathfrak{a}^d)$ such that $\mathfrak{a}^d(\theta) = \mathfrak{a}(\mathfrak{m}^*(\theta))$, and the agent finds it optimal to report his type truthfully under $\mathcal{M}^d$.

---

7. To obtain more general results, assume that $\mathfrak{M}$ contains all possible (probabilistic) convex combinations over its elements, so that $\mathfrak{m} \in \mathfrak{M}$ in fact represents a probability distribution over $\mathfrak{M}$.
8. The participation decision may also take place in the second period when the message space contains a message to this effect.

Mechanism Design

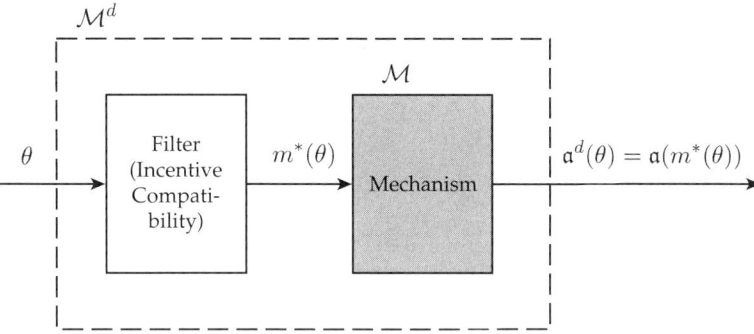

Figure 5.4
Direct mechanism $\mathcal{M}^d$ and the revelation principle.

Proof The proof is trivial. Since under mechanism $\mathcal{M} = (\mathfrak{M}, \mathfrak{a})$ the agent finds it optimal to report $\mathfrak{m}^*(\theta)$, it needs to be the case that

$$\mathfrak{m}^*(\theta) \in \arg\max_{\mathfrak{m} \in \mathfrak{M}} U(\mathfrak{a}(\mathfrak{m}), \theta) \tag{5.16}$$

for $\theta \in \Theta$. If the principal sets $\mathfrak{a}^d(\theta) = \mathfrak{a}(\mathfrak{m}^*(\theta))$, then clearly it is optimal for the agent to send $\mathfrak{m}^d(\theta) = \theta$ in the new mechanism $\mathcal{M}^d$, which is therefore a direct revelation mechanism. ∎

The revelation principle implies that in her search for an optimal mechanism the principal can limit herself without loss of generality to *direct* (i.e., truth-telling) mechanisms (Figure 5.4). The fact that the principal is able to commit to a certain mechanism is essential for the revelation principle to work. Commitment to a mechanism allows the principal to promise to the agent that indeed a revelation mechanism is applied so that truthful messages are incentive-compatible.[9]

Definition 5.2 A direct mechanism $(\Theta, \mathfrak{a})$ is *implementable* if the allocation function $\mathfrak{a} : \Theta \to \mathbb{R}^2$ satisfies the agent's incentive-compatibility (or truth-telling) constraint, namely, if

$$U(t(\theta), x(\theta), \theta) \geq U(t(\hat{\theta}), x(\hat{\theta}), \theta), \tag{5.17}$$

for all $\theta, \hat{\theta} \in \Theta$.

---

9. In environments with *renegotiation*, where the principal is unable to *commit* to a revelation mechanism, the revelation principle fails to apply, and it may become optimal for the principal to select an indirect mechanism. Indirect mechanisms can also help to ensure that allocations are unique (strong implementation); by contrast, the revelation principle ensures only that a truthful action is among the agent's most preferred actions (weak implementation). For more details see Palfrey and Srivastava (1993).

Note that the direct mechanisms in proposition 5.1 are implementable. Relation (5.17) is just a restatement of (5.16). The following analysis restricts attention to differentiable mechanisms, that is, mechanisms in which the allocation function $\mathfrak{a} = (t, x)$ is differentiable (and all the relevant sets are convex).

Assumption 5.2 (Sorting/Spence-Mirrlees Condition) The marginal rate of substitution between the agent's consumption input (the good) and money is monotonic in the agent's type,

$$\frac{\partial}{\partial \theta}\left(-\frac{U_x(t,x,\theta)}{U_t(t,x,\theta)}\right) \geq 0. \tag{5.18}$$

For *quasilinear* preferences, which can be represented by a net utility function of the form $U(x,\theta) - t$ (instead of $U(t,x,\theta)$), the sorting condition (5.18) amounts to requiring increasing differences in (x, θ). In that case, inequality (5.18) reduces to (5.6) or, given differentiability, to the familiar supermodularity condition $U_{x\theta} \geq 0$. The following result provides a useful characterization of an implementable direct mechanism.

Proposition 5.2 (Implementation Theorem) The direct mechanism $(\Theta, \mathfrak{a})$ with $\mathfrak{a} = (t,x) : \Theta \to \mathbb{R}^2$ twice differentiable is implementable if and only if for all $\theta \in \Theta$,

$$U_x(t(\theta), x(\theta), \theta) \, \dot{x}(\theta) + U_t(t(\theta), x(\theta), \theta) \, \dot{t}(\theta) = 0 \tag{5.19}$$

and

$$\dot{x}(\theta) \geq 0. \tag{5.20}$$

Proof $\Rightarrow$: Consider an agent of type $\theta \in \Theta$ and a direct mechanism $(\Theta, \mathfrak{a})$. In choosing his message $\mathfrak{m}(\theta)$, the agent solves

$$\mathfrak{m}(\theta) \in \arg\max_{\hat{\theta} \in \Theta} U(t(\hat{\theta}), x(\hat{\theta}), \theta),$$

for which the first-order necessary optimality condition can be written as

$$U_x(t(\hat{\theta}), x(\hat{\theta}), \theta) \, \dot{x}(\hat{\theta}) + U_t(t(\hat{\theta}), x(\hat{\theta}), \theta) \, \dot{t}(\hat{\theta}) = 0. \tag{5.21}$$

Hence, any direct mechanism must necessarily satisfy (5.21) for $\hat{\theta} = \theta$, namely, equation (5.19) for all $\theta \in \Theta$. The necessary optimality condition (5.21) becomes sufficient if, in addition the corresponding second-order condition,

$$U_{xx}(\dot{x})^2 + 2U_{xt}\dot{x}\dot{t} + U_{tt}(\dot{t})^2 + U_x\ddot{x} + U_t\ddot{t} \leq 0, \tag{5.22}$$

Mechanism Design

is satisfied at a $\hat{\theta}$ that solves (5.21). At a truth-telling optimum, relation (5.22) needs to be satisfied for $\hat{\theta} = \theta$. Differentiating equation (5.19) with respect to θ (note that it holds for all $\theta \in \Theta$) yields

$$(U_{xx}\dot{x} + U_{xt}\dot{t} + U_{x\theta})\dot{x} + U_x\ddot{x} + (U_{xt}\dot{x} + U_{tt}\dot{t} + U_{t\theta})\dot{t} + U_t\ddot{t} = 0,$$

so the second-order condition (5.22) becomes

$$U_{x\theta}\dot{x} + U_{t\theta}\dot{t} \geq 0,$$

or equivalently, using the fact that by (5.19) $\dot{t} = -U_x\dot{x}/U_t$,

$$U_t \dot{x} \frac{\partial}{\partial \theta}\left(\frac{U_x}{U_t}\right) \geq 0.$$

Since by assumption 5.1 the agent's utility decreases in his transfer to the principal, $U_t < 0$, one obtains by assumption 5.2 that necessarily $\dot{x} > 0$ on Θ. $\Leftarrow$: In order to demonstrate that (5.19) and (5.20) are sufficient for the direct mechanism $(\Theta, (t, x))$ to be implementable, one must show that (5.17) in definition 5.2 holds for all $\theta, \hat{\theta} \in \Theta$. If one sets

$$\hat{U}(\hat{\theta}, \theta) = U(t(\hat{\theta}), x(\hat{\theta}), \theta),$$

then the first-order and second-order optimality conditions can be written in the form $\hat{U}_1(\theta, \theta) = 0$ and $\hat{U}_{11}(\theta, \theta) \leq 0$.[10] If $\hat{\theta} \leq \theta$, then

$$\hat{U}(\theta, \theta) - \hat{U}(\hat{\theta}, \theta) = \int_{\hat{\theta}}^{\theta} \hat{U}_1(\vartheta, \theta) d\vartheta \qquad (5.23)$$

$$= \int_{\hat{\theta}}^{\theta} U_t(t(\vartheta), x(\vartheta), \theta) \left(\frac{U_x(t(\vartheta), x(\vartheta), \theta)}{U_t(t(\vartheta), x(\vartheta), \theta)}\dot{x}(\vartheta) + \dot{t}(\vartheta)\right) d\vartheta.$$

From (5.19) and (5.20), together with assumption 5.2, one obtains that for $\vartheta \leq \theta$,

$$0 = \frac{U_x(t(\vartheta), x(\vartheta), \vartheta)}{U_t(t(\vartheta), x(\vartheta), \vartheta)}\dot{x}(\vartheta) + \dot{t}(\vartheta) \geq \frac{U_x(t(\vartheta), x(\vartheta), \theta)}{U_t(t(\vartheta), x(\vartheta), \theta)}\dot{x}(\vartheta) + \dot{t}(\vartheta).$$

By assumption 5.1, $U_t < 0$, so that the right-hand side of (5.23) is nonnegative, $\hat{U}(\theta, \theta) \geq \hat{U}(\hat{\theta}, \theta)$. If $\hat{\theta} > \theta$, then (5.23) still holds. With the integration bounds reversed, assumptions 5.1 and 5.2 lead to the same conclusion, namely, that the right-hand side of (5.23) is nonnegative. In

---

10. $\hat{U}_j$ (resp. $\hat{U}_{jj}$) denotes the partial derivative of $\hat{U}$ (resp. $\hat{U}_j$) with respect to its first argument.

other words, the direct mechanism $(\Theta, (t, x))$ is by (5.17) implementable, which completes the proof. ∎

The implementation theorem directly implies a *representation* of all implementable direct mechanisms, which the principal can use to find an optimal screening mechanism:

1. Choose an arbitrary nondecreasing schedule $x(\theta)$ for the agent's consumption good.

2. For the $x(\,\cdot\,)$ in the last step, find a transfer schedule $t(\theta) - t_0$ by solving the differential equation (5.19). Note that this transfer schedule is determined only up to a constant t_0. The constant t_0 can be chosen so as to satisfy the agent's *participation constraint*,

$$U(t(\theta), x(\theta), \theta) \geq 0, \quad \forall \theta \in [\theta_0, 1], \tag{5.24}$$

where $\theta \in [0, 1]$ is the lowest participating type.[11]

3. Invert the schedule in step 1, that is, find $\varphi(x) = \{\theta \in \Theta : x(\theta) = x\}$, which may be set-valued. The (unit) price schedule as a function of the consumption choice, $\tau(x) \in t(\varphi(x))$, can then be written in the form

$$\tau(x) = \begin{cases} t(\theta) & \text{for some } \theta \in \varphi(x) \neq \emptyset, \\ \infty & \text{if } \varphi(x) = \emptyset. \end{cases} \tag{5.25}$$

This transformation is sometimes referred to as the *taxation principle*. Note that whenever $\varphi(x)$ is set-valued, different types obtain the same price for the same consumption choice. This is called *bunching*, since different types are batched together. Bunching occurs for neighboring types when $\dot{x}(\theta) = 0$ on an interval of positive length (see remark 5.1).

5.4 Nonlinear Pricing

Consider again the sales manager's decision problem (see section 5.1), but this time the manager (principal) assumes that potential buyers have types θ distributed in the continuous type space $\Theta = [0, 1]$ with differentiable cumulative distribution function $F : \Theta \to [0, 1]$ (and density $f = \dot{F}$). The principal wishes to find an optimal allocation function $(t, x) : \Theta \to \mathbb{R}^2$ so as to maximize her expected payoff,

---

11. Assumption 5.1 implies that if type $\theta \in (0, 1)$ decides to participate in the principal's mechanism, then (as a consequence of $U_\theta > 0$) all types in Θ larger than θ also decide to participate. Thus, the set of all participating types must be of the form $\Theta_0 = [\theta_0, 1] \subset \Theta$ for some $\theta_0 \in [0, 1]$.

Mechanism Design

$$\int_{\Theta_0} V(t(\theta), x(\theta), \theta) \, dF(\theta), \tag{5.26}$$

subject to the implementability conditions (5.19) and (5.20) in proposition 5.2 and to the participation constraint (5.24), where $\Theta_0 = [\theta_0, 1]$ is the set of participating types. The principal's payoff function $V : \mathbb{R}^2 \times \Theta \to \mathbb{R}$ is assumed continuously differentiable and satisfies the following assumption, which ensures that the principal likes money, dislikes providing the attribute (at an increasing rate), and for any type does not mind providing the zero bundle, that is, a zero attribute at zero price (assuming that her outside payoff is zero).

Assumption 5.3 $V_{xx}, V_x < 0 < V_t$; $V(0, \theta) \geq 0, \forall \theta \in \Theta$.

With the control $u(t) = \dot{x}(t)$, to find a screening mechanism that maximizes her expected payoff (5.26), subject to (5.19),(5.20), and (5.24), the principal can solve the optimal control problem (OCP)

$$J(u) = \int_{\theta_0}^{1} V(t(\theta), x(\theta), \theta) f(\theta) \, d\theta \longrightarrow \max_{u(\cdot), (t_0, x_0, \theta_0)}, \tag{5.27}$$

$$\dot{t}(\theta) = -\frac{U_x(t(\theta), x(\theta), \theta)}{U_t(t(\theta), x(\theta), \theta)} u(\theta), \qquad t(\theta_0) = t_0, \tag{5.28}$$

$$\dot{x}(\theta) = u(\theta), \qquad x(\theta_0) = x_0, \tag{5.29}$$

$$0 \leq U(t_0, x_0, \theta_0), \tag{5.30}$$

$$u(\theta) \in [0, \bar{u}], \qquad \forall \theta, \tag{5.31}$$

$$\theta \in [\theta_0, 1], \tag{5.32}$$

where $\bar{u} > 0$ is a large (finite but otherwise arbitrary) control constraint. Problem (5.27)–(5.32) is a general OCP of the form (3.35)–(3.39), with assumptions A1–A5 and conditions S, B, and C (see section 3.4) satisfied. Let

$$H(t, x, \theta, u, \psi) = V(t, x, \theta) f(\theta) - \frac{U_x(t, x, \theta)}{U_t(t, x, \theta)} \psi_t u + \psi_x u$$

be the corresponding Hamiltonian, where $\psi = (\psi_t, \psi_x)$ is the adjoint variable. Thus, the Pontryagin maximum principle (PMP) in proposition 3.5 can be used to formulate necessary optimality conditions for the principal's mechanism design problem.

Proposition 5.3 (Optimal Screening Contract) Let assumptions 5.1–5.3 be satisfied, and let $(t^*(\theta), x^*(\theta), t_0^*, x_0^*, \theta_0^*)$, $\theta \in [\theta_0^*, 1]$, be an optimal solution to the principal's screening problem (5.27)–(5.32), with $u^*(\theta) \equiv \dot{x}^*(\theta)$. Then there exist a multiplier $\lambda \in \mathbb{R}$ and an absolutely continuous function $\psi = (\psi_t, \psi_x) : [\theta_0^*, 1] \to \mathbb{R}^2$ such that the following optimality conditions are satisfied.

- Adjoint equation

$$\dot{\psi}_t(\theta) = \frac{\partial}{\partial t} \frac{U_x(t^*(\theta), x^*(\theta), \theta)}{U_t(t^*(\theta), x^*(\theta), \theta)} \psi_t(\theta) u^*(\theta) - V_t(t^*(\theta), x^*(\theta), \theta) f(\theta), \quad (5.33)$$

$$\dot{\psi}_x(\theta) = \frac{\partial}{\partial x} \frac{U_x(t^*(\theta), x^*(\theta), \theta)}{U_t(t^*(\theta), x^*(\theta), \theta)} \psi_t(\theta) u^*(\theta) - V_x(t^*(\theta), x^*(\theta), \theta) f(\theta), \quad (5.34)$$

for all $\theta \in [\theta_0^*, 1]$.

- Transversality

$$\psi(\theta_0^*) = -\lambda U_{(t,x)}(t_0^*, x_0^*, \theta_0^*) \quad \text{and} \quad \psi(1) = 0. \quad (5.35)$$

- Maximality

$$\forall \theta \in [\theta_0^*, 1]: \ u^*(\theta) \neq 0 \Rightarrow \psi_x(\theta) = \frac{U_x(t^*(\theta), x^*(\theta), \theta)}{U_t(t^*(\theta), x^*(\theta), \theta)} \psi_t(\theta). \quad (5.36)$$

- Endpoint optimality

$$\lambda \geq 0, \quad \lambda U(t_0^*, x_0^*, \theta_0^*) = 0, \quad (5.37)$$

$$\lambda U_\theta(t_0^*, x_0^*, \theta_0^*) = V(t_0^*, x_0^*, \theta_0^*) f(\theta_0^*). \quad (5.38)$$

- Nontriviality

$$|\lambda| + \|\psi(\theta)\| \neq 0, \quad \forall \theta \in [\theta_0^*, 1]. \quad (5.39)$$

- Envelope condition

$$V(t^*(\theta), x^*(\theta), \theta) f(\theta) = -\int_\theta^1 H_\theta(t^*(s), x^*(s), s, u^*(s), \psi(s)) \, ds, \quad (5.40)$$

for all $\theta \in [\theta_0^*, 1]$.

Proof The conditions obtain by applying the PMP in proposition 3.5 to the OCP (5.27)–(5.32). ∎

Mechanism Design

Remark 5.1 (Quasilinear Payoffs) If both the agent's and the principal's payoff functions are quasilinear in money, that is, if

$$U(t, x, \theta) = \hat{U}(x, \theta) - t \quad \text{and} \quad V(t, x, \theta) = \hat{V}(x, \theta) + t,$$

for some appropriate functions $\hat{U}, \hat{V}$ (so that all assumptions on U, V remain satisfied), then some optimality conditions in proposition 5.3 can be simplified. For example, (5.33) and (5.35) directly yield

$$\psi_t(\theta) = 1 - F(\theta), \qquad \forall \theta \in [\theta_0^*, 1]. \tag{5.41}$$

The maximality condition (5.36), together with assumption 5.1, then implies that

$$\psi_x(\theta) = -(1 - F(\theta)) \hat{U}_x(x^*(\theta), \theta) \leq 0, \qquad \forall \theta \in [\theta_0^*, 1]. \tag{5.42}$$

Assuming $\lambda \neq 0$, one obtains from the endpoint condition (5.37)

$$t_0^* = \hat{U}(x_0^*, \theta_0^*), \tag{5.43}$$

that is, the lowest participating type obtains zero surplus. From the previous relation, the transversality condition (5.35), and the endpoint-optimality condition (5.38) it can be concluded that

$$\hat{U}(x_0^*, \theta_0^*) + \hat{V}(x_0^*, \theta_0^*) - \frac{1 - F(\theta_0^*)}{f(\theta_0^*)} \hat{U}_\theta(x_0^*, \theta_0^*) = 0. \tag{5.44}$$

The term on the left-hand side of (5.44) is the virtual surplus evaluated for the lowest participating type. Combining (5.41) and (5.42) with the maximality condition (5.36) in proposition 5.3 yields

$$u^*(\theta) \neq 0 \Rightarrow \hat{U}_x(x^*(\theta), \theta) + \hat{V}_x(x^*(\theta), \theta)$$

$$- \frac{1 - F(\theta)}{f(\theta)} \hat{U}_{x\theta}(x^*(\theta), \theta) = 0, \tag{5.45}$$

which is consistent with maximizing the virtual surplus

$$S(x, \theta) = \hat{U}(x, \theta) + \hat{V}(x, \theta) - \frac{1 - F(\theta)}{f(\theta)} \hat{U}_\theta(x, \theta) \tag{5.46}$$

with respect to x, where the term $W = U + V = \hat{U} + \hat{V}$ corresponds to the actual surplus in the system, and the term $((1 - F)/f)\hat{U}_\theta$ represents the social loss due to the asymmetric information. Note also that

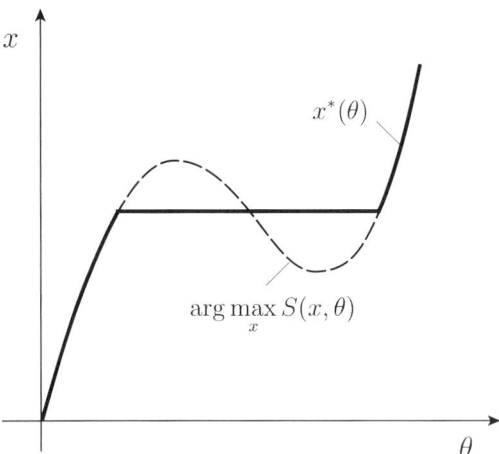

Figure 5.5
Ironing of the optimal attribute schedule and bunching of types.

$S = \hat{U} + \hat{V} - h^c \hat{U}_\theta$, where

$$h^c(\theta) \equiv \frac{1 - F(\theta)}{f(\theta)}$$

is the complementary (or inverse) hazard rate. As in the two-type case of section 5.2, the lowest type ends up without any surplus (full surplus extraction at the bottom), and the highest type obtains an efficient attribute, maximizing $\hat{W}(x, 1)$ (no distortion at the top), since the social loss vanishes for $\theta = 1$ (by virtue of the fact that $h^c(1) = 0$). Last, it is important to ask what happens when the lower control bound becomes binding, that is, when $u^*(\theta) = \dot{x}^*(\theta) = 0$. In that case, the now constant attribute schedule implies that an interval of types is treated in the same way by the principal (the types obtain the same attribute in return for the same transfer). This is called bunching (figure 5.5). The corresponding ironing procedure dates to Mussa and Rosen (1978). □

The following example illustrates how to construct an optimal nonlinear pricing scheme using an optimal screening contract.

Example 5.2 (Nonlinear Pricing) Consider the sales manager's problem of finding an optimal pricing scheme $\tau(x)$ for selling a quantity $x \geq 0$ of her product, when her cost of providing that amount to an agent is $C(x) = cx^2/2$, where $c > 0$ is a given cost parameter. When the agent

Mechanism Design

buys the quantity x at the price t, his net utility is

$$U(t,x,\theta) = \theta x - t,$$

where type $\theta \in \Theta = [0,1]$ belongs to the agent's private information. The manager (principal), with net payoff $V(t,x,\theta) = t - C(x)$, believes that the different agent types are uniformly distributed on Θ, so that $F(\theta) \equiv \theta$. Since both the principal's and the agent's net payoffs are quasilinear in money, one can use the conditions in remark 5.1 to determine the optimal price-quantity schedule $(t^*, x^*) : \Theta \to \mathbb{R}^2$. Indeed, the virtual surplus in (5.46) becomes

$$S(x,\theta) = \theta x - cx^2/2 - (1-\theta)x,$$

so, using (5.45),

$$\forall \theta \in [\theta_0^*, 1]: \quad \dot{x}^*(\theta) > 0 \Rightarrow x^*(\theta) = \frac{[2\theta - 1]_+}{c} \in \arg\max_{x \geq 0} S(x,\theta).$$

Relations (5.43) and (5.44), together with the adjoint equation (5.34) and transversality condition (5.35), can be used to determine the endpoint data $(t_0^*, x_0^*, \theta_0^*)$ (and the Lagrange multiplier λ). Indeed, (5.34) and (5.35) are equivalent to

$$t_0^* - \theta_0^* x_0^* = (2\theta_0^* - 1)x_0^* - c(x_0^*)^2/2 = 0. \tag{5.47}$$

From (5.34)–(5.35) and the endpoint-optimality conditions (5.37)–(5.38) one obtains $\lambda = (t_0^* - c(x_0^*)^2/2)/x_0^*$ and

$$\psi_x(\theta_0^*) = -(1-\theta_0^*)\theta_0^* = -\frac{t_0^* - c(x_0^*)^2/2}{x_0^*}\theta_0^*. \tag{5.48}$$

But (5.47)–(5.48) imply that $(t_0^*, x_0^*, \theta_0^*) = 0$, so by the state equation (5.28) it is

$$t^*(\theta) = \int_0^\theta \vartheta \dot{x}^*(\vartheta)\,d\vartheta = \frac{1}{c}\left[\theta^2 - \frac{1}{4}\right]_+, \quad \forall \theta \in [0,1].$$

Using the taxation principle (5.25) to eliminate the type parameter from the optimal price-quantity schedule $(t^*(\theta), x^*(\theta))$, $\theta \in \Theta$, the sales manager's optimal nonlinear pricing scheme therefore becomes

$$\tau^*(x) = \begin{cases} (2x + cx^2)/4 & \text{if } x \in [0, 1/c], \\ \infty & \text{otherwise.} \end{cases}$$

The corresponding second-best payoffs for the agent and the principal are

$$U(t^*(\theta), x^*(\theta), \theta) \equiv \frac{([\theta - (1/2)]_+)^2}{c} \quad \text{and}$$

$$V(t^*(\theta), x^*(\theta), \theta) \equiv \frac{[8\theta - 4\theta^2 - 3]_+}{4c}.$$

The total second-best welfare and virtual surplus are therefore

$$W(x^*(\theta), \theta) \equiv \frac{[2\theta - 1]_+}{2c} \quad \text{and} \quad S(x^*(\theta), \theta) \equiv \frac{2([\theta - (1/2)]_+)^2}{c},$$

respectively, where $W(x, \theta) = \theta x - cx^2/2$ and $S(x, \theta) = (1 - \theta)x$. As in the two-type model in section 5.2, one obtains full surplus extraction at the bottom of the type space (for all $\theta \in [0, 1/2]$) and no distortion at the top because the highest type obtains the welfare-maximizing quality without any social loss (for $\theta = 1$: $x^*(1) = 1/c \in \arg\max_{x \geq 0} W(x, \theta)$).[12]
□

5.5 Notes

The implementation theorem is ascribed to Mirrlees (1971); the version presented here is by Guesnerie and Laffont (1984). The treatment of mechanism design in sections 5.3 and 5.4 is inspired by Laffont (1989). Lancaster (1966) first realized, "The good, per se, does not give utility to the consumer; it possesses characteristics, and these characteristics give rise to utility .... In general a good will possess more than one characteristic, and many characteristics will be shared by more than one good" (65).

These Lancasterian characteristics are referred to as product attributes, and naturally products contain a number of different such attributes which, facing a heterogeneous consumer base of unknown types,

---

12. If the principal maximizes social surplus instead of her profits, then (by replacing V with $U + V$ in all optimality conditions) the first-best price-quantity schedule is $(t^{FB}(\theta), x^{FB}(\theta)) = (3[\theta^2 - (1/9)]_+/(2c), [3\theta - 1]_+/c)$, leading to the nonlinear pricing scheme $\tau^{FB}(x) = (2x + cx^2)/6$ for $x \in [0, 2/c]$ and $\tau^{FB}(x) = \infty$ otherwise. This notion of first-best retains the agent's autonomy, so the optimization is carried out subject to the incentive-compatibility constraints as formulated in the implementation theorem (proposition 5.2). It is interesting that the first-best leads to an overprovision of quantity for high types, compared to a full-information command-and-control solution, which is also sometimes meant by the term first-best (for $\theta = 1$: $x^{FB}(1) = 2/c > 1/c$).

Mechanism Design

allows a monopolist (the principal) to screen the agents. The product attributes can be used as instruments in the screening process. The screening problem was first examined as such by Stiglitz (1975) (in the context of mitigating adverse selection in a job market). Using multiple instruments to screen consumers of one-dimensional type was considered by Matthews and Moore (1987), and the inverse case of a single instrument (price) given consumers of multidimensional types by Laffont, Maskin, and Rochet (1987), among others. This line of work on nonlinear pricing originates with Mussa and Rosen (1978), based on methods developed earlier by Mirrlees (1971) in the context of optimal income taxation; they treated the case for consumers of a single characteristic and single vertical-attribute products. Wilson (1993) and Armstrong (1996) provided generalizations for fully nonlinear pricing models in the multiproduct case. A multidimensional screening model generalizing these approaches was advanced by Rochet and Choné (1998). Rochet and Stole (2003) provided an excellent overview of recent results. Weber (2005b) generalized the screening problem to allow for externalities between different agent types.

For a more general overview of mechanism design, see Hurwicz and Reiter (2006). Williams (2008) focused on mechanisms with differentiable allocation functions using methods from differential geometry.

5.6 Exercises

5.1 (Screening) Assume that you are the product manager for a company that produces digital cameras. You have identified two consumer types $\theta \in \Theta = \{\theta_L, \theta_H\}$, "amateurs" ($\theta_L$) and "professionals" ($\theta_H$), where $\theta_H > \theta_L > 0$. Based on the results of a detailed survey of the two groups you find that the choice behavior of a type-θ consumer can be represented approximately by the utility function

$$U(t, x, \theta) = \theta(1 - (1-x)^2)/2 - t,$$

where t is the price of a camera and x is an internal (scalar) quality index that you have developed, which orders the different camera models of the company. A consumer's utility is zero if he does not buy. Assume that the production cost of a digital camera of quality x is $C(x) = cx$, where $c \in (0, 1)$. The survey also showed that the proportion of professionals among all consumers is equal to $\mu \in (0, 1)$.

a. If your company has production capacity for only a single camera model, which can only be sold at a single price, determine the profit-maximizing price and quality, (t^m, x^m), of that product. Under what conditions can you avoid shutdown, so that both amateurs and professionals would end up buying this product? What are the company's profits, Π^m?

b. If you could perfectly distinguish amateurs from professionals, what would be the profit-maximizing menu of products, $\{(\hat{t}_i, \hat{x}_i)\}_{i \in \{L,H\}}$, to offer? Determine the associated *first-best* level of profits, $\hat{\Pi}$.

c. If you cannot distinguish between consumer types, what is the optimal *second-best* menu of products, $\{(t_i^*, x_i^*)\}_{i \in \{L,H\}}$? Determine your company's second-best profits Π^* in that case.

d. Determine the optimal nonlinear menu of products,

$\{(t^*(\theta), x^*(\theta))\}_{\theta \in \Theta}$,

for the case where $\Theta = [0,1]$ and all types are equally likely.

5.2 (Nonlinear Pricing with Congestion Externality) Consider a continuum of agents (or agent types), indexed by $\theta \in \Theta = [0,1]$ and distributed with the continuous probability density $f(\theta) = \dot{F}(\theta) > 0$ (where F is the associated cumulative distribution function). Each agent θ likes to consume bandwidth $x(\theta) \in [0,1]$; and he also cares about the aggregate bandwidth consumption

$$y = \int_0^1 x(\theta) \, dF(\theta).$$

If agent θ has to pay $t(\theta)$ for his bandwidth consumption, then his net utility is

$U(t, x, y, \theta) = \theta x(1 - \alpha y) - t,$

where $\alpha \in [0,1]$ describes the degree to which the agent is affected by the congestion externality. Any agent can also choose to consume nothing, in which case his net utility is zero. A principal[13] would like to construct an optimal nonlinear screening contract $(t, x) : \Theta \to \mathbb{R} \times [0,1]$ so as to maximize her expected profit

---

13. Assume that the principal knows only the distribution of agent types but cannot distinguish them other than by offering them a menu of options in the form of a screening contract.

Mechanism Design

$$\bar{V}(t,x) = \int_0^1 \left(t(\theta) - \frac{c(x(\theta))^2}{2} \right) dF(\theta),$$

where $c > 1$ is a cost parameter.

a. Formulate the principal's mechanism design problem as an OCP.

b. Find the optimal screening contract $(t^*(\theta), x^*(\theta))$, $\theta \in \Theta$.

c. Using the taxation principle, convert the optimal screening contract of part b into an optimal price-quantity schedule $\tau^*(x)$, $x \in [0,1]$, that the principal can actually advertise.

d. Analyze and interpret the dependence of your solution on $\alpha \in [0,1]$. What is the effect of the congestion externality?

Appendix A: Mathematical Review

This appendix provides a loose collection of definitions and key results from mathematics that are used in the main text. In terms of notation, ∃ is often used for "there exist(s)," ∀ for "for all," and $\dot{\forall}$ for "for almost all." Also, "a.e." stands for "almost everywhere" and "a.a." for "almost all" or "almost always," typically leaving out a set of Lebesgue measure zero from consideration when saying "almost." The abbreviation "s.t." is sometimes used instead of "subject to." The set $\mathbb{R}$ is the set of all real numbers; $\mathbb{C} = \mathbb{R} + i\mathbb{R}$ is the set of all complex numbers (where $i = \sqrt{-1}$);[1] $\mathbb{R}_+ = [0, \infty)$ is the set of all nonnegative real numbers; and $\mathbb{R}_{++} = (0, \infty)$ is the set of all positive real numbers. Consider two vectors, $x, y \in \mathbb{R}^n$, where $n \geq 2$ is an integer, so that $x = (x_1, \ldots, x_n)$ and $y = (y_1, \ldots, y_n)$. Then $x \geq y$ if $x_j \geq y_j$ for all $j \in \{1, \ldots, n\}$; $x > y$ if $x \geq y$ and at least one component of x is strictly greater than the corresponding component of y; and $x \gg y$ if $x_j > y_j$ for all $j \in \{1, \ldots, n\}$. For example, $\mathbb{R}^n_+ = \{x \in \mathbb{R}^n : x \geq 0\}$, $\mathbb{R}^n_{++} = \{x \in \mathbb{R}^n : x \gg 0\}$, and $\mathbb{R}^n_+ \setminus \{0\} = \{x \in \mathbb{R}^n : x > 0\}$.

A.1 Algebra

Consider the n-dimensional Euclidean space $\mathbb{R}^n$. The *scalar product* of two vectors $x = (x_1, \ldots, x_n)$ and $y = (y_1, \ldots, y_n)$ in $\mathbb{R}^n$ is defined as

$$\langle x, y \rangle = \sum_{i=1}^{n} x_i y_i.$$

With the scalar-product notation, the Euclidean norm $\|x\|$ of x can be expressed as $\|x\| = \sqrt{\langle x, x \rangle}$. The vectors $x^1, \ldots, x^k \in \mathbb{R}^n$ are said to be *linearly dependent* if there exists a nonzero vector $\lambda = (\lambda_1, \ldots, \lambda_k)$

---

1. One of the most beautiful relations in mathematics is *Euler's identity*: $e^{i\pi} + 1 = 0$.

such that $\sum_{j=1}^{k} \lambda_j x^j = 0$; otherwise, the vectors are called *linearly independent*.

An $m \times n$ matrix $A = [a_{ij}]_{i,j=1}^{m,n}$ is a rectangular array of real numbers, arranged in m rows and n columns. If the number of rows equals the number of columns (i.e., $m = n$), then the matrix A is *square*. The sum of two $m \times n$ matrices $A = [a_{ij}]_{i,j=1}^{m,n}$ and $B = [b_{ij}]_{i,j=1}^{m,n}$ is given by $A + B = [a_{ij} + b_{ij}]_{i,j=1}^{m,n}$ and is obtained by summing the corresponding elements in A and B. The product of a matrix A with a *scalar* $\alpha \in \mathbb{R}$ is given by $\alpha A = [\alpha a_{ij}]_{i,j=1}^{m,n}$. The product of that matrix with an $n \times l$ matrix $B = [b_{ij}]_{i,j=1}^{n,l}$ is given by the $m \times l$ matrix $AB = [\sum_{k=1}^{n} a_{ik} b_{kj}]_{i,j=1}^{m,l}$.[2] For convenience, the index notation is dropped when the context is clear. The *transpose* of the $m \times n$ matrix $A = [a_{ij}]$ is the $n \times m$ matrix $A' = [a'_{ij}]$ with $a'_{ij} = a_{ji}$ for all i, j. If A is square (i.e., $m = n$) and $A' = A$, it is called *symmetric*. The square matrix A has an *inverse*, denoted by A^{-1}, if the product AA^{-1} is equal to the *identity matrix I*, which is square and such that all its entries are zero except for the diagonal entries, which are equal to 1. Such an inverse exists if and only if $A = [a_{ij}]_{i,j=1}^{n}$ is *nonsingular*, that is, if all its row vectors $(a_{ij})_{j=1}^{n}, i \in \{1, \ldots, n\}$, are linearly independent. In other words, the square matrix is nonsingular if and only if the equation $Ax = 0$ (where the vector $x = (x_1, \ldots, x_n)'$ is interpreted as an $n \times 1$ matrix) has the unique solution $x = 0$.

The *rank* of an $m \times n$ matrix A is equal to the maximum number of its row vectors that are linearly independent. The matrix is called *of full rank* if its *rank is maximal*, that is, equal to $\min\{m, n\}$. Thus, a square matrix A is nonsingular if and only if it is of full rank. An alternative criterion can be formulated in terms of the *determinant*, which is defined by the recursive *Laplace expansion rule*, for any fixed $j \in \{1, \ldots, n\}$,

$$\det A = \sum_{i=1}^{n} a_{ij} A_{ij} \left(= \sum_{i=1}^{n} a_{ji} A_{ji} \right),$$

where A_{ij} is the (sub-)determinant of matrix A after its ith row and its jth column have been removed. The determinant of a 1×1 matrix is equal to its only element. Then, the square matrix A is nonsingular if and only if its determinant is nonzero.

Right-multiplying the row vector $x = (x_1, \ldots, x_n)'$ with the $n \times n$ matrix A yields the *linear transformation Ax*. Of particular interest are the

---

2. Matrix multiplication is not commutative.

eigenvectors $v = (v_1, \ldots, v_n) \neq 0$, for which there exists a scalar λ (which is generally a complex number) such that $Av = \lambda v$. That means that the linear transformation Av of an eigenvector leaves the vector essentially unchanged, except for a possible expansion ($|\lambda| \geq 1$) or contraction ($|\lambda| \leq 1$) of v (together with a possible 180-degree rotation). Accordingly, each *eigenvalue* λ of A is such that $Av = \lambda v$ for some eigenvector v. The last relation is equivalent to the matrix equation $(A - \lambda I)v = 0$, which has a nonzero solution v if and only if the matrix $A - \lambda I$ is singular, so that the *characteristic equation*,

$$\det(A - \lambda I) = 0,$$

is satisfied. The eigenvalues of A are the solutions of the last equation, which correspond to the roots of an nth degree polynomial. A *root* of an nth order polynomial $p(\lambda) = \alpha_0 + \alpha_1 \lambda + \alpha_2 \lambda^2 + \cdots + \alpha_n \lambda^n$ is such that $p(\lambda) = 0$.

Proposition A.1 (Fundamental Theorem of Algebra) Let $\alpha = (\alpha_1, \ldots, \alpha_n) \in \mathbb{R}^n \setminus \{0\}$. Then the polynomial $p(\lambda) = \sum_{i=k}^{n} \alpha_k \lambda^k$ has exactly n (complex) roots, $\lambda_1, \ldots, \lambda_n \in \mathbb{C}$.

Proof See Hungerford (1974, 265–267). ∎

Thus, the eigenvalues of A are in general complex numbers (i.e., elements of $\mathbb{C}$). Note also that A is singular if and only if one of its eigenvalues is zero. If A is nonsingular, with eigenvalues $\lambda_1, \ldots, \lambda_n$, then the eigenvalues of its inverse A^{-1} are given by $1/\lambda_1, \ldots, 1/\lambda_n$, whereas the eigenvalues of its transpose A' are the same as those of A.

A symmetric square matrix A is called *positive semidefinite* if

$$x'Ax \geq 0, \qquad \forall x \in \mathbb{R}^n \setminus \{0\}.$$

If the previous inequality is strict, then A is called *positive definite*. Similarly, A is called *negative (semi)definite* if $-A$ is positive (semi)definite. A symmetric matrix has real eigenvalues; if the matrix is also positive (semi)definite, then its eigenvalues are positive (resp., nonnegative).

A.2 Normed Vector Spaces

A *set* $\mathcal{S}$ is a collection of elements, which can be numbers, actions, outcomes, or any other objects. The set of nonnegative integers, $\{0, 1, 2, \ldots\}$, is denoted by $\mathbb{N}$. A *field* $\mathbb{F} = (\mathcal{S}, +, \cdot)$ is a set $\mathcal{S}$ together with the binary operations of addition, $+: \mathcal{S} \times \mathcal{S} \to \mathcal{S}$, and multiplication, $\cdot: \mathcal{S} \times \mathcal{S} \to \mathcal{S}$, with the following *field properties* for all $a, b, c \in \mathcal{S}$:

- *Commutativity.* $a + b = b + a$ and $a \cdot b = b \cdot a$.
- *Associativity.* $(a + b) + c = a + (b + c)$ and $(a \cdot b) \cdot c = a \cdot (b \cdot c)$.
- *Distributivity.* $a \cdot (b + c) = a \cdot b + a \cdot c$.
- *Additive identity.* There is a *zero* element, $0 \in \mathcal{S}$, such that $a + 0 = a$, independent of which $a \in \mathcal{S}$ is chosen.
- *Additive inverse.* There exists (in $\mathcal{S}$) an *additive inverse* of a, denoted by $-a$, such that $a + (-a) = 0$.
- *Multiplicative identity.* There is a *one* element, $1 \in \mathcal{S}$, such that $a \cdot 1 = a$, independent of which $a \in \mathcal{S}$ is chosen.
- *Multiplicative inverse.* There exists (in $\mathcal{S}$) a *multiplicative inverse* of $a \neq 0$, denoted by $1/a$, such that $a \cdot (1/a) = 1$.
- *Closure.* $a + b \in \mathcal{S}$ and $a \cdot b \in \mathcal{S}$.

The set $\mathbb{R}$ of all real numbers and the set $\mathbb{C}$ of all complex numbers together with the standard addition and multiplication are fields. For simplicity $\mathbb{R}$ and $\mathbb{C}$ are often used instead of $(\mathbb{R}, +, \cdot)$ and $(\mathbb{C}, +, \cdot)$. A *vector space* or *linear space* (over a field $\mathbb{F} = (\mathcal{S}, +, \cdot)$) is a set $\mathcal{X}$ of objects, called *vectors*, together with the binary operations of *(vector) addition*, $+: \mathcal{X} \times \mathcal{X} \to \mathcal{X}$, and *(scalar) multiplication*, $\cdot: \mathcal{S} \times \mathcal{X} \to \mathcal{X}$, such that the following *vector-space properties* are satisfied:

- The vector addition is commutative and associative.
- There is a *zero vector* (referred to as *origin*), $0 \in \mathcal{X}$, such that $x + 0 = x$ for all $x \in \mathcal{X}$.
- For any $x \in \mathcal{X}$ there is an *additive inverse* $-x \in \mathcal{X}$ such that $x + (-x) = 0$.
- For any *scalars* $a, b \in \mathcal{S}$ and all vectors $x, y \in \mathcal{X}$ it is $a \cdot (b \cdot x) = (a \cdot b) \cdot x$ (*associativity*) as well as $a \cdot (x + y) = a \cdot x + b \cdot y$ and $(a + b) \cdot x = a \cdot x + b \cdot x$ (*distributivity*).
- If 1 is an additive identity in $\mathcal{S}$, then $1 \cdot x = x$ for all $x \in \mathcal{X}$.

In this book attention is restricted to the fields $\mathbb{F} \in \{\mathbb{R}, \mathbb{C}\}$ of the real and complex numbers. "Vector space $\mathcal{X}$" usually means a vector space $(\mathcal{X}, +, \cdot)$ over the fields $(\mathbb{R}, +, \cdot)$ or $(\mathbb{C}, +, \cdot)$; in the latter case it might be referred to an "complex vector space $\mathcal{X}$."

Example A.1 As prominent examples of vector spaces consider first the set $\mathbb{R}^n$ of real n-vectors (for any given integer $n \geq 1$), second the set of all

Appendix A

sequences $\{x^0, x^1, \ldots\} = \{x^k\}_{k=0}^{\infty}$ formed of elements $x^k \in \mathbb{R}^n$, and third the set $C^0([a,b], \mathbb{R}^n)$ of continuous functions $f : [a,b] \to \mathbb{R}^n$ for given real numbers a, b with $a < b$.[3] □

A *norm* on a vector space $\mathcal{X}$ is a real-valued function $\|\cdot\|$ on $\mathcal{X}$ such that for all vectors $x, y \in \mathcal{X}$ and any scalar $a \in \mathbb{F}$ the following *norm properties* hold:

- *Positive definiteness.* $\|x\| \geq 0$, and $\|x\| = 0$ if and only if $x = 0$.
- *Positive homogeneity.* $\|ax\| = |a|\|x\|$.
- *Triangular inequality.* $\|x+y\| \leq \|x\| + \|y\|$.

The ordered pair $(\mathcal{X}, \|\cdot\|)$ is termed a *normed vector space* (or *normed linear space*). The set $\mathcal{Y} \subseteq \mathcal{X}$ is a *(linear) subspace* of $\mathcal{X}$ (or more precisely, of $(\mathcal{X}, \|\cdot\|)$) if $x, y \in \mathcal{Y} \Rightarrow ax + by \in \mathcal{Y}$, for any scalars a and b.

Example A.2 (1) $\ell_p^n = (\mathbb{F}^n, \|\cdot\|_p)$ is a normed vector space, where for any number p with $1 \leq p \leq \infty$ the *p-norm* on the vector space $\mathbb{F}^n$ is defined by

$$\|a\|_p = \begin{cases} \left(\sum_{i=1}^n |a_i|^p\right)^{1/p} & \text{if } 1 \leq p < \infty, \\ \max\{|a_1|, \ldots, |a_n|\} & \text{if } p = \infty, \end{cases}$$

for all $a = (a_1, \ldots, a_n) \in \mathbb{F}^n$.

(2) The collection of all sequences $x = \{x^k\}_{k=0}^{\infty}$ with $x^k \in \mathbb{F}^n$, $k \geq 0$, for which with *p*-norm

$$\|x\|_p = \begin{cases} \left(\sum_{i=1}^n \|x^k\|^p\right)^{1/p} & \text{if } 1 \leq p < \infty, \\ \sup\{\|x^k\| : k \in \mathbb{N}\} & \text{if } p = \infty, \end{cases} < \infty$$

is a normed vector space (referred to as ℓ_p).

(3) Let η be a positive measure on a σ-algebra Σ of subsets of a nonempty measurable set $\mathcal{S}$. For any p with $1 \leq p \leq \infty$ the set $\mathbf{L}_p(\mathcal{S}, \Sigma, \eta)$ of Lebesgue-integrable functions f, which are such that

$$\|x\|_p = \begin{cases} \left(\int_{\mathcal{S}} \|f(x)\|^p d\eta(x)\right)^{1/p} & \text{if } 1 \leq p < \infty \\ \inf\{c > 0 : \eta\left(\{x \in \mathcal{S} : \|f(x)\| > c\}\right) = 0\} & \text{if } p = \infty \end{cases} < \infty,$$

is a normed vector space.

---

3. The concept of a function is introduced in section A.3.

(4) Let $\Omega \subset \mathbb{R}^n$ be a nonempty compact (i.e., closed and bounded) set. The space $C^0(\Omega, \mathbb{R}^n)$ of continuous functions $f : \Omega \to \mathbb{R}^n$ with the maximum norm

$$\|f\|_\infty = \max_{x \in \Omega} \|f(x)\|,$$

where $\|\cdot\|$ is a suitable norm on $\mathbb{R}^n$, is a normed vector space. $\square$

Remark A.1 (Norm Equivalence) One can show that all norms on $\mathbb{F}^n$ are equivalent in the sense that for any two norms $\|\cdot\|$ and $|\cdot|$ on $\mathbb{F}^n$ there exist positive constants $\alpha, \beta \in \mathbb{R}_{++}$ such that $\alpha\|x\| \leq |x| \leq \beta\|x\|$, for all $x \in \mathbb{F}^n$. $\square$

A sequence $x = \{x^k\}_{k=0}^\infty$ of elements of a normed vector space $\mathcal{X}$ converges to a *limit* $\bar{x}$, denoted by $\lim_{k \to \infty} x^k = \bar{x}$, if for any $\varepsilon > 0$ there exists an integer $N = N(\varepsilon) > 0$ such that

$$k \geq N \Rightarrow \|x^k - \bar{x}\| \leq \varepsilon.$$

It follows immediately from this definition that the limit of a sequence (if it exists) must be unique. If any convergent sequence $\{x^k\}_{k=0}^\infty \subset \mathcal{X}$ has a limit that lies in $\mathcal{X}$, then $\mathcal{X}$ is called *closed*. The *closure* $\bar{\mathcal{X}}$ of a set $\mathcal{X}$ is the union of $\mathcal{X}$ and all its *accumulation points*, that is, the limits of any convergent (sub)sequences $\{x_k\}_{k=0}^\infty \subset \mathcal{X}$. The set $\mathcal{X}$ is *open* if for any $x \in \mathcal{X}$ there exists $\varepsilon > 0$ so that the ε-ball $\mathcal{B}_\varepsilon(x) = \{\hat{x} : \|\hat{x} - x\| < \varepsilon\}$ is a subset of $\mathcal{X}$.

Let $\mathcal{X}$ be a nonempty subset of a normed vector space. A family of open sets is an *open cover* of $\mathcal{X}$ if their union contains $\mathcal{X}$. The set $\mathcal{X}$ is called *compact* if from any open cover of $\mathcal{X}$ one can select a *finite* number of sets (i.e., a finite subcover) that is also an open cover of $\mathcal{X}$. It turns out that in a finite-dimensional vector space this definition leads to the following characterization of compactness:

$\mathcal{X}$ is compact $\Leftrightarrow$ $\mathcal{X}$ is closed and bounded.[4]

Compact sets are important for the construction of optimal solutions because every sequence of elements of such sets has a convergent subsequence.

Proposition A.2 (Bolzano-Weierstrass Theorem) Every sequence $x = \{x^k\}_{k=0}^\infty$ in a compact subset $\mathcal{X}$ of a normed vector space has a convergent subsequence (with limit in $\mathcal{X}$).

---

4. The set $\mathcal{X}$ is *bounded* if there exists $M > 0$ such that $\|x\| \leq M$ for all $x \in \mathcal{X}$.

Proof Let $\mathcal{S} = \bigcup_{k=0}^{\infty}\{x^k\} \subset \mathcal{X}$ be the range of the sequence x. If $\mathcal{S}$ is finite, then there exist indices $k_1 < k_2 < k_3 < \cdots$ and $y \in \mathcal{S}$ such that $x^{k_1} = x^{k_2} = x^{k_3} = \cdots = y$. The subsequence $\{x^{k_j}\}_{j=1}^{\infty}$ converges to $y \in \mathcal{X}$. If $\mathcal{S}$ is infinite, then it contains a limit point, y.[5] Indeed, if it did not contain a limit point, then each $\hat{y} \in \mathcal{S}$ must be an isolated point of $\mathcal{S}$ (see footnote 6), that is, there exists a collection of open balls, $\{\mathcal{B}_{\hat{y}}\}_{\hat{y}\in\mathcal{S}}$, which covers $\mathcal{S}$ and which is such that $\mathcal{B}_{\hat{y}} \cap \mathcal{S} = \{\hat{y}\}$ for all $\hat{y} \in \mathcal{S}$. But then $\mathcal{X}$ cannot be compact, since $\mathcal{S} \subset \mathcal{X}$ and an open cover of $\mathcal{X}$ can be found that does not have a finite subcover of $\mathcal{X}$. It is therefore enough to select a subsequence $\{x^{k_j}\}_{j=1}^{\infty}$ of x such that $k_1 < k_2 < \cdots$ and $x^{k_j} \in \{\hat{y} \in \mathcal{S} : \|\hat{y} - y\| < 1/j\}$ for all $j \geq 1$, which implies that $x^{k_j} \to y$ as $j \to \infty$, completing the proof. ∎

The sequence $\{x^k\}_{k=0}^{\infty}$ is a *Cauchy sequence* if for any $\varepsilon > 0$ there exists an integer $K = K(\varepsilon) > 0$ such that

$$k, l \geq K \Rightarrow \|x^k - x^l\| \leq \varepsilon.$$

Note that any Cauchy sequence is bounded, since for $\varepsilon = 1$ there exists $K > 0$ such that $\|x^k - x^K\| \leq 1$ for all $k \geq K$, which implies that

$$\|x^k\| = \|x^k - x^K + x^K\| \leq \|x^k - x^K\| + \|x^K\| \leq 1 + \|x^K\|,$$

for all $k \geq K$. On the other hand, any convergent sequence x is a Cauchy sequence, since (using the notation of the earlier definition of limit)

$$k, l \geq K(\varepsilon) = N\left(\frac{\varepsilon}{2}\right) \Rightarrow \|x^k - x^l\| \leq \|x^k - \bar{x} + \bar{x} - x^l\| \leq \|x^k - \bar{x}\|$$
$$+ \|\bar{x} - x^l\| \leq \varepsilon.$$

A normed vector space $\mathcal{X}$ in which the converse is always true, that is, in which every Cauchy sequence converges, is called *complete*. A complete normed vector space is also referred to as a *Banach space*.

Example A.3 Let $1 \leq p \leq \infty$.

(1) The normed vector space ℓ_p^n is complete. Indeed, if $\{a^k\}_{k=0}^{\infty}$ is a Cauchy sequence, then it is bounded. By the Bolzano-Weierstrass theorem any bounded sequence in $\mathbb{F}^n$ contains a convergent subsequence $\{\hat{a}^k\}_{k=0}^{\infty}$, the limit of which is denoted by $\bar{a}$. Thus, for any $\varepsilon > 0$ there exists $\hat{N}(\varepsilon) > 0$ such that $\|\hat{a}^k - \bar{a}\| \leq \varepsilon$ for all $k \geq \hat{N}(\varepsilon)$. This in

---

5. A point y is a *limit point* of $\mathcal{S}$ if for any $\varepsilon > 0$ the set $\{\hat{x} \in \mathcal{S} : \|\hat{x} - x\| < \varepsilon\}$ contains a point $\hat{y} \in \mathcal{S}$ different from y.

turn implies that the original sequence $\{a^k\}_{k=0}^{\infty}$ converges to $\bar{a}$, since by construction

$$k \geq k_0 = \max\left\{K\left(\frac{\varepsilon}{2}\right), \hat{N}\left(\frac{\varepsilon}{2}\right)\right\} \Rightarrow \|a^k - \bar{a}\| \leq \|a^k - \hat{a}^{k_0}\|$$
$$+ \|\hat{a}^{k_0} - \bar{a}\| \leq \frac{\varepsilon}{2} + \frac{\varepsilon}{2} = \varepsilon.$$

Because of the norm equivalence in $\mathbb{F}^n$ (see remark A.1) it is unimportant which norm $\|\cdot\|$ is considered.

(2) One can as well show (see, e.g., Luenberger 1969, 35–37) that the other spaces introduced in example A.2 (namely, ℓ_p, $\mathbf{L}_p(\mathcal{S}, \Sigma, \eta)$, and $C^0(\Omega, \mathbb{R}^n)$) are also complete. □

Proposition A.3 (Banach Fixed-Point Theorem) (Banach 1922) Let Ω be a closed subset of a Banach space $\mathcal{X}$, and let $f : \Omega \to \Omega$ be a contraction mapping in the sense that

$$\|f(\hat{x}) - f(x)\| \leq K\|\hat{x} - x\|, \quad \forall \hat{x}, x \in \Omega,$$

for some $K \in [0, 1)$. Then there exists a unique $x^* \in \Omega$ for which

$$f(x^*) = x^*.$$

Moreover, the fixed point x^* of f can be obtained by the method of successive approximation, so that starting from any $x^0 \in \Omega$ and setting $x^{k+1} = f(x^k)$ for all $k \geq 0$ implies that

$$\lim_{k \to \infty} x^k = \lim_{k \to \infty} f^k(x^0) = x^*.$$

Proof Let $x^0 \in \Omega$ and let $x^k = f(x^k)$ for all $k \geq 0$. Then the sequence $\{x^k\}_{k=0}^{\infty} \subset \Omega$ is a Cauchy sequence, which can be seen as follows. For any $k \geq 0$ the difference of two subsequent elements of the sequence is bounded by a fraction of the difference of the first two elements, since

$$\|x^{k+1} - x^k\| = \|f(x^k) - f(x^{k-1})\| \leq K\|x^k - x^{k-1}\| \leq \cdots \leq K^k\|x^1 - x^0\|.$$

Thus, for any $l \geq 1$ it is

$$\|x^{k+l} - x^k\| \leq \|x^{k+l} - x^{k+l-1}\| + \|x^{k+l-1} - x^{k+l-2}\| + \cdots + \|x^{k+1} - x^k\|$$
$$\leq (K^{k+l-1} + K^{k+l-2} + \cdots + K^k)\|x^1 - x^0\|$$
$$\leq K^k \sum_{\kappa=0}^{\infty} K^\kappa \|x^1 - x^0\| = \frac{K^k}{1-K}\|x^1 - x^0\| \to 0 \quad \text{as } k \to \infty.$$

Appendix A

Hence, by completeness of the Banach space $\mathcal{X}$, it is $\lim_{k\to\infty} x^k = x^* \in \mathcal{X}$. In addition, since Ω is by assumption closed, it is also $x^* \in \Omega$. Note also that because

$$\|f(x^*) - x^*\| \le \|f(x^*) - x^k\| + \|x^k - x^*\| \le K\|x^* - x^{k-1}\| + \|x^k - x^*\| \to 0,$$

as $k \to \infty$, one can conclude that $f(x^*) = x^*$, that is, x^* is indeed a fixed point of the mapping f. Now if $\hat{x}^*$ is another fixed point, then

$$\|\hat{x}^* - x^*\| = \|f(\hat{x}^*) - f(x^*)\| \le K\|\hat{x}^* - x^*\|,$$

and necessarily $\hat{x}^* = x^*$, since $K < 1$. This implies uniqueness of the fixed point and concludes the proof. ∎

The Banach fixed-point theorem is sometimes also referred to as *contraction mapping principle*. It is used in chapter 2 to establish the existence and uniqueness of the solution to a well-posed initial value problem (IVP) (see proposition 2.3). The following example illustrates how the contraction mapping principle can be used to establish the existence and uniqueness of a Nash equilibrium in a game of complete information.

Example A.4 (Uniqueness of a Nash Equilibrium) Consider a two-player static game of complete information in the standard normal-form representation Γ (see section 4.2.1), with action sets $\mathcal{A}^i = [0, 1]$ and twice continuously differentiable payoff functions $U^i : [0, 1]^2 \to [0, 1]$ for $i \in \{1, 2\}$. If $0 < r^i(a^{-i}) < 1$ is a best response for player i, it satisfies

$$U^i_{a^i}(r^i(a^{-i}), a^{-i}) = 0.$$

Differentiation with respect to a^{-i} yields

$$\frac{dr^i(a^{-i})}{da^{-i}} = -\frac{U^i_{a^i a^{-i}}(r^i(a^{-i}), a^{-i})}{U^i_{a^i a^i}(r^i(a^{-i}), a^{-i})}.$$

Set $r(a) = (r^1(a^2), r^2(a^1))$; then any fixed point $a^* = (a^{1*}, a^{2*})$ of r is a Nash equilibrium. Note that r maps the set of strategy profiles $\mathcal{A} = [0, 1]^2$ into itself. Let $\hat{a}, a$ be two strategy profiles. Then by the mean-value theorem (see proposition A.14),

$$\|r(\hat{a}) - r(a)\|_1 = |r^1(\hat{a}^2) - r^1(a^2)| + |r^2(\hat{a}^1) - r^2(a^1)|$$
$$\le \max\{L^1, L^2\}(|\hat{a}^1 - a^1| + |\hat{a}^2 - a^2|)$$
$$= \max\{L^1, L^2\}\|\hat{a} - a\|_1,$$

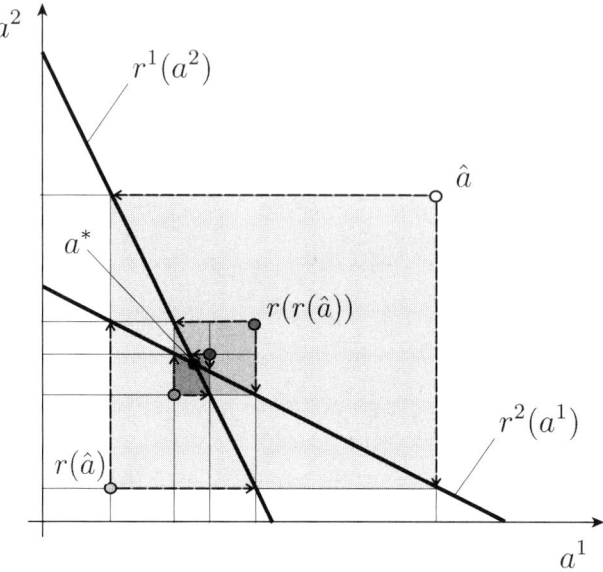

Figure A.1
Uniqueness of a Nash equilibrium by virtue of the Banach fixed-point theorem (see example A.4).

where $\|\cdot\|_1$ is the ℓ_1-norm on the vector space $\mathbb{R}^2$ (see example A.2), and

$$L^i = \max_{(a^i, a^{-i}) \in \mathcal{A}} \left| \frac{U^i_{a^i a^{-i}}(a^i, a^{-i})}{U^i_{a^i a^i}(a^i, a^{-i})} \right|, \quad \forall i \in \{1, 2\}.$$

Thus, if $L^1, L^2 < 1$, then r is a contraction mapping, and by the Banach fixed-point theorem there exists a unique Nash equilibrium $a^* = r(a^*)$, that is, a fixed point of r, and this equilibrium can be found starting at an arbitrary strategy profile $\hat{a} \in \mathcal{A}$ by successive approximation: iterative application of r. That is, $r(r(r(\cdots r(\hat{a}) \cdots))) \to a^*$ for any $\hat{a} \in \mathcal{A}$. Figure A.1 illustrates the successive-approximation procedure. Note that this example can be generalized to games with $N > 2$ players. □

A.3 Analysis

Let $\mathcal{X}$ and $\mathcal{Y}$ be nonempty subsets of a normed vector space. A relation that maps each element $x \in \mathcal{X}$ to an element $f(x) = y \in \mathcal{Y}$ is called a *function*, denoted by $f : \mathcal{X} \to \mathcal{Y}$. The sets $\mathcal{X}$ and $\mathcal{Y}$ are then called the *domain* and *co-domain* of f, respectively. If $\mathcal{X}$ is a normed vector space containing functions, then f is often referred to as a *functional* (or *operator*). The function is *continuous at a point* $x \in \mathcal{X}$ if for any $\varepsilon > 0$ there exists a $\delta > 0$ such that [6]

$$\forall \hat{x} \in \mathcal{X}: \quad \|\hat{x} - x\| < \delta \Rightarrow \|f(\hat{x}) - f(x)\| < \varepsilon.$$

The function is *continuous* (*on* $\mathcal{X}$) if it is continuous at any point of $\mathcal{X}$. Note that sums, products, and compositions of continuous functions are also continuous.[7] A useful alternative characterization of continuity is as follows: a function $f : \mathcal{X} \to \mathcal{Y}$ is continuous on its *domain* $\mathcal{X}$ if and only if the *preimage* of any open subset $\mathcal{G} \subset \mathcal{Y}$, denoted by $f^{-1}(\mathcal{G}) \equiv \{x \in \mathcal{X} : f(x) \in \mathcal{G}\}$, is open. Continuous transformations, which are continuous mappings, are important in topology. A closed curve in $\mathbb{R}^2$, for example, is a continuous transformation of a unit circle. A weaker notion than continuity is measurability. The function f is *measurable* if the preimage of any measurable set is a measurable set.[8] The measurable function f is *essentially bounded* on the measurable set $\mathcal{G}$ if there exists a constant $M > 0$ such that $\{x \in \mathcal{G} : \|f(x)\| > M\}$ is of (Lebesgue) measure zero.

The following is a classic result in topology, which is used in the proof of the Poincaré-Bendixson theorem (see proposition 2.13).

Proposition A.4 (Jordan Curve Theorem) (Jordan 1909) Every simple (i.e., self-intersection-free) closed curve in $\mathbb{R}^2$ divides the plane into two disjunct pieces, the inside and the outside.

Proof See Armstrong (1983, 112–114).[9] ∎

The next result characterizes the existence of a convergent sequence of functions, which are all defined on a compact set. It is therefore similar to the Bolzano-Weierstrass theorem (see proposition A.2) for sequences defined on compact sets.

---

6. The definition of continuity is automatically satisfied at an *isolated point* of $\mathcal{X}$, i.e., at a point $x \in \mathcal{X}$ for which there exists $\varepsilon > 0$ such that $\{\hat{x} \in \mathcal{X} : \|\hat{x} - x\| < \varepsilon\} = \{x\}$. Thus, any function is continuous on a finite set.
7. The same holds true for quotients f/g of continuous real-valued functions f and g as long as g never vanishes.
8. For a brief introduction to measure theory and Lebesgue integration, see, e.g., Kirillov and Gvishiani (1982).
9. Guillemin and Pollack (1974, 85–89) provide an interesting outline for the proof of the Jordan-Brouwer separation theorem, which can be viewed as a generalization of the Jordan curve theorem to higher dimensions.

Proposition A.5 (Arzelà-Ascoli Theorem) Let $\mathcal{X}$ be a compact subset of a normed vector space, and let $\mathcal{F}$ be a family of functions $f : \mathcal{X} \to \mathbb{R}^n$. Then every sequence $\{f_k\}_{k=0}^\infty \subset \mathcal{F}$ contains a uniformly convergent subsequence if and only if $\mathcal{F}$ is uniformly bounded and equicontinuous.[10]

Proof See, for example, Zorich (2004, II, 398–399). ∎

If $x = (x_1, \ldots, x_n)$ is an interior point of $\mathcal{X} \subset \mathbb{R}^n$, then the function f is *(partially) differentiable with respect to (the ith coordinate)* x_i if the limit, called partial derivative,

$$f_{x_i}(x) \equiv \lim_{\delta \to 0} \frac{f(x + \delta e_i) - f(x)}{\delta} \quad \left(= \frac{\partial f(x)}{\partial x_i} \right),$$

exists, where $e_i \in \mathbb{R}^n$ is the ith Euclidean unit vector (such that $\langle x, e_i \rangle = x_i$). The function is called *differentiable at x* if its partial derivatives with respect to all coordinates exist, and it is called *differentiable* (on an open set $\mathcal{X}$) if it is differentiable at any point $x \in \mathcal{X}$. The *Jacobian (matrix)* of $f = (f_1, \ldots, f_m)' : \mathcal{X} \to \mathbb{R}^m$ is the matrix of partial derivatives,

$$f_x(x) = \left[\frac{\partial f_i(x)}{\partial x_j} \right]_{i,j=1}^{n,m}.$$

If the Jacobian is itself (componentwise) differentiable, then the tensor $f_{xx}(x)$ of second derivatives is called the *Hessian*. In particular, for $m = 1$ it is

$$f_{xx}(x) = \left[\frac{\partial^2 f(x)}{\partial x_i \partial x_j} \right]_{i,j=1}^{n,n}.$$

This Hessian matrix is symmetric if it is continuous, a statement also known as *Schwarz's theorem* (or *Clairaut's theorem*).

Example A.5 (Total Derivative) The total derivative of a differentiable function $F : \mathbb{R} \to \mathbb{R}^n$ is the differential of $F(t)$ with respect to the independent variable t: it is given by $F_t(t) = \lim_{\delta \to 0} \left(F(t + \delta) - F(t) \right) / \delta$, analogous to relation (2.1) in chapter 2, and it is usually denoted by $\dot{F}(t)$. □

---

10. The family of functions is *uniformly bounded* (on $\mathcal{X}$) if the union of images $\bigcup_{f \in \mathcal{F}} f(\mathcal{X})$ is bounded. It is *equicontinuous* (on $\mathcal{X}$) if for any $\varepsilon > 0$ there exists $\delta > 0$ such that $\|\hat{x} - x\| < \delta \Rightarrow \|f(\hat{x}) - f(x)\| < \varepsilon$, for all $f \in \mathcal{F}$ and all $\hat{x}, x \in \mathcal{X}$. A sequence of functions $\{f_k\}_{k=0}^\infty$ (all with the same domain and co-domain) *converges uniformly* to the limit function f (with the same domain and co-domain), denoted by $f_k \rightrightarrows f$, if for any $\varepsilon > 0$ there exists $K > 0$ such that $\|f_k - f\| < \varepsilon$ for all $k \geq K$ on the entire domain.

Appendix A

The inverse operation of differentiation is integration.[11] Let t_0, T with $t_0 < T$ be given real numbers. In order for a function $F : [t_0, T] \to \mathbb{R}^n$ to be representable as an integral of its derivative $\dot{F} = f$ it is not necessary that f be continuous, only that it be integrable.

Proposition A.6 (Fundamental Theorem of Calculus) If the real-valued function F is equal to the anti-derivative of the function f a.e. on the interval $[t_0, T]$, namely,

$$\dot{F}(t) = f(t), \quad \forall\, t \in [t_0, T],$$

and f is essentially bounded, then

$$F(T) - F(t_0) = \int_{t_0}^{T} f(t)\, dt.$$

Remark A.2 (Absolute Continuity) A function $F : [t_0, T] \to \mathbb{R}^n$ is *absolutely continuous* (on $[t_0, T]$) if for any $\varepsilon > 0$ there exists $\delta > 0$ such that for any pairwise disjoint intervals $\mathcal{I}_k \subset [t_0, T], k \in \{1, 2, \ldots, N\}$ (allowing for countably many intervals, so that possibly $N = \infty$) it is

$$\sum_{k=1}^{N} \operatorname{diam} \mathcal{I}_k < \delta \Rightarrow \sum_{k=1}^{N} \operatorname{diam} f(\mathcal{I}_k) < \varepsilon,$$

where the diameter of a set S is defined as the largest Euclidean distance between two points in S, $\operatorname{diam} S = \sup\{\|\hat{x} - x\| : \hat{x}, x \in S\}$. Intuitively, a function is therefore absolutely continuous if the images $f(\mathcal{I}_k)$ together stay small whenever the preimages $\mathcal{I}_k$ stay small. One can show that an absolutely continuous function F is differentiable with $\dot{F} = f$ a.e., and

$$F(t) = F(t_0) + \int_{t_0}^{t} f(t)\, dt, \quad \forall\, t \in [t_0, T].$$

The set of all absolutely continuous functions defined on $[t_0, T]$ is given by the *Sobolev space* $\mathbf{W}_{1,\infty}[t_0, T]$. The total derivatives of functions in $\mathbf{W}_{1,\infty}[t_0, T]$ are then elements of the normed vector space of essentially bounded integrable functions, $\mathbf{L}_\infty[t_0, T]$ (see example A.2). □

---

[11]. A brief summary of the theory of Lebesgue integration (or alternatively, measure theory) is provided by Kirillov and Gvishiani (1982, ch. 2); for more detail, see, e.g., Lang (1993).

Continuity and differentiability of functions at a point can be sufficient for local solvability of implicit equations. The following is a key result in nonlinear analysis.

Proposition A.7 (Implicit Function Theorem) Let $\mathcal{X}, \mathcal{Y}, \mathcal{Z}$ be normed vector spaces (e.g., $\mathbb{R}^m$, $\mathbb{R}^n$, and $\mathbb{R}^l$) such that $\mathcal{Y}$ is also complete. Suppose that $(\hat{x}, \hat{y}) \in \mathcal{X} \times \mathcal{Y}$ and let

$$\mathcal{B}_\varepsilon(\hat{x}, \hat{y}) = \{(x, y) \in \mathcal{X} \times \mathcal{Y} : \|\hat{x} - x\| + \|\hat{y} - y\| < \varepsilon\}$$

define an ε-neighborhood of the point $(\hat{x}, \hat{y})$. Assume that the function $F : \mathcal{W} \to \mathcal{Z}$ is such that $F(\hat{x}, \hat{y}) = 0$, $F(\cdot)$ is continuous at $(\hat{x}, \hat{y})$, and F is differentiable on $\mathcal{W}$ with a derivative that is continuous at $(\hat{x}, \hat{y})$. If the linear mapping $y \mapsto z = F_y(\hat{x}, \hat{y}) y$ is invertible, then there exist an open set $\mathcal{U} \times \mathcal{V}$, which contains $(\hat{x}, \hat{y})$, and a function $f : \mathcal{U} \to \mathcal{V}$, such that (1) $\mathcal{U} \times \mathcal{V} \subset \mathcal{W}$; (2) $\forall (x, y) \in \mathcal{U} \times \mathcal{V} : F(x, y) = 0 \Leftrightarrow y = f(x)$; (3) $\hat{y} = f(\hat{x})$; and (4) $f(\cdot)$ is continuous at $\hat{x}$.

Proof See, for example, Zorich (2004, vol. 2, 97–99). ∎

Remark A.3 (Implicit Smoothness) It is notable that the smoothness properties of the original function F in proposition A.7 at the point $(\hat{x}, \hat{y})$ imply the same properties of the implicit function f at $\hat{x}$ (where $f(\hat{x}) = \hat{y}$). For example, if F is continuous (resp., r-times continuously differentiable) in a neighborhood of $(\hat{x}, \hat{y})$, then f is continuous (resp., r-times continuously differentiable) in a neighborhood of $\hat{x}$. □

The implicit function theorem can be used to prove the following result, which is useful in many practical applications that involve implicit differentiation.

Proposition A.8 (Inverse Function Theorem) Let $\mathcal{X}, \mathcal{Y}$ be normed vector spaces, such that $\mathcal{Y}$ is complete. Assume that $\mathcal{G} \subset \mathcal{Y}$ is an open set that contains the point $\hat{y} \in \mathcal{Y}$, and that $g : \mathcal{G} \to \mathcal{X}$ is a differentiable function such that the derivative g_y is continuous at $\hat{y}$. If the linear transformation $y \mapsto g_y(\hat{y}) y$ is invertible, then there exist open sets $\mathcal{U}$ and $\mathcal{V}$ with $(g(\hat{y}), \hat{y}) \in \mathcal{U} \times \mathcal{V}$, such that (1) $\mathcal{U} \times \mathcal{V} \subset \mathcal{X} \times \mathcal{Y}$ and the restriction of g to $\mathcal{V}$ (i.e., the function $g : \mathcal{V} \to \mathcal{U}$) is bijective,[12] and its inverse $f : \mathcal{U} \to \mathcal{V}$ is continuous on $\mathcal{U}$ and differentiable at $\hat{x} = g(\hat{y})$, and $f_x(\hat{x}) = (g_y(\hat{y}))^{-1}$.

---

12. The function $g : \mathcal{V} \to \mathcal{U}$ is called *bijective* if it is both *injective*, i.e., $y \neq \hat{y}$ implies that $g(y) \neq g(\hat{y})$, and *surjective*, i.e., its image $g(\mathcal{V})$ is equal to its co-domain $\mathcal{U}$.

Appendix A

Analogous to the statements in remark A.3, local smoothness properties of g (e.g., continuity, r-fold differentiability) imply the corresponding local smoothness properties of f. This section concludes by providing the proof of a well-known inequality[13] that is used in chapter 2 (see proof of proposition 2.5).

Proposition A.9 (Gronwall-Bellman Inequality) Let $\alpha, \beta, x : [t_0, T] \to \mathbb{R}$ be continuous functions such that $\beta(t) \geq 0$ for all $t \in [t_0, T]$, given some $t_0, T \in \mathbb{R}$ with $t_0 < T$. If

$$x(t) \leq \alpha(t) + \int_{t_0}^{t} \beta(s) x(s) \, ds, \qquad \forall t \in [t_0, T], \tag{A.1}$$

then also

$$x(t) \leq \alpha(t) + \int_{t_0}^{t} \alpha(s) \beta(s) \exp\left[\int_{s}^{t} \beta(\theta) \, d\theta\right] ds, \qquad \forall t \in [t_0, T]. \tag{A.2}$$

Proof Let $y(t) = \int_{t_0}^{t} \beta(s) x(s) \, ds$. Then by assumption

$$\Delta(t) \equiv \alpha(t) + y(t) - x(t) \geq 0. \tag{A.3}$$

Moreover, $\dot{y}(t) = \beta(t) x(t) = \beta(t) \left(\alpha(t) + y(t) - \Delta(t) \right)$, yielding a linear IVP (with variable coefficients),

$$\dot{y} - \beta(t) y = (\alpha(t) - \Delta(t)) \beta(t), \qquad y(t_0) = 0.$$

Thus, the Cauchy formula in proposition 2.1 and inequality (A.3) together imply that

$$y(t) = \int_{t_0}^{t} (\alpha(s) - \Delta(s)) \beta(s) \exp\left[\int_{s}^{t} \beta(\theta) \, d\theta\right] ds$$

$$\leq \int_{t_0}^{t} \alpha(s) \beta(s) \exp\left[\int_{s}^{t} \beta(\theta) \, d\theta\right] ds, \tag{A.4}$$

for all $t \in [t_0, T]$. By (A.1) it is $x(t) \leq \alpha(t) + y(t)$, which, using inequality (A.4), implies (A.2). ∎

Remark A.4 (Simplified Gronwall-Bellman Inequality) (1) The statement of the inequalities in proposition A.9 simplifies when $\alpha(t) \equiv \alpha_0 \in \mathbb{R}$, in which case

---

13. This result is ascribed to Gronwall (1919) and Bellman (1953).

$$x(t) \leq \alpha_0 + \int_{t_0}^{t} \beta(s) x(s) \, ds, \quad \forall t \in [t_0, T],$$

implies that

$$x(t) \leq \alpha_0 \exp\left[\int_{t_0}^{t} \beta(\theta) \, d\theta\right], \quad \forall t \in [t_0, T].$$

(ii) If in addition $\beta(t) \equiv \beta_0 \geq 0$, then

$$x(t) \leq \alpha_0 + \beta_0 \int_{t_0}^{t} x(s) \, ds, \quad \forall t \in [t_0, T],$$

implies that

$$x(t) \leq \alpha_0 \exp[\beta_0(t - t_0)], \quad \forall t \in [t_0, T]. \quad \square$$

A.4 Optimization

In neoclassical economics, optimal choice usually corresponds to an agent's selecting a decision x from a set $\mathcal{X} \subset \mathbb{R}^n$ of feasible actions so as to maximize his objective function $f : \mathcal{X} \to \mathbb{R}$. This amounts to solving

$$\max_{x \in \mathcal{X}} f(x). \tag{A.5}$$

The following result provides simple conditions that guarantee the existence of solutions to this maximization problem.

Proposition A.10 (Weierstrass Theorem) Let $\mathcal{X} \subset \mathbb{R}^n$ be a compact set. Any continuous function $f : \mathcal{X} \to \mathbb{R}$ takes on its extrema (i.e., its minimum and maximum) in $\mathcal{X}$, that is, there exist constants $m, M \in \mathbb{R}$ and points $\underline{x}, \bar{x} \in \mathcal{X}$ such that

$$m = \min\{f(x) : x \in \mathcal{X}\} = f(\underline{x}) \quad \text{and} \quad M = \max\{f(x) : x \in \mathcal{X}\} = f(\bar{x}).$$

Proof See Bertsekas (1995, 540–541). ∎

If the objective function f is differentiable at an extremum, then there is a simple (first-order) necessary optimality condition.[14]

---

[14]. If f is concave and the domain $\mathcal{X}$ is convex, then by the *Rademacher theorem* (Magaril-Il'yaev and Tikhomirov 2003, 160) f is almost everywhere differentiable and its set-valued subdifferential $\partial f(x)$ exists at any point of x. The Fermat condition at a point $\hat{x}$ where f is not differentiable becomes $0 \in \partial f(\hat{x})$. For example, when $n = 1$ and $f(x) = -|x|$, then f attains its maximum at $\hat{x} = 0$ and $0 \in \partial f(\hat{x}) = [-1, 1]$.

Appendix A 247

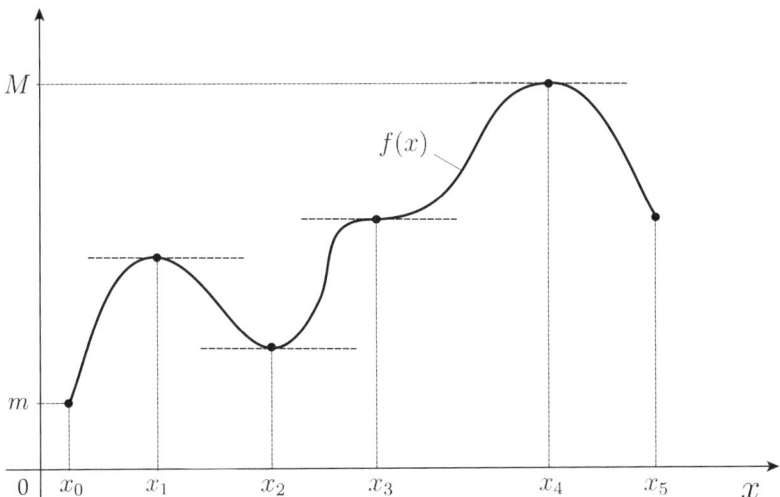

Figure A.2
The continuous function $f : [x_0, x_5] \to \mathbb{R}$ achieves its minimum m at the boundary point x_0 and its maximum M at the inner point x_4. The points x_1, x_2, x_3 and x_4 are critical points of f satisfying the condition in Fermat's lemma. The points x_1 and x_2 are local extrema.

Proposition A.11 (Fermat's Lemma) If the real-valued function $f : \mathbb{R}^n \to \mathbb{R}$ is differentiable at an interior extremum $\hat{x}$, then its first derivative vanishes at that point, $f_x(\hat{x}) = 0$.

Proof Since the function $f(x)$ is by assumption differentiable at the interior extremum $\hat{x}$, it is

$$f(\hat{x} + \Delta) - f(\hat{x}) = \langle f_x(\hat{x}), \Delta \rangle + \langle \rho(\hat{x}; \Delta), \Delta \rangle = \langle f'(\hat{x}) + \rho(\hat{x}; \Delta), \Delta \rangle,$$

where $\rho(\hat{x}; \Delta) \to 0$ as $\Delta \to 0$. If $f_x(\hat{x}) \neq 0$, then for small $\|\Delta\|$ the left-hand side of the last relation is sign-definite (since $\hat{x}$ is a local extremum), whereas the scalar product on the right-hand side can take on either sign, depending on the orientation of Δ, which yields a contradiction. Hence, necessarily $f_x(\hat{x}) = 0$. ∎

Figure A.2 illustrates the notion of extrema for a real-valued function on a compact interval. The extrema m and M, guaranteed to exist by the Weierstrass theorem, are taken on at the boundary and in the interior of the domain, respectively. The next two propositions are auxiliary results to establish the mean-value theorem for differentiable real-valued functions.

Proposition A.12 (Rolle's Theorem) Let $f : [a,b] \to \mathbb{R}$ be a continuous real-valued function, differentiable on the open interval (a,b), where $-\infty < a < b < \infty$. Suppose further that $f(a) = f(b)$. Then there exists a point $\hat{x} \in (a,b)$ such that $f_x(\hat{x}) = 0$.

Proof Since f is continuous on the compact set $\Omega = [a,b]$, by the Weierstrass theorem there exist points $\underline{x}, \bar{x} \in \Omega$ such that $f(\underline{x})$ and $f(\bar{x})$ are the extreme values of f on Ω. If $f(\underline{x}) = f(\bar{x})$, then f must be constant on Ω. If $f(\underline{x}) \neq f(\bar{x})$, then $\underline{x}$ or $\bar{x}$ must lie in (a,b), since $f(a) = f(b)$. That point is denoted by $\hat{x}$, and Fermat's lemma implies that $f_x(\hat{x}) = 0$. ∎

Proposition A.13 (Lagrange's Theorem) Let $f : [a,b] \to \mathbb{R}$ be a continuous real-valued function, differentiable on the open interval (a,b), where $-\infty < a < b < \infty$. Then there exists a point $\hat{x} \in (a,b)$ such that

$$f(b) - f(a) = f_x(\hat{x})(b - a).$$

Proof The function $\hat{f}$, with

$$\hat{f}(x) = f(x) - \frac{f(b) - f(a)}{b - a}(x - a)$$

for all $x \in [a,b]$, satisfies the assumptions of Rolle's theorem. Hence, there exists a point $\hat{x} \in (a,b)$ such that

$$\hat{f}_x(\hat{x}) = f_x(\hat{x}) - \frac{f(b) - f(a)}{b - a} = 0,$$

which completes the argument. ∎

Proposition A.14 (Mean-Value Theorem) Let $f : \mathcal{D} \to \mathbb{R}$ be a differentiable real-valued function, defined on the domain $\mathcal{D} \subset \mathbb{R}^n$. Assume that a closed line segment with endpoints $\hat{x}$ and x lies in $\mathcal{D}$, i.e., $\theta\hat{x} + (1-\theta)x \in \mathcal{D}$ for all $\theta \in [0,1]$. Then there exists $\lambda \in (0,1)$ such that

$$f(\hat{x}) - f(x) = \langle f_x(\lambda\hat{x} + (1-\lambda)x), \hat{x} - x \rangle.$$

Proof The function $F : [0,1] \to \mathbb{R}$, with

$$F(\theta) = f(\theta\hat{x} + (1-\theta)x),$$

for all $\theta \in [0,1]$, satisfies the assumptions of Lagrange's theorem. Thus, there exists $\lambda \in (0,1)$ such that

$$f(\hat{x}) - f(x) = F(1) - F(0) = F_\theta(\lambda) = \langle f_x(\lambda\hat{x} + (1-\lambda)x), \hat{x} - x \rangle,$$

which concludes the proof. ∎

Appendix A

Consider the choice problem discussed at the outset of this section, but the now continuously differentiable objective function and the choice set depend on an exogenous parameter, leading to the following *parameterized optimization problem*:

$$\max_{x \in \mathcal{X}(p)} f(x,p), \tag{A.6}$$

where $p \in \mathbb{R}^m$ is the problem parameter and

$$\mathcal{X}(p) = \{x \in \mathbb{R}^n : g(x,p) \leq 0\} \tag{A.7}$$

is the parameterized choice set, with $g : \mathbb{R}^{n+m} \to \mathbb{R}^k$ a continuously differentiable constraint function. The resulting maximized objective function (also termed value function) is

$$F(p) = \max_{x \in \mathcal{X}(p)} f(x,p). \tag{A.8}$$

Some of the most important insights in economics derive from an analysis of the comparative statics (changes) of solutions with respect to parameter movements. To quantify the change of the value function $F(p)$ with respect to p, recall the standard Lagrangian formalism for solving constrained optimization problems. In the *Lagrangian*

$$\mathcal{L}(x,p,\lambda) = f(x,p) - \langle \lambda, g(x,p) \rangle, \tag{A.9}$$

$\lambda \in \mathbb{R}^k$ is the *Lagrange multiplier* (similar to the adjoint variable in the Hamiltonian framework in chapter 3). The idea for solving the constrained maximization problem (A.6) is to relax the constraint and charge for any violation of these constraints using the vector λ of shadow prices. The *Karush-Kuhn-Tucker (necessary optimality) conditions*[15] become

$$\mathcal{L}_x(x,p,\lambda) = f_x(x,p) - \lambda' g_x(x,p) = 0, \tag{A.10}$$

together with the *complementary slackness condition*

$$g_j(x,p) > 0 \Rightarrow \lambda_j = 0, \quad \forall j \in \{1,\ldots,k\}. \tag{A.11}$$

Assuming a differentiable solution $x(p), \lambda(p)$ one obtains

$$F_p(p) = f_x(x(p),p) x_p(p) + f_p(x(p),p). \tag{A.12}$$

---

15. For the conditions to hold, the maximizer $x(p)$ needs to satisfy some constraint qualification, e.g., that the active constraints are positively linearly independent (Mangasarian-Fromovitz conditions). This type of regularity condition is used in assumptions A4 and A5 (see section 3.4) for the endpoint constraints and the state-control constraints of the general finite-horizon optimal control problem.

From the first-order necessary optimality condition (A.10),

$$f_x(x(p),p) = \sum_{j=1}^{k} \lambda_j(p) g_{j,x}(x(p),p). \tag{A.13}$$

The complementary slackness condition (A.11), on the other hand, implies that $\langle \lambda(p), g(x(p),p) \rangle = 0$. Differentiating this relation with respect to p yields

$$\sum_{j=1}^{k} \left[\lambda_{j,p}(p) g_j(x(p),p) + \lambda_j(p) \left(g_{j,x}(x(p),p) x_p(p) + g_{j,p}(x(p),p) \right) \right] = 0,$$

so that, using (A.13),

$$f_x(x(p),p) x(p) = -\sum_{j=1}^{k} \left[\lambda_{j,p}(p) g_j(x(p),p) + \lambda_j(p) g_{j,p}(x(p),p) \right].$$

Hence, the expression (A.12) becomes

$$F_p(p) = f_p(x(p),p) + \sum_{j=1}^{k} \lambda_j(p) g_{j,p}(x(p),p) = \mathcal{L}_p(x(p),p,\lambda(p)).$$

This proves the *envelope theorem*, which addresses changes of the value function with respect to parameter variations.[16]

Proposition A.15 (Envelope Theorem) Let $x(p)$ (with Lagrange multiplier $\lambda(p)$) be a differentiable solution to the parameterized optimization problem (A.6), and let $F(p)$ in (A.8) be the corresponding value function. Then $F_p(p) = \mathcal{L}_p(x(p),p,\lambda(p))$.

In the special case where no constraint is binding at a solution $x(p)$ of (A.6), the envelope theorem simply states that $F_p(p) = f_p(x(p),p)$, that is, at the optimal solution the slope of the value function with respect to a change in the parameter p is equal to the slope of the objective function with respect to p.

In general, the solution to the constrained optimization problem (A.6) is set-valued for each parameter p. The behavior of this set

$$X(p) = \arg\max_{x \in X(p)} f(x,p)$$

---

16. A more general version of this theorem was proposed by Milgrom and Segal (2002).

Appendix A   251

as a function of p cannot be described using the standard continuity and smoothness concepts.

A *set-valued function* $\varphi : \mathcal{X} \rightrightarrows \mathcal{Y}$ maps elements of the normed vector space $\mathcal{X}$ to subsets of the normed vector space $\mathcal{Y}$. The function is called *upper semicontinuous* at a point $\hat{x} \in \mathcal{X}$ if for each open set $\mathcal{B}$ with $\varphi(\hat{x}) \subset \mathcal{B}$ there exists a neighborhood $\mathcal{G}(\hat{x})$ such that

$$x \in \mathcal{G}(\hat{x}) \Rightarrow \varphi(x) \subset \mathcal{B}.$$

The following result establishes the regularity of the solution to (A.6) and its value function $F(p)$ in (A.8) with respect to changes in the parameter p.

Proposition A.16 (Berge Maximum Theorem) (Berge 1959) Assume that the functions f and g in (A.6)–(A.7) are continuous. Then the set of solutions $X(p)$ is nonempty, upper semicontinuous, and compact-valued. Furthermore, the value function $F(p)$ in (A.8) is continuous.

Proof   See Berge (1959; pp. 115–117 in 1963 English translation).   ∎

The Berge maximum theorem is a useful tool to verify the plausibility of solutions to parameterized optimization problems. First, it states that the solution set $X(p)$ cannot really make jumps, but by its upper semi-continuity can add distant solutions only via indifference. For example, if $X(p) \subseteq \{0, 1\}$ for all p, then along any path from $\check{p}$ to $\hat{p}$ in the parameter space with $x(\check{p}) = \{0\}$ and $X(\hat{p}) = \{1\}$ there will be at least one point p along the way where $X(p) = \{0, 1\}$ and the agent is indifferent between the two actions. The Berge maximum theorem is used together with the following well-known fixed-point theorem to establish the existence of a Nash equilibrium of a finite game in proposition 4.1.

Proposition A.17 (Kakutani Fixed-Point Theorem) (Kakutani 1941) Let $\mathcal{S} \neq \emptyset$ be a convex compact subset of $\mathbb{R}^n$ (where $n \geq 1$). Let $F : \mathcal{S} \rightrightarrows \mathcal{S}$ be an upper semicontinuous set-valued mapping with the property that $F(x)$ is nonempty and convex for all $x \in \mathcal{S}$. Then F has a fixed point.

Proof   See Aubin (1998, 154).   ∎

A.5   Notes

This mathematical review has established (with a few exceptions) only those results used in the main text. No effort has been made to provide any kind of completeness. A good source for linear algebra is Strang (2009) and for more advanced algebra, Hungerford (1974). There are many classic texts on analysis and calculus, such as Rudin (1976) and Apostol (1974). A modern treatment is provided by Zorich (2004) in two volumes. More advanced topics of nonlinear analysis and functional analysis are discussed in Kolmogorov and Fomin (1957), Dunford and Schwartz (1958), Aubin and Ekeland (1984), Aubin (1998), and Schechter (2004). Border (1985) covers the application of fixed-point theorems in economics, while Granas and Dugundji (2003) provide a comprehensive overview of mathematical fixed-point theory. For an introduction to optimization, see Luenberger (1969), Bertsekas (1995), and Brinkhuis and Tikhomirov (2005). More powerful results in optimization are obtained when the considered problems are convex (see, e.g., Boyd and Vandenberghe 2004).

Appendix B: Solutions to Exercises

B.1 Numerical Methods

The optimal control problems considered in this book assume that an explicit model of a dynamic system is available. The exercises at the end of chapters 2–5 provide examples of how structural insights can be obtained from analytical solutions. In many practical situations, because of the complexity of the problem at hand, it will not be possible to obtain such analytical solutions, in which case one has to rely on numerical approximations for specific parametrizations of the optimal control problem.

There are two main classes of methods to solve optimal control problems: indirect and direct. An *indirect method* relies on necessary optimality conditions provided by the Pontryagin maximum principle (PMP) to generate solution candidates, for example, by solving a two-point boundary value problem or by forward-simulating the Hamiltonian system consisting of system equation and adjoint equation for different initial values of the adjoint variable in the direction of steepest ascent of the objective functional. A *direct method* approximates trajectories by using functions that are indexed by finite-dimensional parameters and in this way converts the infinite-dimensional optimal control problem into a finite-dimensional optimization problem (of finding the best parameter values), which can be solved using commercially available nonlinear programming packages. Figure B.1 provides an overview.

Indirect Methods Given that the vast majority of the analytical solutions to optimal control problems in this book are obtained using the necessary conditions provided by the PMP, it is natural to expect that there exist successful implementations of this approach. Ignoring any

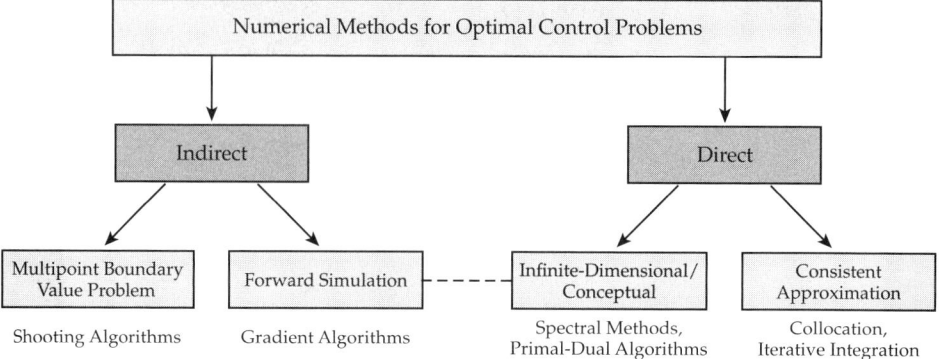

Figure B.1
Overview of numerical solution methods.

constraints, the maximum principle in proposition 3.4 for the simplified finite-horizon optimal control problem leads to a *(two-point) boundary value problem* (BVP), that is, an ordinary differential equation (ODE) that specifies endpoint conditions at both ends of the time interval $[t_0, T]$. The split boundary conditions make the solution of a BVP much more difficult than the integration of a standard initial value problem (IVP), especially when the time interval is large. To make things concrete, consider the simple optimal control problem (3.9)–(3.12) in section 3.3.1, and assume that there is a unique feedback law $u = \chi(t, x, \psi)$ such that the maximality condition

$$H(t, x, \chi(t, x, \psi), \psi) = \max_{u \in \mathcal{U}} H(t, x, u, \psi)$$

is satisfied. Then the Hamiltonian system (3.10)–(3.11) and (3.26)–(3.27) is a two-point BVP that can be written in the form

$$\dot{x}(t) = \hat{H}_\psi(t, x(t), \psi(t)), \qquad x(t_0) = x_0, \qquad (\text{B.1})$$

$$\dot{\psi}(t) = -\hat{H}_x(t, x(t), \psi(t)), \qquad \psi(T) = 0, \qquad (\text{B.2})$$

where

$$\hat{H}(t, x, \psi) \equiv H(t, x, \chi(t, x, \psi), \psi).$$

A numerical *shooting algorithm* tries to solve the BVP (B.1)–(B.2) by relaxing one of the boundary conditions, converting it to an initial condition instead, for instance, by imposing $\psi(t_0) = \alpha$ instead of $\psi(T) = 0$, yielding the IVP

Appendix B

$$\dot{x}(t) = \hat{H}_\psi(t, x(t), \psi(t)), \qquad x(t_0) = x_0, \tag{B.3}$$

$$\dot{\psi}(t) = -\hat{H}_x(t, x(t), \psi(t)), \qquad \psi(t_0) = \alpha, \tag{B.4}$$

with $(\hat{x}(t, \alpha), \hat{\psi}(t, \alpha))$, $t \in [t_0, T]$, as solution trajectories for different values of the parameter $\alpha = \hat{\alpha}$. Provided the BVP (B.1)–(B.2) possesses a solution, it may be obtained by finding $\hat{\alpha}$ such that [1]

$$\hat{\psi}(T, \hat{\alpha}) = 0.$$

The underlying reason that shooting methods work for well-posed systems is the continuous dependence of solutions to ODEs on the initial conditions (see proposition 2.5 and footnote 25 in section 3.6.1). The main drawback of using the BVP (B.1)–(B.2) to determine solution candidates for an optimal control problem is that the success invariably depends on the quality of the initial guess for the parameter α and the norm of the discretization grid (which has the same dimensionality as the costate ψ). For more details on shooting methods for solving two-point BVPs, see, for example, Roberts and Shipman (1972), Fraser-Andrews (1996), and Sim et al. (2000).

Consider now an alternative to a regular shooting method, which is useful especially in cases where $|T - t_0|$ is large, so that small perturbations of the parameter α from the unknown value $\hat{\alpha}$ can cause a large deviation of $\hat{\psi}(T, \alpha)$ from $\hat{\psi}(T, \hat{\alpha}) = 0$. For any intermediate time $\tau \in [t_0, T]$, the parameterized IVP (B.3)–(B.4) can be forward-simulated on the interval $[t_0, \tau]$ by selecting the trajectory $(x(t, \alpha), \psi(t, \alpha))$ (i.e., selecting the parameter $\alpha = \hat{\alpha}(\tau)$) that maximizes $\hat{z}(\tau, \alpha)$, which is obtained from the solution trajectory $\hat{z}(t, \alpha)$, $t \in [t_0, \tau]$, of the IVP

$$\dot{z}(t) = \hat{h}(t, x(t, \alpha), \psi(t, \alpha)), \qquad z(t_0) = 0,$$

where $\hat{h}(t, x, \psi) \equiv h(t, x, \chi(t, x, \psi))$. In this setting, the function $\hat{z}(\tau, \alpha)$ represents an approximation of the value of the objective functional when the optimal control problem is truncated to the interval $[t_0, \tau]$. Figure B.2 provides the intuition of this method, where at each $\tau \in (t_0, T)$ the parameter α is selected based on the largest attainable value of the objective function at $t = \tau$. By Bellman's principle of optimality the concatenation of optimal policies for all τ describes a solution to the optimal control problem over the entire time interval $[t_0, T]$. For more details on this *gradient method* of steepest ascent (or steepest descent

---

1. Similarly, it is possible to consider the terminal condition $x(T) = \beta$ instead of the initial condition $x(t_0) = x_0$, and then determine trajectories $(\hat{x}(t, \beta), \hat{\psi}(t, \beta))$, to determine $\beta = \hat{\beta}$ such that $\hat{x}(t_0, \hat{\beta}) = x_0$.

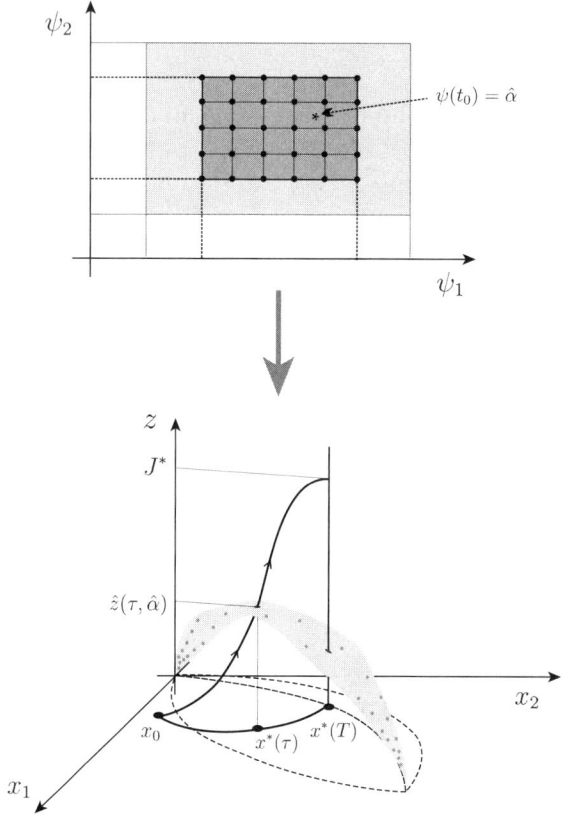

Figure B.2
Forward simulation of Hamiltonian system (B.3)–(B.4) with parameter α selected based on steepest ascent of $\hat{z}(t, \alpha)$.

for the minimization problems commonly considered in engineering) see, for instance, Lee and Markus (1967, app. A). Indirect methods may lack robustness because a numerical solution of the Hamiltonian BVP may be possible only if the initial guess of the parameter α is sufficiently close to its actual value. Note also that the necessary optimality conditions provided by the PMP are satisfied by state-control trajectories that maximize (resp. minimize) the objective functional, are only local extrema, or (as saddle points) represent no extrema at all.

Direct Methods In contrast to the indirect methods discussed earlier, direct methods attempt to maximize the objective functional, subject to the system equation and state-control constraints. For this, a direct method converts the (by its very nature) infinite-dimensional optimal control problem to a finite-dimensional nonlinear programming problem. The class of conceptual algorithms achieves this by using finite-dimensional families of admissible controls (e.g., piecewise constant functions), but typically retaining the use of infinite-dimensional operations such as integration and differentiation.[2] These operations are usually fraught with numerical errors, which may limit the accuracy of the methods.[3] Discretization techniques to approximate the original problem arbitrarily closely using sufficiently many discretization points are referred to as consistent approximation methods. So-called Galerkin methods are based on selecting appropriate collocation points (or knots), at which a discretized version of the optimal control problem is solved, usually respect to a finite set $\mathcal{Z}$ of basis functions $u(t; z)$, $z \in \mathcal{Z}$. The optimal control problem is then transcribed in terms of the finite-dimensional parameter vector $z \in \mathcal{Z}$ as a nonlinear optimization problem of the form

$$F(z) \longrightarrow \max_z, \tag{B.5}$$

$$G(z) = 0, \tag{B.6}$$

$$H(z) \le 0, \tag{B.7}$$

where the functions G and H encapsulate the state equation, endpoint constraints, and state-control constraints of an optimal control problem. The mix of ODEs and algebraic equations constraining the path leads to differential algebraic equations (DAEs), which can be solving using pseudospectral methods (transforming the problem into a different space, for example, by using the Laplace transform) or iterative integration techniques. An implementation of a collocation method based on a Runge-Kutta integration method of piecewise-polynomial functions was realized as a MATLAB toolbox, Recursive Integration Optimal Trajectory Solver (RIOTS), by Schwartz et al. (1997). Implementations of the pseudospectral methods are provided by the MATLAB toolboxes PROPT by TOMLAB (Rutquist and Edvall 2009) and DIDO (Ross and Fahroo 2003).

---

2. For references to such methods, see, e.g., Schwartz (1996).
3. Numerical errors can sometimes be avoided using more sophisticated methods such as the algorithmic differentiation technique implemented by TOMLAB in the optimal control toolbox PROPT for MATLAB.

Remark B.1 (Approximate Dynamic Programming) In a discrete-time framework and using the Hamilton-Jacobi-Bellman (HJB) equation (or the Bellman equation for infinite-horizon problems) there exist approximation methods usually referred to as approximate dynamic programming techniques. For more details, see, for example, Bertsekas (1996), Sutton and Barto (1998), and Powell (2007). □

B.2 Ordinary Differential Equations

2.1 (Growth Models)

a. (Generalized Logistic Growth) The ODE $\dot{x} = f(t, x)$ can be rewritten as a Bernoulli ODE of the form

$$\frac{d}{dt}\left(\frac{x}{\bar{x}}\right) - \alpha\gamma\left(\frac{x}{\bar{x}}\right) + \alpha\gamma\left(\frac{x}{\bar{x}}\right)^{(1/\gamma)+1} = 0,$$

so that the standard substitution $\varphi = -\gamma(x/\bar{x})^{-1/\gamma}$ leads to the linear first-order ODE

$$\dot{\varphi} + \alpha\varphi = -\alpha\gamma.$$

Taking into account the initial condition $\varphi(t_0) = -\gamma(x_0/\bar{x})^{-1/\gamma}$, the Cauchy formula yields

$$\varphi(t) = -\gamma\left(\frac{x_0}{\bar{x}}\right)^{-1/\gamma} e^{-\alpha(t-t_0)} - \alpha\gamma \int_{t_0}^{t} e^{-\alpha(t-s)} ds$$

$$= -\gamma\left[\left(\frac{x_0}{\bar{x}}\right)^{-1/\gamma} e^{-\alpha(t-t_0)} + (1 - e^{-\alpha(t-t_0)})\right],$$

so that

$$x(t) = \frac{\bar{x}}{\left[1 + \left(\frac{\bar{x}^{1/\gamma} - x_0^{1/\gamma}}{x_0^{1/\gamma}}\right) e^{-\alpha(t-t_0)}\right]^{\gamma}}, \quad t \geq t_0,$$

solves the original ODE with initial condition $x(t_0) = x_0 > 0$.

b. (Gompertz Growth) Taking the limit for $\gamma \to \infty$ on the right-hand side of the ODE for generalized logistic growth in part a, one obtains, using l'Hôpital's rule,[4]

---

4. In its simplest form, *l'Hôpital's rule* states, for two real-valued functions $f, g : \mathbb{R} \to \mathbb{R}$ and given $\hat{t} \in \mathbb{R}$, that if $\lim_{t \to \hat{t}} f(t) = g(t) \in \{0, \pm\infty\}$, then $\lim_{t \to \hat{t}} f(t)/g(t) = \lim_{t \to \hat{t}} \dot{f}(t)/\dot{g}(t)$, provided that the functions $f(\cdot)$ and $g(\cdot)$ are differentiable at $t = \hat{t}$.

Appendix B

$$\lim_{\gamma \to \infty} \alpha\gamma \left(1 - \left(\frac{x}{\bar{x}}\right)^{1/\gamma}\right) x = \alpha x \lim_{\gamma \to \infty} \frac{1 - (x/\bar{x})^{1/\gamma}}{1/\gamma}$$

$$= \alpha x \lim_{\gamma \to \infty} \frac{(1/\gamma^2)(x/\bar{x})^{1/\gamma} \ln(x/\bar{x})}{-1/\gamma^2}$$

$$= \alpha x \ln(\bar{x}/x).$$

Thus, Gompertz growth can be viewed as a limit of generalized logistic growth, for $\gamma \to \infty$. The ODE for Gompertz growth is separable, so that by direct integration

$$\ln\left(\ln\left(\frac{x}{\bar{x}}\right)\right) - \ln\left(\ln\left(\frac{x_0}{\bar{x}}\right)\right) = \int_{x_0}^{x} \frac{d\xi}{\xi \ln(\xi/\bar{x})} = -\alpha(t - t_0),$$

for all $t \geq t_0$, and therefore

$$x(t) = \bar{x}\left(\frac{x_0}{\bar{x}}\right)^{e^{-\alpha(t-t_0)}}, \quad t \geq t_0,$$

solves the IVP for Gompertz growth with initial condition $x(t_0) = x_0 > 0$.

c. (Bass Diffusion) Using the separability of $f(t, x)$, it is[5]

$$\frac{\bar{x}}{\alpha\bar{x} + \beta} \ln\left(\frac{\alpha x + \beta}{\alpha x_0 + \beta} \cdot \frac{\bar{x} - x_0}{\bar{x} - x}\right) = \int_{x_0}^{x} \frac{d\xi}{(1 - (\xi/\bar{x}))(\alpha\xi + \beta)}$$

$$= \int_{t_0}^{t} \rho(s) \, ds \equiv R(t),$$

or equivalently,

$$\frac{\alpha x + \beta}{\bar{x} - x} = \frac{\alpha x_0 + \beta}{\bar{x} - x_0} \exp\left[\left(\alpha + \frac{\beta}{\bar{x}}\right) R(t)\right].$$

Hence,

$$x(t) = \bar{x} - \frac{\alpha\bar{x} + \beta}{\alpha + \frac{\alpha x_0 + \beta}{\bar{x} - x_0} \exp\left[\left(\alpha + \frac{\beta}{\bar{x}}\right) R(t)\right]}, \quad t \geq t_0,$$

solves the Bass diffusion IVP with initial condition $x(t_0) = x_0$, provided that $0 < x_0 < \bar{x}$.

d. (Estimation of Growth Models) The preceding models are fitted to the total shipments of compact discs in the United States from 1985 to 2008,

---

5. The integral on the left-hand side is computed by a partial fraction expansion.

(see table B.1). At time $t_k = 1985 + k$, $k \in \{0, 1, \ldots, 23\}$, the number $x_k = x(t_k)$ of compact discs shipped (net, after returns) corresponds to the change $y_k \approx \dot{x}(t_k) = f(t_k, x_k) + \varepsilon_k$, where ε_k is the residual at time t_k. The model parameters are determined (e.g., using a numerical spreadsheet solver) so as to minimize the sum of squared residuals,

$$\text{SSR} = \sum_{k=0}^{23} \varepsilon_k^2 = \sum_{k=0}^{23} (y_k - f(t_k, x_k))^2.$$

Table B.2 reports the parameter values together with estimates for the standard error, $\text{SE} = \sqrt{\text{SSR}/d}$, where the number of degrees of freedom d is equal to the number of data points (24) minus the number of free model parameters, minus 1. The corresponding diffusion curves are depicted in figure B.3.

2.2 (Population Dynamics with Competitive Exclusion)

a. Consider a linear transformation of the form

$$\tau = \gamma t, \qquad x_1(\tau) = \frac{\xi_1(t)}{A_1}, \qquad x_2(\tau) = \frac{\xi_2(t)}{A_2},$$

which, when substituted in the original system of ODEs, yields

Table B.1
Total Shipments of Compact Discs in the United States, 1985–2008 (millions)

Year	1985	1986	1987	1988	1989	1990	1991	1992
Units	22.6	37.6	62.4	103.7	172.4	286.5	343.8	412.6

Year	1993	1994	1995	1996	1997	1998	1999	2000
Units	495.4	662.1	722.9	778.9	753.1	847.0	938.9	942.5

Year	2001	2002	2003	2004	2005	2006	2007	2008
Units	881.9	803.3	745.9	720.5	696.0	616.0	511.1	384.7

Sources: Recording Industry Association of America; Kurian and Chernov (2007, 323).

Table B.2
Model Parameters That Minimize SSR in Exercise 2.1d

Model	Parameters	SE
Generalized logistic growth	$\alpha = 0.1792$, $\gamma = 2.9059$, $\bar{x} = 15915.8$	40.78
Gompertz growth	$\alpha = 0.1494$, $\bar{x} = 16073.7$	73.16
Bass diffusion	$\alpha = 0.2379$, $\beta = 77.9227$, $\bar{x} = 14892.6$	49.48

Appendix B

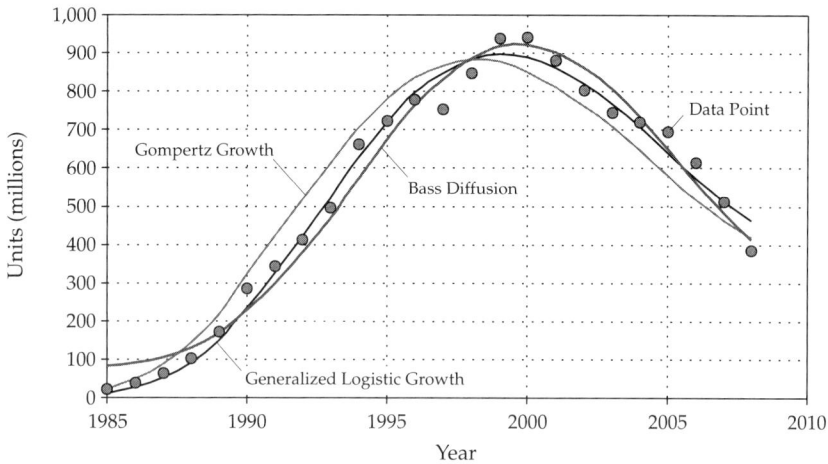

(a)

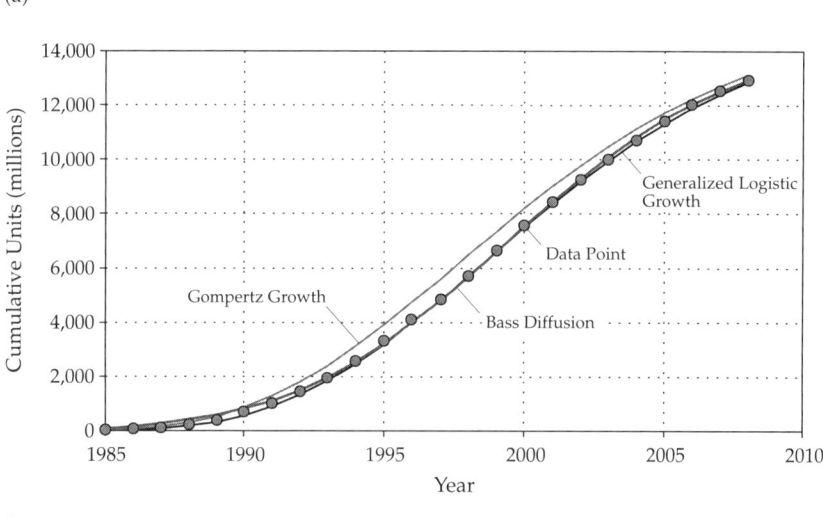

(b)

Figure B.3
Approximation of (a) shipments, and (b) cumulative shipments, of compact discs using different growth models.

$$\gamma A_1 \dot{x}_1(\tau) = a_1 A_1 x_1(\tau) \left(1 - \frac{A_1}{\bar{\xi}_1} x_1(\tau) - \frac{A_2 b_{12}}{\bar{\xi}_1} x_2(\tau)\right),$$

$$\gamma A_2 \dot{x}_2(\tau) = a_2 A_2 x_2(\tau) \left(1 - \frac{A_2}{\bar{\xi}_2} x_2(\tau) - \frac{A_1 b_{21}}{\bar{\xi}_2} x_1(\tau)\right).$$

Hence, for $\gamma = a_1$, $A_1 = \bar{\xi}_1$, and $A_2 = \bar{\xi}_2$, one obtains the desired system of ODEs, with $\beta_{12} = \frac{\bar{\xi}_2}{\bar{\xi}_1} b_{12}$, $\beta_{21} = \frac{\bar{\xi}_1}{\bar{\xi}_2} b_{21}$, and $\alpha = \frac{a_2}{a_1}$. The coefficient β_{12} determines how x_2 affects growth or decay of x_1 depending on the sign of $1 - x_{10} - \beta_{12} x_{20}$. Similarly, α and $\alpha \beta_{21}$ determine the relative growth or decay of x_2 as a function of x_2 and x_1, respectively, based on the sign of $1 - x_{20} - \beta_{21} x_{10}$.

b. The steady states (or equilibria) of the system are the roots of the system function $f(x)$. Indeed, it is $f(\bar{x}) = 0$ if and only if

$$\bar{x} \in \left\{ (0,0), (0,1), (1,0), \left(\frac{1 - \beta_{12}}{1 - \beta_{12} \beta_{21}}, \frac{1 - \beta_{21}}{1 - \beta_{12} \beta_{21}}\right) \right\}.$$

From this it can be concluded that there is a (unique) equilibrium in the strict positive quadrant $\mathbb{R}^2_{++}$ if (provided that $\beta_{12} \beta_{21} \neq 1$)

$$\min\{(1 - \beta_{12})(1 - \beta_{12} \beta_{21}), (1 - \beta_{21})(1 - \beta_{12} \beta_{21})\} > 0.$$

Figure B.4 depicts the corresponding values of β_{12} and β_{21}.

c. Using the linearization criterion one can examine, based on the eigenvalues of the Jacobian matrix $f_x(\bar{x})$, the local stability properties of any equilibrium point $\bar{x} \in \mathbb{R}^2_+$ (as determined in part b). Indeed, it is

$$f_x(x) = \begin{pmatrix} 1 - 2x_1 - \beta_{12} x_2 & -\beta_{12} x_1 \\ -\alpha \beta_{21} x_2 & \alpha(1 - 2x_2 - \beta_{21} x_1) \end{pmatrix}.$$

Thus

- $\bar{x} = 0$ is unstable, since

$$f_x(\bar{x}) = \begin{pmatrix} 1 & 0 \\ 0 & \alpha \end{pmatrix}$$

has the positive eigenvalues $\lambda_1 = 1$ and $\lambda_2 = \alpha$.
- $\bar{x} = (0, 1)$ is stable if $\beta_{12} > 1$, since

$$f_x(\bar{x}) = \begin{pmatrix} 1 - \beta_{12} & 0 \\ -\alpha \beta_{21} & -\alpha \end{pmatrix}$$

Appendix B

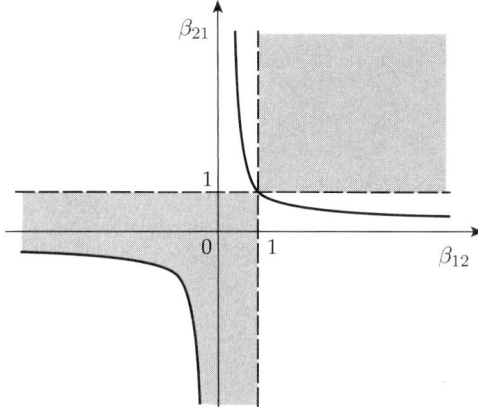

Figure B.4
Parameter region (shaded) that guarantees a positive equilibrium for the system of exercise 2.2.

has the negative eigenvalues $\lambda_1 = -\alpha$ and $\lambda_2 = 1 - \beta_{12}$.

- $\bar{x} = (1, 0)$ is stable if $\beta_{21} > 1$, since

$$f_x(\bar{x}) = \begin{pmatrix} -1 & -\beta_{12} \\ 0 & \alpha(1 - \beta_{21}) \end{pmatrix}$$

has the negative eigenvalues $\lambda_1 = -1$ and $\lambda_2 = \alpha(1 - \beta_{21})$.

It was shown in part b when β_{12}, β_{21} are positive, that $\bar{x} = \left(\frac{1-\beta_{12}}{1-\beta_{12}\beta_{21}}, \frac{1-\beta_{21}}{1-\beta_{12}\beta_{21}}\right)$ exists as a positive equilibrium if and only if $(\beta_{12} - 1)(\beta_{21} - 1) > 0$, that is, both parameters are either greater or smaller than 1. The equilibrium $\bar{x}$ is stable if the eigenvalues of $f_x(\bar{x})$,

$$\lambda_i \in \left\{ \frac{(\beta_{12} - 1) + \alpha(\beta_{21} - 1) \pm \sqrt{a^2 - 4\alpha b}}{2(1 - \beta_{12}\beta_{21})} \right\}, \quad i \in \{1, 2\},$$

where $a = \beta_{12} - 1 + \alpha(\beta_{21} - 1)$ and $b = (1 - \beta_{12}\beta_{21})(\beta_{12} - 1)(\beta_{21} - 1)$, have negative real parts (i.e., for $\beta_{12}, \beta_{21} < 1$).

d. Figure B.5 depicts phase diagrams for the cases where either $\beta_{12}, \beta_{21} > 1$ or $0 < \beta_{12}, \beta_{21} < 1$. In the unstable case a *separatrix* divides the state space into initial (nonequilibrium) states from which the system converges either to exclusion of the first population with limit state $(0, 1)$ or to exclusion of the second population with limit state $(1, 0)$. Similarly, in

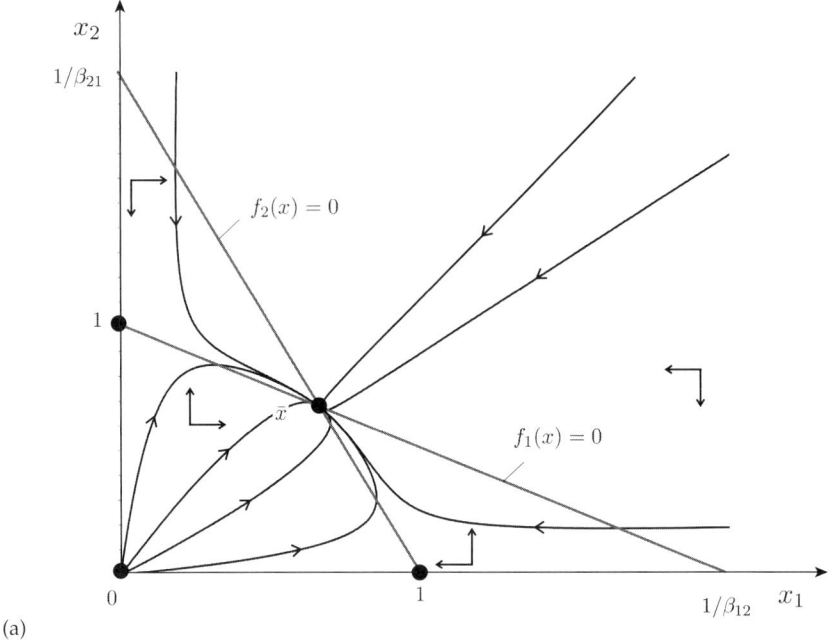

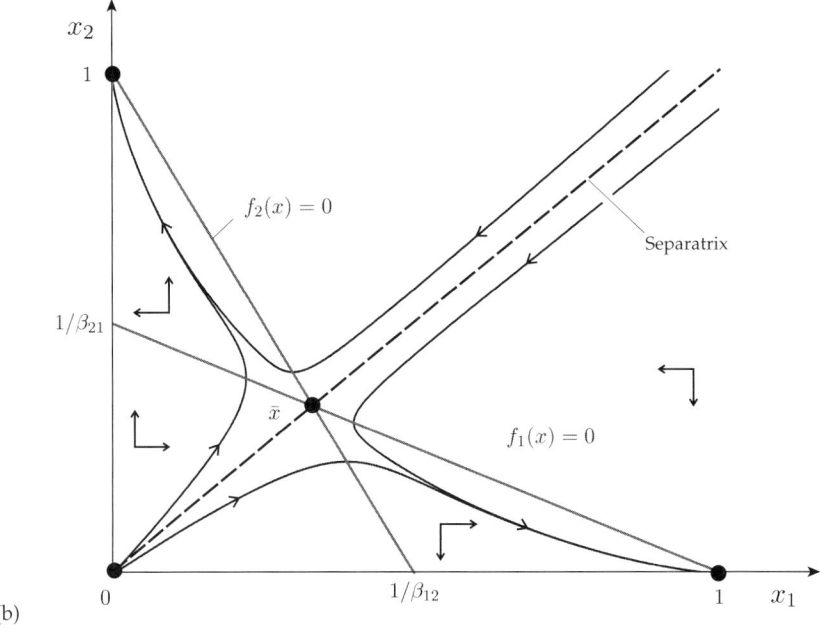

Figure B.5
Phase diagrams for the system of exercise 2.2d: (*a*) stable case where $\beta_{12}\beta_{21} < 1$, and (*b*) unstable case where $\beta_{12}\beta_{21} > 1$.

Appendix B

the absence of a positive steady state, that is, when $(\beta_{12} - 1)(\beta_{21} - 1) < 0$, the system trajectory $x(t)$ converges either to $(0, 1)$ (for $\beta_{12} > 1$) or to $(1, 0)$ (for $\beta_{21} > 1$) as $t \to \infty$.

e. The nature of the population coexistence (in the sense of competition versus cooperation) is determined by the existence of a stable positive steady state. As seen in parts c and d, coexistence is possible if $0 < \beta_{12}, \beta_{21} < 1$. As a practical example, one can think of the state variables x_1 and x_2 as the installed bases for two firms that produce two perfectly substitutable products. For almost all initial values, either the first or the second firm will monopolize the market.

2.3 (Predator-Prey Dynamics with Limit Cycle)

a. As in exercise 2.2, a steady state $\bar{x}$ of the system is such that $f(\bar{x}) = 0$, which is equivalent to

$$\bar{x} \in \left\{ (0,0), (1,0), \left(\frac{a - \sqrt{b}}{2}, \frac{a - \sqrt{b}}{2} \right), \left(\frac{a + \sqrt{b}}{2}, \frac{a + \sqrt{b}}{2} \right) \right\},$$

where $a = 1 - \beta - \delta$ and $b = a^2 + 4\beta$. Note that since $\beta, \delta > 0$, only the last solution is positive.

b. Using the linearization criterion, one can determine the stability of the positive steady state by examining the eigenvalues λ_1, λ_2 of the system matrix

$$f_x(\bar{x}) = - \begin{pmatrix} \left(1 - \frac{\delta \bar{x}_1}{(\bar{x}_1 + \beta)^2}\right) \bar{x}_1 & \frac{\delta \bar{x}_1}{\bar{x}_1 + \beta} \\ -\alpha & \alpha \end{pmatrix}.$$

This matrix is asymptotically stable if $\text{Re}(\lambda_1), \text{Re}(\lambda_2) < 0$, that is, if

$$\text{trace}(f_x(\bar{x})) = \lambda_1 + \lambda_2 = -\left(1 - \frac{\delta \bar{x}_1}{(\bar{x}_1 + \beta)^2}\right) \bar{x}_1 - \alpha < 0,$$

and

$$\det(f_x(\bar{x})) = \lambda_1 \lambda_2 = \alpha \left(1 + \frac{\beta \delta}{(\bar{x}_1 + \beta)^2}\right) \bar{x}_1 > 0.$$

The second inequality is always satisfied, which means that both eigenvalues always have the same sign. Thus, in the generic case where the eigenvalues are nonzero the system is either asymptotically stable or unstable, depending on whether the first inequality holds or not. Figure B.6 shows the (α, β, δ)-region for which $f_x(\bar{x})$ is asymptotically stable. In this region, it is $\delta \geq 1/2$.

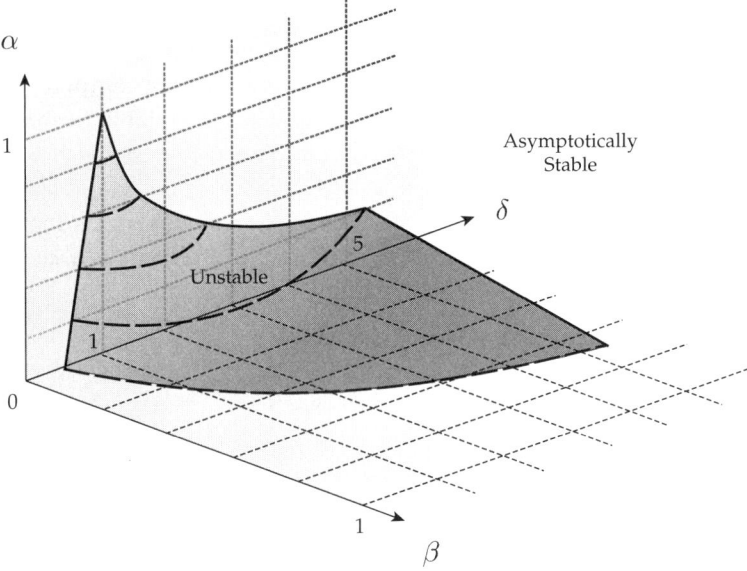

Figure B.6
Parameter region that guarantees asymptotic stability for the unique positive equilibrium of exercise 2.3b.

c. Consider a parameter tuple (α, β, δ) such that the system is unstable, as determined in part b, for instance, $\alpha = \beta = 1/5$ and $\delta = 2$ (see figure B.6). Given the system function $f(x) = (f_1(x), f_2(x))$, the two nullclines are given by the set of states x for which $f_1(x) = 0$ and $f_2(x) = 0$, respectively. These nullclines divide the state space into four regions where the direction of the state trajectories is known. Note that the positive equilibrium $\bar{x}$ is located at the intersection of these nullclines. Since both eigenvalues of the system have positive real parts, no trajectory starting at a point x_0 different from $\bar{x}$ can converge to $\bar{x}$. Thus, if there is a nonempty compact invariant set $\Omega \subset \mathbb{R}^2_{++}$, such that any initial condition $x(0) = x_0 \in \Omega$ implies that $x(t) = \phi(t, x_0) \in \Omega$ for all $t \geq 0$, then by the Poincaré-Bendixson theorem the set Ω must contain a nontrivial limit cycle.[6] Figure B.7 shows that the invariant set Ω does indeed exist, and a formal proof (which would consist in explicitly checking the direction of the vector field that generates the flow of the system at the boundary of Ω) can be omitted.

---

6. By Lemma 2.4 the positive limit set $\mathcal{L}^+_{x_0}$ of x_0 is nonempty, compact, and invariant.

Appendix B

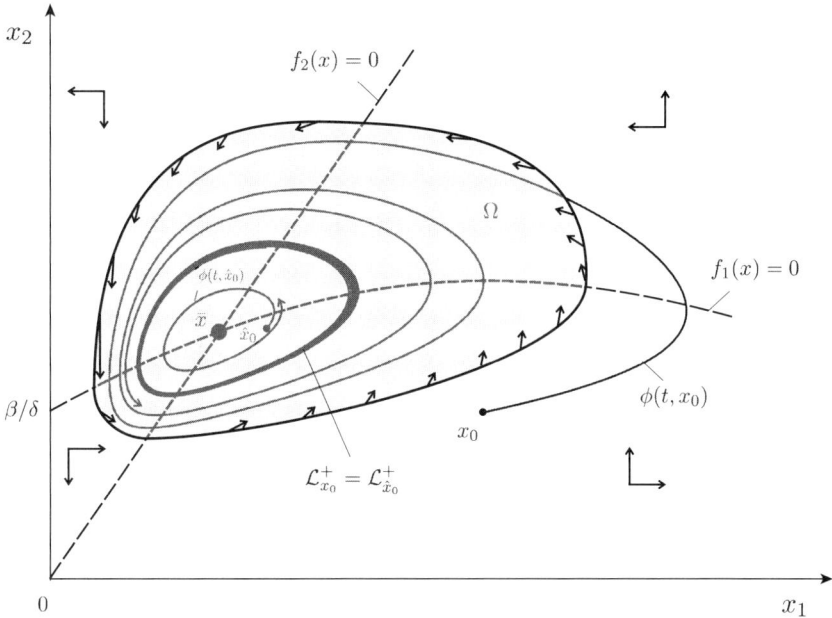

Figure B.7
Invariant set Ω and (stable) limit cycle for the system of exercise 2.3c when the positive equilibrium $\bar{x}$ is unstable.

d. In the Lotka-Volterra model (see example 2.8), the system trajectory starts at a limit cycle for any given initial condition, whereas in this model the system trajectory approaches a certain limit cycle as $t \to \infty$. Thus, in the Lotka-Volterra model, a slight change in the initial conditions results in a consistently different trajectory, whereas the model considered here exhibits a greater robustness with respect to the initial condition because of the convergence of trajectories to a zero-measure set of limit points.

2.4 (Predator-Prey Dynamics and Allee Effect)

Part 1: System Description

a. Consider, for example, the functions

$$g(x_1) = \exp[-\alpha(x - \hat{x}_1)^2] \quad \text{and} \quad h(x_2) = \sqrt{1 + \beta x_2},$$

which satisfy the given assumptions for any desired $\hat{x}_1 > 0$ as long as $0 < \alpha \leq \bar{\alpha} = \frac{1}{2(\hat{x}_1)^2}$ and $\beta > 0$. The classical Lotka-Volterra predator-prey system obtains by setting $\alpha = \beta = 0$.

The right-hand sides of the two ODEs are zero, if $x_2 = g(x_1)$ and $x_1 = h(x_2)$, respectively. These relations describe the nullclines of the system, the intersection of which determines the equilibria of the system (see part 2). Depending on the particular parametrization of the functions $g(\cdot)$ and $h(\cdot)$, the nullclines may intersect in $\mathbb{R}^2_{++}$ either before or after $\hat{x}_1$. For example, using the parametrization for $\hat{x}_1 = 5$, $\alpha = \bar{\alpha} = 1/50$, we have that for $\beta = 1/10$ the only positive equilibrium is at $\bar{x} \approx (1.0359, 0.7303)$, and for $\beta = 1$ it is $\bar{x} \approx (1.3280, 0.7636)$.

Figure B.8 depicts the phase diagrams for $\beta \in \{1/10, 1\}$. In contrast to the classical Lotka-Volterra system, depending on the parameter values one may obtain either a stable (for $\beta = 1$) or an unstable (for $\beta = 1/10$) positive steady state.

b. Since $g(x_1)$ is strictly increasing on $[0, \hat{x}_1]$, the relative growth rate of x_1 (i.e., $g(x_1) - x_2$) increases with the population size x_1 and decreases as x_1 passes $\hat{x}_1$, finally approaching zero. This phenomenon is known as the (weak) *Allee effect* and does not appear in the classical Lotka-Volterra model, where $g(x_1) \equiv 1$ is constant.

Part 2: Stability Analysis

c. From figure B.8 it is apparent that the origin is not a stable equilibrium. In order to prove this claim formally, consider the linearized system

$$\dot{x} = f_x(0) x = \begin{bmatrix} g(0) & 0 \\ 0 & -h(0) \end{bmatrix} x.$$

The system matrix $f_x(0)$ has the eigenvalues $g(0)$ and $-h(0)$, the first of which is by assumption positive. This implies that the origin is unstable.

d. If $\bar{x} = (\bar{x}_1, \bar{x}_2) \in \mathbb{R}^2_{++}$ is an equilibrium, then it must be a positive solution of

$x_2 = g(x_1),$

$x_1 = h(x_2).$

Concerning existence, note first that $h(\cdot)$ is by assumption increasing and concave, so that its (increasing and convex) inverse $h^{-1}(\cdot)$ exists. Thus, $x_2 = h^{-1}(x_1)$, and a solution can be obtained, provided that

$h^{-1}(x_1) = g(x_1).$

Appendix B

(a)
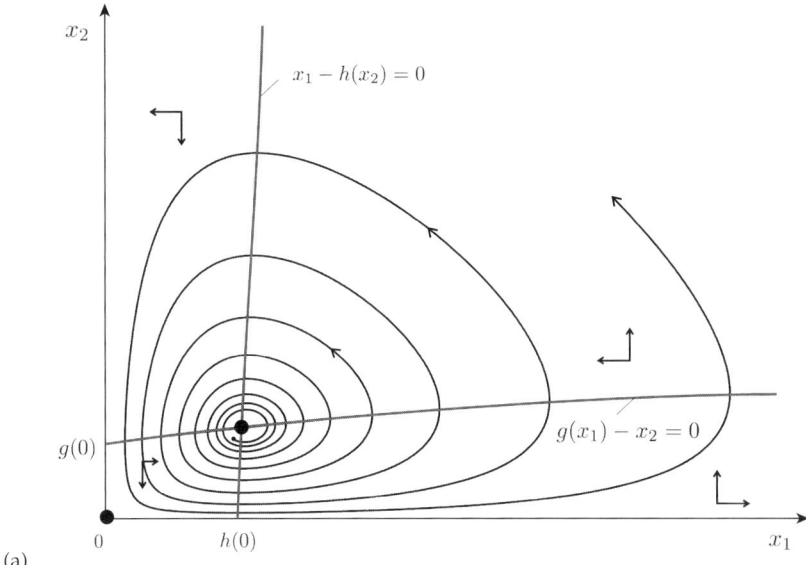

(b)
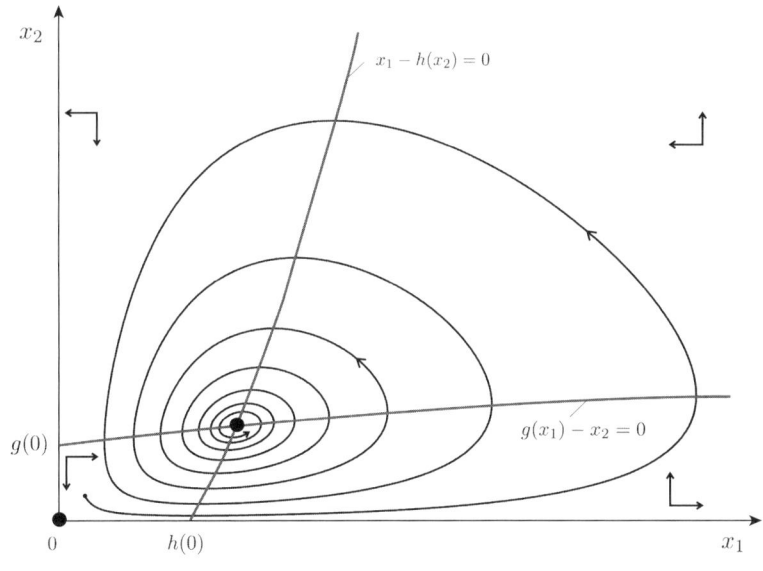

Figure B.8
Predator-prey dynamics of exercise 2.4: (*a*) unstable equilibrium ($\beta = 0.1$), and (*b*) stable equilibrium ($\beta = 1$).

But the previous equation possesses a solution $\bar{x}_1 > h(0)$, since $h^{-1}(h(0)) = 0 < g(x_1)$ for all $x_1 \geq h(0)$, and at the same time $h^{-1}(x_1)$ is increasing while $\lim_{x_1 \to \infty} g(x_1) = 0$ by assumption. Hence, a positive equilibrium $\bar{x} \gg 0$ exists. Concerning uniqueness, note that if the smallest solution $\bar{x}_1 \leq \hat{x}_1$, then since $g(x_1)$ is concave by assumption, the function $h(g(x_1))$ is increasing and concave. Therefore the fixed-point problem $x_1 = h(g(x_1)) = (h \circ g)(x_1)$ has a unique solution on $[h(0), \hat{x}_1]$ (e.g., by the Banach fixed-point theorem (proposition A.3), since the concave function $h \circ g$ is a contraction mapping on the compact domain $[h(0), \hat{x}_1]$). On the other hand, if the smallest solution $\bar{x}_1 > \hat{x}$, there cannot be any further solutions because $h^{-1}(x_1)$ is increasing and $g(x_1)$ decreasing for $x_1 > \hat{x}_1$.

The linearization criterion provides sufficient conditions for the stability of the positive equilibrium $\bar{x}$. The linearized system is of the form

$$\dot{x} = f_x(\bar{x}) x = \begin{bmatrix} \bar{x}_1 g'(\bar{x}_1) & -\bar{x}_1 \\ \bar{x}_2 & -\bar{x}_2 h'(\bar{x}_2) \end{bmatrix} x.$$

The system matrix $f_x(\bar{x})$ is Hurwitz if (using the equilibrium relations)

$$\frac{g'(\bar{x}_1)}{g(\bar{x}_1)} < \frac{h'(\bar{x}_2)}{h(\bar{x}_2)} \quad \text{and} \quad g'(\bar{x}_1) h'(\bar{x}_2) < 1,$$

which is sufficient for the (local asymptotic) stability of the nonlinear system $\dot{x} = f(x)$ at $x = \bar{x}$. Note that both inequalities are satisfied if $\bar{x}_1 > \hat{x}_1$. On the other hand, if

$$\max \left\{ \frac{g'(\bar{x}_1)}{g(\bar{x}_1)} - \frac{h'(\bar{x}_2)}{h(\bar{x}_2)}, g'(\bar{x}_1) h'(\bar{x}_2) - 1 \right\} > 0,$$

then one of the eigenvalues of $f_x(\bar{x})$ has a positive real part, which implies by the linearization criterion that $\bar{x}$ must be an unstable equilibrium of the nonlinear system $\dot{x} = f(x)$.

Remark B.2 For the particular parametrization of g and h in exercise 2.4a, we obtain that $g'(x_1)/g(x_1) = 2\alpha(\hat{x}_1 - x_1)$ and $h'(x_2)/h(x_2) = \frac{2}{\beta(1+\beta x_2)}$. When $\alpha = 1/50$, $\hat{x}_1 = 5$, then the conditions yield instability of the unique positive equilibrium for $\beta = 1/5$ and stability for $\beta \in \{1, 100\}$. □

Part 3: Equilibrium Perturbations and Threshold Behavior

e. As in part d, the fact that $\bar{x}_1 > \hat{x}_1$ implies that the positive equilibrium $\bar{x}$ is stable. Furthermore, since $x_0 > \bar{x}$, we have $\dot{x}_1 < 0$, and the system will start going back toward $\bar{x}_1$. Depending on the value of x_2, the system may or may not converge to $\bar{x}$ without passing by $\hat{x}_1$. In the

Appendix B

situation where the state trajectory passes over $\hat{x}_1$, which happens when $\|x_0 - \bar{x}\|$ is large enough, $\dot{x}_1$ starts to decrease on $[0, \hat{x}_1]$. This will continue until x_2 becomes less than $g(x_1)$ which makes $\dot{x}_1$ positive. Therefore, passing $\hat{x}_1$ the state trajectory will cause the system to orbit around the steady state until it converges to $\bar{x}$. When the state trajectory does not pass $\hat{x}_1$, it converges to $\bar{x}$ immediately.

f. If the conditions given in part d are satisfied, that is, $\bar{x}$ is a stable steady state, and if $\|x_0 - \bar{x}\|$ is small enough, by the definition of a stable steady state, the state trajectory will converge to $\bar{x}$.[7] However, if $\bar{x}$ is not stable, since $\mathcal{X} = \mathbb{R}^2_{++}$ is an invariant set for the system, there exists a nonempty compact invariant subset Ω of $\mathcal{X}$ that does not include any steady state. Consequently, by the Poincaré-Bendixson theorem there exists a limit cycle to which the state trajectory converges.

Part 4: Interpretation and Verification

g. One can easily show that the relative growth rate of population x_1 increases for $x_1 < \hat{x}_1$ and decreases for $x_1 > \hat{x}_1$. Furthermore, the larger the population x_2, the slower the growth rate of x_1. On the other hand, the relative growth rate of population x_2 increases with x_1. Consequently, one can interpret x_1 as a platform for the growth of x_2. For example, if x_1 represents the number of highways in a state and x_2 denotes the number of cars in that state, then the more highways, the greater the number of cars. As the number of highways increases, the need for additional highways will diminish. This reflects the property of $g(x_1)$ which decreases when $x_1 > \hat{x}_1$. Moreover, having a large number of cars increases the highway depreciation through overuse, which justifies the factor $-x_2$ in the relative growth rate of x_1.

h. See parts a and d.

B.3 Optimal Control Theory

3.1 (Controllability and Golden Rule)

a. Since one can always consider the control γu_2 instead of u_2, it is possible to set $\gamma = 1$, without any loss of generality. Moreover, using the substitution

$$\hat{u}_2 = \frac{\gamma u_2}{1 - x_2} \in [0, 1],$$

---

[7]. If the perturbation is large, the state trajectory may not converge to $\bar{x}$. In this case, it is possible to have a limit cycle to which the state trajectory converges.

the system dynamics remain unchanged, since

$$D(x, u_2) = \max\{0, 1 - x_2 - \gamma u_2\}(\alpha_2 x_1 + \alpha_3 x_2)$$
$$= (1 - x_2)(\alpha_2 x_1 + \alpha_3 x_2)(1 - \hat{u}_2) \equiv \hat{D}(x_2, \hat{u}_2).$$

Remark B.3 The objective functional in exercise 3.1d with the transformed control variable $\hat{u} = (\hat{u}_1, \hat{u}_2)$, where $\hat{u}_1 = u_1$, becomes

$$\hat{J}(\hat{u}) = \gamma \int_0^\infty e^{-rt}((1 - x_2(t))\,\hat{u}_2(t)\,\hat{D}(x, \hat{u}_2(t)) - \hat{c}\,\hat{u}_1(t))\,dt,$$

where $\hat{c} = c/\gamma$. □

b. With the new control variable $\hat{u} \in \hat{\mathcal{U}} = [0, \bar{u}_1] \times [0, 1]$ as in part a, the system is described by the ODE

$$\dot{x} = \begin{bmatrix} -\alpha_1 x_1 + \hat{u}_1 \\ \hat{D}(x, \hat{u}_2) - \beta x_2 \end{bmatrix} \equiv f(x, \hat{u}), \quad (B.8)$$

with initial state $x_0 = (x_{10}, x_{20}) \in (0, \bar{u}_1/\alpha_1) \times (0, 1)$. The components v_1, v_2 of the vector field on the right-hand side of (B.8) are contained in $[v_{1\min}, v_{1\max}]$ and $[v_{2\min}, v_{2\max}]$, respectively, where $v_{1\min} = -\alpha_1 x_1$, $v_{1\max} = -\alpha_1 x_1 + \bar{u}_1$, $v_{2\min} = -\beta x_2$, $v_{2\max} = (1 - x_2)(\alpha_2 x_1 + \alpha_3 x_2) - \beta x_2$. The bounding box $[v_{1\min}, v_{1\max}] \times [v_{2\min}, v_{2\max}]$ describes the vectogram $f(x, \hat{\mathcal{U}})$. Thus, taking into account the possibility of control, the dynamics of the system are captured by the differential inclusion

$$\dot{x} \in f(x, \hat{\mathcal{U}}).$$

Given a set-valued vector field on the right-hand side, useful phase diagrams are those at the corner points of the bounding box, notably where $f \in \{(v_{1\min}, v_{2\min}), (v_{1\max}, v_{2\max})\}$. Figures B.9a and B.9b depict phase diagrams for the cases where $\beta < \alpha_2$ and $\beta \geq \alpha_3$, respectively.

c. Let $\bar{x}_1 = \bar{u}_1/\alpha_1$. To determine a set $\mathcal{C}$ of controllable states, first note that since $\dot{x}_1 \in [v_{1\min}, v_{1\max}] = [-\alpha_1 x, \alpha_1 x + \bar{u}_1]$, the advertising effect x_1 can be steered from any $x_{10} \in (0, \bar{x}_1)$ to any $\hat{x}_1 \in (0, \bar{x}_1)$ in finite time, independent of x_2 and $\hat{u}_2$. One can do so in the fastest manner using a constant control $\hat{u}_1 \in \{0, \bar{u}_1\}$ such that $(\hat{x}_1 - x_{10})\,\hat{u}_1 \geq 0$, which (using the Cauchy formula) yields the trajectory

$$x_1(t) = x_{10} e^{-\alpha_1 t} + \frac{\hat{u}_1}{\alpha_1}(1 - e^{-\alpha_1 t}), \quad \forall t \in [0, T_1],$$

Appendix B

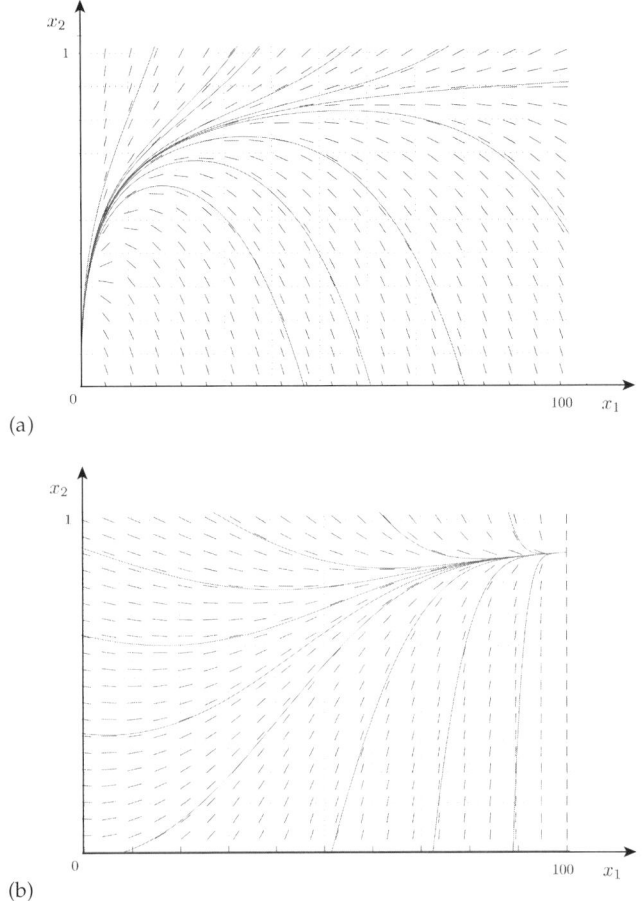

(a)

(b)

Figure B.9
Phase diagram of the system of exercise 3.1 with parameters $(\alpha_1, \alpha_2, \alpha_3, \beta, \gamma) = (1, .05, .1, .6)$: (a) $\hat{u}(t) \equiv (0, 0)$; (b) $\hat{u}(t) \equiv (\bar{u}_1, 0)$; (c) $\hat{u}(t) \equiv (0, 1)$; and (d) $\hat{u}(t) \equiv (\bar{u}_1, 1)$.

where $T_1 = (\frac{1}{\alpha_1}) \ln \left(\frac{\alpha_1 x_{10} - \hat{u}_1}{\alpha_1 \hat{x}_1 - \hat{u}_1} \right)$. Regarding the controllability of x_2, because of the negative drift term $-\beta x_2$ the set of installed-base values $\hat{x}_2$ that can be reached from a given state $x_{20} \in (0, 1)$ in finite time is limited. Indeed,

$$v_{2\max} \geq 0 \Leftrightarrow (1 - x_2)(\alpha_2 x_1 + \alpha_3 x_2) \geq \beta x_2 \Leftrightarrow x_2 \leq \bar{x}_2(x_1),$$

where

$$\bar{x}_2(x_1) = \frac{\alpha_3 - \alpha_2 x_1 - \beta + \sqrt{(\alpha_3 - \alpha_2 x_1 - \beta)^2 + 4\alpha_2 \alpha_3 x_1}}{2\alpha_3}.$$

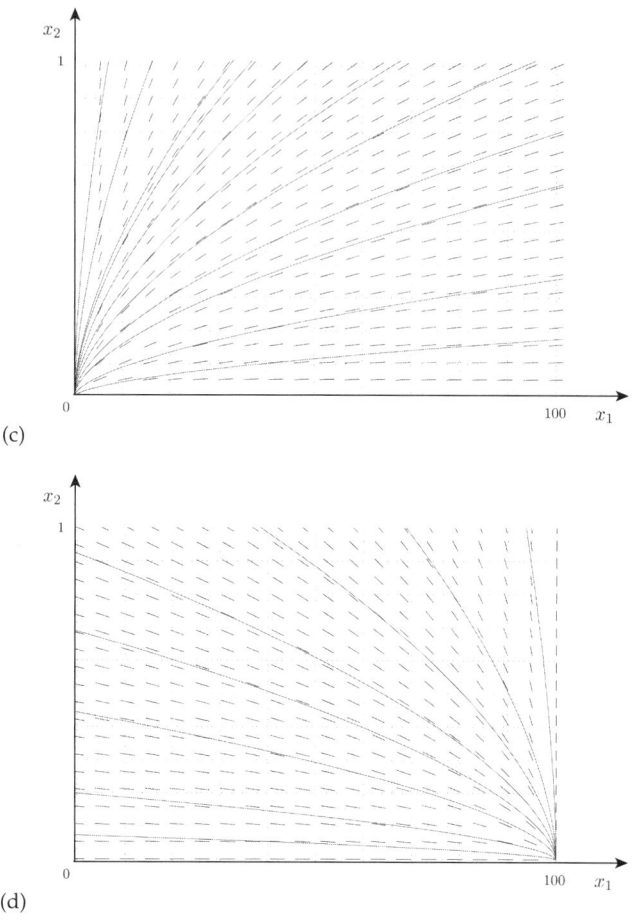

(c)

(d)

Figure B.9
(*continued*)

Thus, if the target state $\hat{x} = (\hat{x}_1, \hat{x}_2)$ is such that $\hat{x}_2 \in (0, \bar{x}_2(\hat{x}_1))$, it is possible to steer the system first to the state $(x_{10}, \hat{x}_2)$ in finite time T_2 by applying the control $\hat{u} = (\alpha x_{10}, 0)$, keeping the marketing effect constant and charging zero price, and then to keep installed base at the target level $\hat{x}_2$ by using $\hat{u}_2 = \beta x_2 (1 - x_2)^{-1} (\alpha_2 x_1 + \alpha_3 x_2)^{-1}$ while adjusting the marketing effect from x_{10} to $\hat{x}_1$ using a constant marketing effort $\hat{u}_1$ (either zero or $\bar{u}_1$) as described earlier.

Hence, it is possible to choose $\mathcal{C}_\varepsilon = \{(x_1, x_2) \in \mathbb{R}_+^2 : \varepsilon \leq x_1 \leq \bar{x}_1 - \varepsilon, \varepsilon \leq x_2 \leq \bar{x}_2(x_1) - \varepsilon\}$ as $\mathcal{C}$, given any sufficiently small $\varepsilon > 0$ (figure B.10).

Appendix B

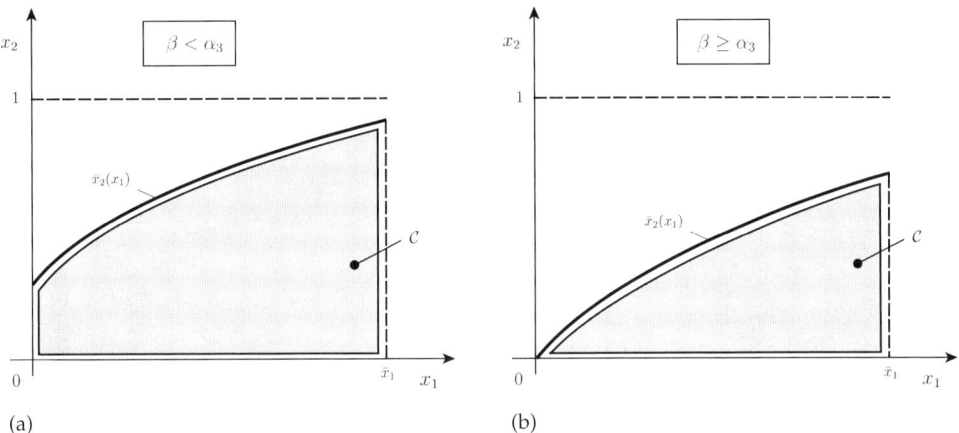

Figure B.10
Set of controllable states $\mathcal{C}$: (a) when $\beta < \alpha_3$, and (b) when $\beta \geq \alpha_3$.

Moreover, the (open) set of states that can be reached from an initial state $x_0 = (x_{10}, x_{20}) \in (0, \bar{x}_1) \times (0, \bar{x}_2(x_{10}))$

$$\mathcal{R} = \{(x_1, x_2) \in \mathbb{R}_+^2 : 0 < x_1 < \bar{x}_1,\ 0 < x_2 < \bar{x}_2(x_1)\}.$$

When the initial state $x(0) = x_0 \gg 0$ is not in $\mathcal{C}$, then the trajectory will enter the invariant set $\mathcal{R}$ in finite time and remain there forever. Note also that $\lim_{\varepsilon \to 0^+} \mathcal{C}_\varepsilon = \mathcal{R}$.

d. Clearly, the system equation (B.8) implies that the set $\mathcal{X} = [0, \bar{x}_1] \times [0, 1]$ is a compact invariant subset of the state space $\mathbb{R}_+ \times [0, 1]$. Therefore, intuitively, the optimal solution for the infinite-horizon profit-maximization problem, if it exists, should approach either a limit cycle or a steady state. The dynamically optimal steady-state solution, given its existence, should satisfy the necessary conditions given by the PMP. The current-value Hamiltonian is

$$\hat{H}(t, x, \hat{u}, \nu) = (1 - x_2)\hat{u}_2 \hat{D}(x, \hat{u}_2) - c\hat{u}_1$$
$$+ \nu_1(-\alpha_1 x_1 + \hat{u}_1) + \nu_2(\hat{D}(x, \hat{u}_2) - \beta x_2),$$

which is maximized with respect to $\hat{u} \in \hat{\mathcal{U}}$ by

$$\hat{u}_1^* \in \begin{cases} \{\bar{u}_1\} & \text{if } \nu_1 > c, \\ [0, \bar{u}_1] & \text{if } \nu_1 = c, \\ \{0\} & \text{otherwise,} \end{cases}$$

and

$$\hat{u}_2^* = \max\left\{0, \min\left\{1, \frac{1}{2}\left(1 - \frac{v_2}{1-x_2}\right)\right\}\right\}.$$

The adjoint equations are

$$\dot{v}_1 = (r + \alpha_1)v_1 - \alpha_2(1-x_2)(1-\hat{u}_2)v_2$$
$$\quad - \alpha_2(1-x_2)^2(1-\hat{u}_2)\hat{u}_2,$$
$$\dot{v}_2 = ((\alpha_2 x_1 + 2\alpha_3 x_2 - \alpha_3)(1-\hat{u}_2) + r + \beta)v_2$$
$$\quad + (1-x_2)(2\alpha_2 x_1 + 3\alpha_3 x_2 - \alpha_3)(1-\hat{u}_2)\hat{u}_2.$$

Provided that the dynamically optimal steady-state solution is interior, by the maximality of the Hamiltonian it is $v_1 = c$ and $\hat{u}_2 = -\frac{v_2}{2(1-\hat{x}_2)}$. From $\dot{x} = 0$ one can conclude $\hat{u}_1 = \alpha_1 \hat{x}_1$ and $\hat{D}(\hat{x}, \hat{u}_2) = \beta \hat{x}_2$, so

$$\alpha_2 \hat{x}_1 = \frac{2\beta - \alpha_3(1 - \hat{x}_2 + v_2)}{1 - \hat{x}_2 + v_2}\hat{x}_2,$$

and by $\dot{v} = 0$,

$$0 = (r + \alpha_1)c - \frac{\alpha_2}{4}(1 - \hat{x}_2 + v_2)^2,$$

$$0 = (r + \beta)v_2 - \frac{\alpha_3}{4}(1 - \hat{x}_2 + v_2)^2 + \beta \hat{x}_2,$$

and thus, after back-substitution, the dynamically optimal steady state (turnpike)

$$\hat{x}_1 = \hat{x}_2\left[\frac{\beta}{\sqrt{\alpha_2 c(r + \alpha_1)}} - \frac{\alpha_3}{\alpha_2}\right]_+,$$

$$\hat{x}_2 = \min\left\{1, \left[\frac{(r+\beta)(1 - 2\sqrt{\frac{c(r+\alpha_1)}{\alpha_2}}) + c(r+\alpha_1)\frac{\alpha_3}{\alpha_2}}{r + 2\beta}\right]_+\right\}.$$

For $(\alpha_1, \alpha_2, \alpha_3, \beta, r, \hat{c}) = (1, .05, .1, .6, .1, .0005)$ it is $\hat{x} = (\hat{x}_1, \hat{x}_2) \approx (47.93, .4264)$.

e. The turnpike state $\hat{x} = (\hat{x}_1, \hat{x}_2)$ in part d is contained in $\mathcal{R}$ and therefore also in $\mathcal{C}_\varepsilon$ for small enough $\varepsilon > 0$, since

$$(1 - \hat{x}_2)(\alpha_2 \hat{x}_1 + \alpha_3 \hat{x}_2) > (1 - \hat{x}_2)(\alpha_2 \hat{x}_1 + \alpha_3 \hat{x}_2)(1 - \hat{u}_2) = \beta \hat{x}_2,$$

Appendix B

as long as the firm charges a positive price $\hat{u}_2 > 0$ (using the argument of part c). The system exhibits a turnpike in the sense that optimal stationary growth is achieved when the system starts at $x_0 = \hat{x}$ and then never leaves that state. Thus, if $x_0 \neq \hat{x}$, it is (at least intuitively) optimal for the decision maker to steer the system to the dynamically optimal steady state $\hat{x}$ as profitably as possible. Even if the decision maker is ignorant about the optimal trajectory, he can obtain close-to-optimal profits by steering the system to $\hat{x}$ quickly as described for part c. Under that strategy, for a small x_0, the decision maker would offer the product initially for free to increase the customer base up to the desired level (penetration pricing) and then compensate for product obsolescence or depreciation (i.e., the term $-\beta x_2$ in the system equation (B.8)) and for the eventually positive price by sufficient advertising (branding).

3.2 (Exploitation of an Exhaustible Resource)

a. To deal with the state-dependent control constraint,

$$c(t) \in [0, \mathbf{1}_{\{x(t) \geq 0\}} \bar{c}],$$

which expresses the requirement that consumption at time $t \in [0, T]$ can be positive only if the capital stock $x(t)$ at that time is positive, it is useful to introduce the endpoint constraint $x(\hat{T}) = 0$, where $\hat{T} \in (0, T]$ is free. The social planner's optimal control problem is therefore

$$J(c) = \int_0^{\hat{T}} e^{-rt} U(c(t)) \, dt \longrightarrow \max_{c(\cdot), \hat{T}}$$

s.t. $\dot{x} = -c,$

$x(0) = x_0, \qquad x(\hat{T}) = 0,$

$c(t) \in [0, \bar{c}], \qquad \forall t \in [0, \hat{T}],$

$\hat{T} \in (0, T].$

b. The current-value Hamiltonian associated with the social planner's optimal control problem in part a is

$$\hat{H}(t, x, c, \nu) = U(c) - \nu c,$$

where $\nu : [0, \hat{T}] \to \mathbb{R}$ is the (absolutely continuous) current-value adjoint variable. Given an optimal terminal time $\hat{T} \in [0, T]$ and an optimal admissible state-control trajectory $(x^*(t), c^*(t))$, $t \in [0, \hat{T}^*]$, the PMP yields the following necessary optimality conditions:

- Adjoint equation

$$\dot{v}(t) = rv(t) - \hat{H}_x(t, x^*(t), c^*(t), v(t)) = rv(t), \qquad \forall t \in [0, \hat{T}^*],$$

so $v(t) = v_0 e^{rt}$, $t \in [0, \hat{T}^*]$, for some constant $v_0 = v(0)$.

- Maximality

$$c^*(t) = \min\{U_c^{-1}(v_0 e^{rt}), \bar{c}\} \in \arg\max_{c \in [0, \bar{c}]} \hat{H}(t, x^*(t), c, v(t)),$$

for all $t \in [0, \hat{T}^*]$.

- Time optimality

$$\hat{H}(\hat{T}^*, x^*(\hat{T}^*), c^*(\hat{T}^*), v(\hat{T}^*)) - \lambda = 0,$$

where $\lambda \geq 0$ is a constant such that $\lambda(T - \hat{T}^*) = 0$.

c. For any given $t \in [0, \hat{T}^*]$ the optimal consumption $c^*(t)$ in part b is obtained (as long as it is interior) from the condition

$$0 = \hat{H}_c(t, x^*(t), c, v(t)) = U_c(c) - v(t).$$

Differentiating this condition for $c = c^*(t)$ with respect to t yields

$$0 = U_{cc}(c^*(t))\dot{c}^*(t) - \dot{v}(t)$$
$$= U_{cc}(c^*(t))\dot{c}^*(t) - rv(t)$$
$$= U_{cc}(c^*(t))\dot{c}^*(t) - rU_c(c^*(t)),$$

for all $t \in [0, \hat{T}^*]$, which is equivalent to the Hotelling rule. The economic interpretation of the rule is that the relative growth $\dot{c}/c$ of consumption c over time is negative. The absolute value of this negative relative consumption growth is proportional to the discount rate r and inversely proportional to society's relative risk aversion (another name for η). This means that when society becomes more risk-averse, its consumption should be more even over time (often termed consumption smoothing). On the other hand, if society is very impatient, that is, when the social discount rate r is very large, then most of the consumption should take place early on.

d. For $U(c) = \ln(c)$, one obtains from the maximality condition in part b that

$$c^*(t) = \min\{e^{-rt}/v_0, \bar{c}\},$$

Appendix B

for all $t \in [0, \hat{T}^*]$. Using both state-endpoint constraints (for $x^*(0)$ and $x^*(\hat{T}^*)$), it is therefore

$$x^*(\hat{T}^*) = x_0 - \int_0^{\hat{T}^*} c^*(t)\, dt = x_0 - \int_0^{\hat{T}^*} \min\{e^{-rt}/v_0, \bar{c}\}\, dt$$

$$= x_0 - \bar{c}t_s - \int_{t_s}^{\hat{T}^*} \frac{e^{-rt}}{v_0}\, dt$$

$$= 0,$$

or equivalently,

$$x_0 = \bar{c}t_s + \frac{1}{rv_0}(e^{-rt_s} - e^{-r\hat{T}^*}) \equiv \bar{c}t_s + \frac{\bar{c}}{r}(1 - e^{-r(\hat{T}^* - t_s)}), \tag{B.9}$$

where $t_s = \min\{\hat{T}^*, [-(1/r)\ln(v_0\bar{c})]_+\} \in [0, \hat{T}^*]$ is a switching time. There are two cases to consider:

1. $\hat{T}^* = T$. Under this condition the time-optimality condition given in part (b) is not active. Consequently, the optimal switching time is given by (B.9) when $\hat{T}^*$ is replaced by T.[8]

2. $\hat{T}^* < T$. Under this condition, consider two possibilities: (i) $c^*(\hat{T}^*) = \bar{c}$. In this situation, $t_s = 0$ and $\hat{T}^* = \frac{1}{r}\ln\left(\frac{\hat{c}}{\hat{c} - x_0 r}\right)$. (ii) $c^*(\hat{T}^*) \neq \bar{c}$. In this situation, the switching time can be determined using the time-optimality condition in part b:

$$U(c^*(\hat{T}^*)) - v(\hat{T}^*)c^*(\hat{T}^*) = \ln(\bar{c}e^{r(\hat{T}^* - t_s)}) - 1 = 0,$$

which implies that $t_s = \hat{T}^* - \frac{1}{r}\ln\left(\frac{e}{\bar{c}}\right)$. Therefore,

$$t_s = \frac{\bar{x}}{\bar{c}} - \frac{e - 1}{re}, \qquad \hat{T}^* = t_s + \frac{1}{r}\ln\left(\frac{e}{\bar{c}}\right).$$

e. Since the Hamiltonian is independent of the state x, the feedback-control law $\mu(t, x)$ cannot depend on x. Thus, $\mu(t, x) = c^*(t)$. That is, no new information or rule can be obtained for the decision maker about the optimal spending policy as a function of the remaining wealth.

f. When $T \to \infty$, the constraint $\hat{T} \leq T$ becomes inactive and everything else in the social planner's optimal control problem remains the same. In other words, only the second case of part d should be considered.

---

8. In this solution, it is assumed that $x_0 < \bar{c}T$. If $x_0 \geq \bar{c}T$, i.e., the initial amount of the resource is more than what can be consumed at the maximum extraction rate, then the optimal solution is given by $c = \bar{c}$.

3.3 (Exploitation of a Renewable Resource)

a. The firm's profit-maximization problem can be written as the following optimal control problem:

$$J(u) = \int_0^T e^{-rt} u(t) F(x(t))\, dt \longrightarrow \max_{u(\cdot)}$$

s.t. $\dot{x} = \alpha(x - \bar{x}) + (1 - u(t))F(x),$

$x(0) = x(T) = x_0,$

$u(t) \in [0, 1].$

Remark B.4 One could remove $\bar{x}$ from the formulation by introducing the new state variable $\xi = x - \bar{x}$; however, this is not without loss of generality, since $G(\xi) = F(\xi + \bar{x})$ is generally equal to zero at $\xi = 0$. □

b. The current-value Hamiltonian is

$$\hat{H}(t, x, u, \nu) = uF(x) + \nu \left(\alpha(x - \bar{x}) + (1 - u)F(x) \right).$$

Given an optimal state-control trajectory $(x^*(t), u^*(t))$, $t \in [0, T]$, the PMP yields the following necessary optimality conditions:

- Adjoint equation

$$\dot{\nu}(t) = -(\alpha - r)\nu(t) - \left(u^*(t) + (1 - u^*(t))\nu(t) \right) F_x(x^*(t)),$$

for all $t \in [0, T]$.

- Maximality

$$u^*(t) \in \arg\max_{u \in [0,1]} \hat{H}(t, x^*(t), u, \nu(t)),$$

so

$$u^*(t) \in \begin{cases} \{1\} & \text{if } \nu(t) > 1, \\ [0, 1] & \text{if } \nu(t) = 1, \\ \{0\} & \text{otherwise,} \end{cases}$$

for all $t \in [0, T]$.

Consider now the corresponding Hamiltonian system,

$\dot{x} = -\alpha(\bar{x} - x),$

$\dot{\nu} = -(\alpha - r)\nu - F_x(x),$

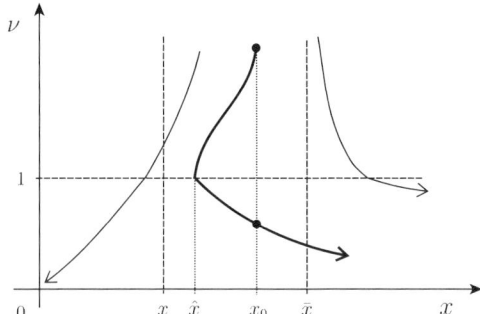

Figure B.11
Optimal state and control trajectories for exploitation of renewable resource.

for $v > 1$, and

$\dot{x} = -\alpha(\bar{x} - x) + F(x),$

$\dot{v} = -(\alpha - r)v - vF_x(x),$

for $v < 1$.[9] A phase diagram is given in figure B.11.[10]

c. Because the Hamiltonian is linear in the control variable, the optimal policy $u^*(t)$ becomes discontinuous whenever the function $v(t) - 1$ changes signs on the interval $(0, T)$. An interior solution $u^*(t) \in (0, 1)$ on a time interval of nonzero measure implies that $v(t) = 1$ on that interval, and thus $\dot{v}(t) = 0$ there. But the adjoint equation in part b would then yield

$-(\alpha - r) - F_x = 0,$

which is an impossibility for $\alpha \geq r$, since $F_x > 0$ by assumption. Thus, $u^*(t) \in \{0, 1\}$ a.e. on $[0, T]$. Furthermore, since $\dot{v} < 0$ in the Hamiltonian system, independent of the control, it is not possible to have more than one control switch. Such a switch, say, at time $t_s \in (0, T)$, must move the control from $u^*(t_s^-) = 1$ to $u^*(t_s^+) = 0$.

Consider now the state trajectory $x^*(t) = \phi_1(t, x_0), t \in [0, t_s]$, when the control $u^*(t) = 1$ is applied and the system is exploited. The Cauchy formula yields that

$x^*(t) = \bar{x} - (\bar{x} - x_0)e^{\alpha t}, \quad \forall t \in [0, t_s],$

---

9. The singular case where $v = 1$ is eliminated in part c.
10. The state $\underline{x}$ in figure B.11 is the lower bound for initial states at which the ecological system can survive on its own without dying out; it is determined as the solution of $\alpha \underline{x} + F(\underline{x}) = \alpha \bar{x}$.

so $\hat{x} = \phi_1(t_s, x_0) = \bar{x} - (\bar{x} - x_0)e^{\alpha t_s}$ is the size of the animal population at the end of the exploitation phase $[0, t_s]$, or equivalently, at the beginning of the regeneration phase $[t_s, T]$. The sustainability condition $x^*(T) = x_0$ can be satisfied if and only if $\phi_0(T - t_s, \hat{x}) = x_0$, where ϕ_0 is the flow of the state equation when a zero control is applied. Clearly, if one defines

$$G(x) = \int_{x_0}^{x} \frac{d\xi}{F(\xi) - \alpha(\bar{x} - \xi)},$$

which is an increasing function in x, then

$$\hat{x} = \phi_0(t_s - T, x_0) = G^{-1}(t_s - T).$$

Hence, the switching time $t_s \in [0, T]$ is determined as the solution of

$$G^{-1}(t_s - T) = \bar{x} - (\bar{x} - x_0)e^{\alpha t_s},$$

provided that a solution to this equation exists.[11]

d. Given the switching time $t_s \in (0, T)$, the optimal policy, as characterized in part c, can be described in words as follows. Exploit the system fully for $t \in [0, t_s]$, harvesting all the population surplus $F(x)$, and subsequently regenerate the population to its original state by ceasing to harvest for $t \in (t_s, T]$. Note that since t_s is independent of r, the optimal state-control trajectory $(x^*(t), u^*(t))$, $t \in [0, T]$, is independent of r. The effect of an increase in α on the switching time can be found by implicitly differentiating the equation that characterizes the switching time. Indeed,

$$-\frac{\partial t_s}{\partial \alpha} \frac{1}{G_x(\hat{x})} = -(\bar{x} - x_0)e^{\alpha t_s}\left(t_s + \alpha \frac{\partial t_s}{\partial \alpha}\right),$$

so

$$\frac{\partial t_s}{\partial \alpha} = \frac{(\bar{x} - x_0)e^{\alpha t_s} G_x(\hat{x})}{1 - (\bar{x} - x_0)e^{\alpha t_s} G_x(\hat{x})}.$$

That is, at least for small α, the switching time is increasing in α. This makes intuitive sense, as the length of the exploitation phase can increase when the population exhibits faster dynamics.

e. Let $F(x) = \sqrt{x}$. Then

---

11. If there is no such solution, then the firm's problem does not have a solution; the system is not sufficiently controllable.

Appendix B

$$G(x) = \int_{x_0}^{x} \frac{d\xi}{\sqrt{\xi - \alpha(\tilde{x} - \xi)}} = \int_{\sqrt{x_0}}^{\sqrt{x}} \frac{2z\,dz}{z - \alpha(\tilde{x} - z^2)}$$

$$= \frac{2}{a} \int_{\sqrt{x_0}}^{\sqrt{x}} \left[\frac{1+a}{1+a+2\alpha z} - \frac{1-a}{1-a+2\alpha z} \right] dz$$

$$= \frac{1}{\alpha a} \left((1+a) \ln \left[\frac{1+a+2\alpha\sqrt{x}}{1+a+2\alpha\sqrt{x_0}} \right] - (1-a) \ln \left[\frac{1-a+2\alpha\sqrt{x}}{1-a+2\alpha\sqrt{x_0}} \right] \right),$$

where $a = \sqrt{1 + 4\alpha^2 \tilde{x}}$, and the switching time t_s solves the fixed-point problem

$$t_s = T + G(\tilde{x} - (\tilde{x} - x_0)e^{\alpha t_s}).$$

For $(\alpha, x_0, \tilde{x}, 2) = (1, 10, 12, 2)$, it is $t_s \approx 0.3551$. For $t \in [0, t_s]$, it is $x^*(t) = 12 - 2e^t$, so $\hat{x} = x^*(t_s) \approx 9.1472$. The optimal control is $u^*(t) = 1_{\{t \leq t_s\}}$ (figure B.12).

3.4 (Control of a Pandemic)

a. The social planner's welfare maximization problem is given by

$$J(u) = -\int_0^T e^{-rt} \left(x(t) + cu^\kappa(t) \right) dt \longrightarrow \max_{u(\cdot)}$$

s.t. $u(t) \in [0, \bar{u}]$,

$\dot{x} = \alpha(1-x)x - ux, \qquad x(0) = x_0,$

$x(t) \geq 0,$

where the finite horizon $T > 0$, the discount rate $r > 0$, the initial state $x_0 \in (0, 1)$, and the constants $\alpha, c, \bar{u} > 0$ are given.

b. The current-value Hamiltonian associated with the social planner's optimal control problem[12] in part a is

$$\hat{H}(t, x, u, \nu) = -x - cu^\kappa + \nu \left(\alpha(1-x)x - ux \right),$$

where $\nu: [0, T] \to \mathbb{R}$ is the (absolutely continuous) current-value adjoint variable. Given an optimal admissible state-control trajectory $(x^*(t), u^*(t))$, $t \in [0, T]$, the PMP yields the following necessary optimality conditions:

For $\kappa = 1$,

---

12. In this problem, assume that $\bar{u} > \alpha$.

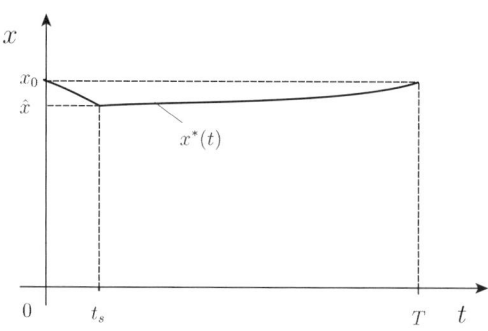
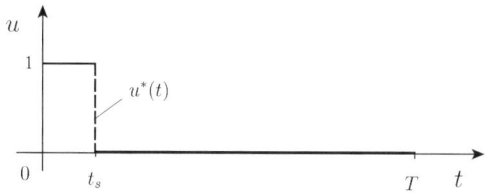

Figure B.12
Phase diagram of Hamiltonian system.

- Adjoint equation

$$\dot{v} = (r + u^* - \alpha + 2\alpha x^*)v + 1.$$

- Maximality

$$u^*(t) \in \begin{cases} \{0\} & \text{if } v(t) > -c/x^*(t), \\ [0, \bar{u}] & \text{if } v(t) = -c/x^*(t), \\ \{\bar{u}\} & \text{otherwise,} \end{cases}$$

for (almost) all $t \in [0, T]$.

- Transversality

$v(T) = 0.$

For $\kappa = 2$,

- Adjoint equation

$$\dot{v} = (r + u^* - \alpha + 2\alpha x^*)v + 1.$$

- Maximality

$$u^*(t) = \max\left\{0, \min\left\{\bar{u}, -\frac{v(t)x^*(t)}{2c}\right\}\right\},$$

Appendix B

for (almost) all $t \in [0, T]$.

- Transversality

$$v(T) = 0.$$

c. Let $\kappa = 1$. Since $-c/x^*(T) < 0$, and since by the transversality condition, it is $v(T) = 0$, it follows from the maximality condition that $u^*(t) = 0$ for $t \in [t_s, T]$ where $t_s \in [0, T)$ is a switching time. By the system equation it is therefore

$$x^*(t) = \frac{x_s}{x_s + (1 - x_s)e^{-\alpha(t-t_s)}}, \qquad \forall\, t \in [t_s, T],$$

where $x_s = x^*(t_s)$. Using the Cauchy formula, the adjoint equation, together with the transversality condition, therefore yields that

$$v(t) = -\int_t^T \exp\left[-\int_t^s (r - \alpha(1 - 2x^*(\varsigma)))\, d\varsigma\right] ds$$

$$= -\int_t^T \left(\frac{x_s + (1 - x_s)e^{-\alpha(t-t_s)}}{x_s + (1 - x_s)e^{-\alpha(s-t_s)}}\right)^2 e^{-(\alpha+r)(s-t)}\, ds,$$

for all $t \in [t_s, T]$. By the definition of the switching time, at $t = t_s$, it is $x^*(t_s) = x_s$ and

$$-\frac{c}{x_s} = v(t_s) = -\int_{t_s}^T \frac{e^{-(\alpha+r)(s-t_s)}\, ds}{(x_s + (1 - x_s)e^{-\alpha(s-t_s)})^2},$$

which directly relates x_s and t_s. Now examine the possibility of a singular control arc, where $v(t) = -c/x^*(t)$ on a time interval. In that case, the system equation yields that $u^* = \alpha(1 - x^*) - (\dot{x}^*/x^*)$, whence the adjoint equation takes the form

$$\dot{v} = rv + \alpha x^* v - \frac{\dot{x}^*}{x^*}v + 1 = rv + \dot{v} + 1 - \alpha c,$$

using the fact that $vx^* = -c$ implies $\dot{v}x^* + v\dot{x}^* = 0$ on the relevant time interval. Thus, necessarily

$$v(t) = -\frac{1 - \alpha c}{r} = \text{const.}$$

on such a singular arc, which in turn also implies that

$$x^*(t) = \frac{cr}{1 - \alpha c} = \text{const.} \quad \text{and} \quad u^*(t) = \alpha\left(1 - \frac{cr}{1 - \alpha c}\right) = \text{const.}$$

Thus, for $(\alpha+r)c > 1$, no singular arc can exist. For $(\alpha+r)c \leq 1$, it is in principle possible to have a singular arc (provided that $\bar{u}$ is large enough), which would then correspond to a temporary turnpike of the system, as state, co-state (adjoint variable), and control are all constant in this scenario.

If, starting at $t=0$, the regulator finds it optimal to apply the constant control $\bar{u}$, the state of the system evolves according to

$$x^*(t) = \frac{x_0\left(1-\frac{\bar{u}}{\alpha}\right)}{x_0 + \left(\left(1-\frac{\bar{u}}{\alpha}\right) - x_0\right)e^{-\alpha\left(1-\frac{\bar{u}}{\alpha}\right)t}}.$$

Provided that $\bar{u} \geq \alpha - r$, this implies via the adjoint equation that $v(t)$ increases, thus approaching the switching point $v(t_s) = -c/x^*(t_s)$ from below. Thus, there can be at most one switching point, with possibly a singular control arc along which the system effectively stays still until switching to the zero control at the end of the trajectory. Let $\hat{t}_s \leq t_s$ be the time at which the system reaches the switching point or singular control arc. Then

$$v(\hat{t}_s) = -\frac{c}{x_s} = -\int_{t_s}^{T} \frac{e^{-(\alpha+r)(s-t_s)}\,ds}{(x_s + (1-x_s)e^{-\alpha(s-t_s)})^2} = v(t_s),$$

and

$$x_s = \frac{x_0\left(1-\frac{\bar{u}}{\alpha}\right)}{x_0 + \left(\left(1-\frac{\bar{u}}{\alpha}\right) - x_0\right)e^{-\alpha\left(1-\frac{\bar{u}}{\alpha}\right)\hat{t}_s}}.$$

Omitting a detailed proof, it is now fairly straightforward to piece the socially optimal policy together. Assume that $\bar{x} \in (0,1)$ is the turnpike state. Then, as long as the time horizon T is long enough, it is optimal to steer the system as fast as possible to $x_s = \bar{x}$ by either applying zero or maximal control, and to stay on the turnpike as long as necessary to make sure that the adjoint variable can move from $-c/\bar{x}$ to 0 using a zero control. If the time horizon is too short to reach the turnpike, then $\hat{t}_s = t_s$, and t_s is determined as solution of

$$\frac{\left(x_0 + \left(\left(1-\frac{\bar{u}}{\alpha}\right) - x_0\right)e^{-\alpha\left(1-\frac{\bar{u}}{\alpha}\right)t_s}\right)c}{x_0\left(1-\frac{\bar{u}}{\alpha}\right)} = \int_{t_s}^{T} \frac{e^{-(\alpha+r)(s-t_s)}\,ds}{(x_s + (1-x_s)e^{-\alpha(s-t_s)})^2},$$

provided that it is positive. Otherwise, it is optimal for the regulator not to intervene at all, that is, to select $u^*(t) \equiv 0$. One can see that the latter

Appendix B

can happen if the intervention cost c is too large, that is, if (considering the previous equation for $x_s = x_0$ and $t_s = 0$)

$$c > \int_0^T \frac{x_0 \, e^{-(\alpha+r)s} \, ds}{(x_0 + (1-x_0)e^{-\alpha s})^2},$$

independent of $\bar{u}$.

Let $\kappa = 2$. Assuming that the intervention limit $\bar{u}$ is not binding on the optimal trajectory, the Hamiltonian system becomes

$$\dot{x}^* = \alpha(1-x^*)x^* + \frac{\nu x^*}{2c},$$

$$\dot{\nu} = (r - \alpha + 2\alpha x^*)\nu - \frac{x^* \nu^2}{2c} + 1.$$

This system of ODEs is highly nonlinear, so one can proceed via a qualitative analysis. The nullclines for $\dot{x} = \dot{\nu} = 0$ are described, for $x \in (0,1)$ and $\nu < 0$, by

$$\nu = -2\alpha c(1-x) \quad \text{and} \quad x = \frac{1 - (\alpha - r)\nu}{(-\nu)\left(2\alpha - \frac{\nu}{2c}\right)}.$$

Both relations are satisfied at a turnpike $(\bar{x}, \bar{\nu})$, of which (because the nullcline for $\dot{\nu} = 0$ does not intersect the axes) there can be none, one, or two values. The first of these can be neglected as an uninteresting measure-zero case because it is not robust to parameter perturbations. Consider the normal case when there are two possible turnpikes. An examination of the phase diagram reveals that only the one with the larger x-value can be reached along monotonic trajectories[13] of the Hamiltonian system in (x, ν)-space. If $x_0 < \bar{x}$, the optimal trajectory (x^*, ν) first approaches the turnpike before increasing to the point $(x_T, 0)$ toward the end of the horizon. The terminal state x_T is implied by the solution to the Hamiltonian system of ODEs, together with the two-point boundary conditions $x(0) = x_0$ and $\nu(T) = 0$. For $x_0 > \bar{x}$ the optimal state trajectory x^* will generally be nonmonotonic in t (as long as T is finite), first approaching $\bar{x}$ and then ending up at $x^*(T) = x_T$.

---

13. When the time horizon $T \to \infty$, the optimal policy becomes independent of time, and the principle of optimality (for this one-dimensional system) therefore implies that the optimal trajectories must be monotonic. (Otherwise, at a given state it would be sometimes optimal to go up and sometimes optimal to go down.)

For both $\kappa = 1$ and $\kappa = 2$, the optimal policy is such that $x_T > \bar{x}$, that is, toward the end of the horizon it is optimal for the social planner to reduce the intervention effort and let the epidemic spread beyond the long-run steady state. In contrast to the linear-cost case, when $\kappa = 2$ the regulator will exercise some intervention effort. In the linear-cost case, the regulator stops all action at the end of any finite time horizon.

d. The dynamically optimal steady state can be be understood as the spread of the epidemic, at which it would be cost-efficient to stabilize the population using a steady healthcare intervention effort. In the case $\kappa = 1$, where the solution can be given explicitly, the steady state $\bar{x} = cr/(1 - \alpha c)$ increases in the infectivity α, the cost c, and the discount rate r. The results are qualitatively equivalent for $\kappa = 2$.

The longer the planning horizon T, the more important this steady state becomes as a goal for public policy, and the implemented healthcare measures can be considered permanent. In the given model, attempting to eradicate the disease completely is never optimal because doing so requires that u be unbounded. Last, if the intervention bound $\bar{u}$ is too small, it may not be possible to reach a dynamically optimal steady state; yet the regulator should try to come as close as possible before (toward the end of the horizon) ceasing or dampening the intervention effort.

e. See also the discussion in part d. It becomes optimal to steer the system in the least-cost way to its turnpike and keep it there indefinitely.

f. Figure B.13 shows the optimal state-control trajectories for some numerical examples. To avoid the numerical problems associated with a singular solution, the case where $\kappa = 1$ is approximated by setting $\kappa = 1.05$ instead. This explains why the trajectories do not exactly attain the turnpike computed in part c.

3.5 (Behavioral Investment Strategies)

Part 1: Optimal Consumption Plan

a. The investor's optimal control problem is

$$J_T(c) = \int_0^T e^{-rt} \ln(c(t))\, dt \longrightarrow \max_{c(\cdot)}$$

s.t. $\dot{x} = \alpha x - c,$

$x(0) = x_0, \qquad x(t) \geq 0,$

$c(t) \in [0, \bar{c}], \qquad \forall t \in [0, T].$

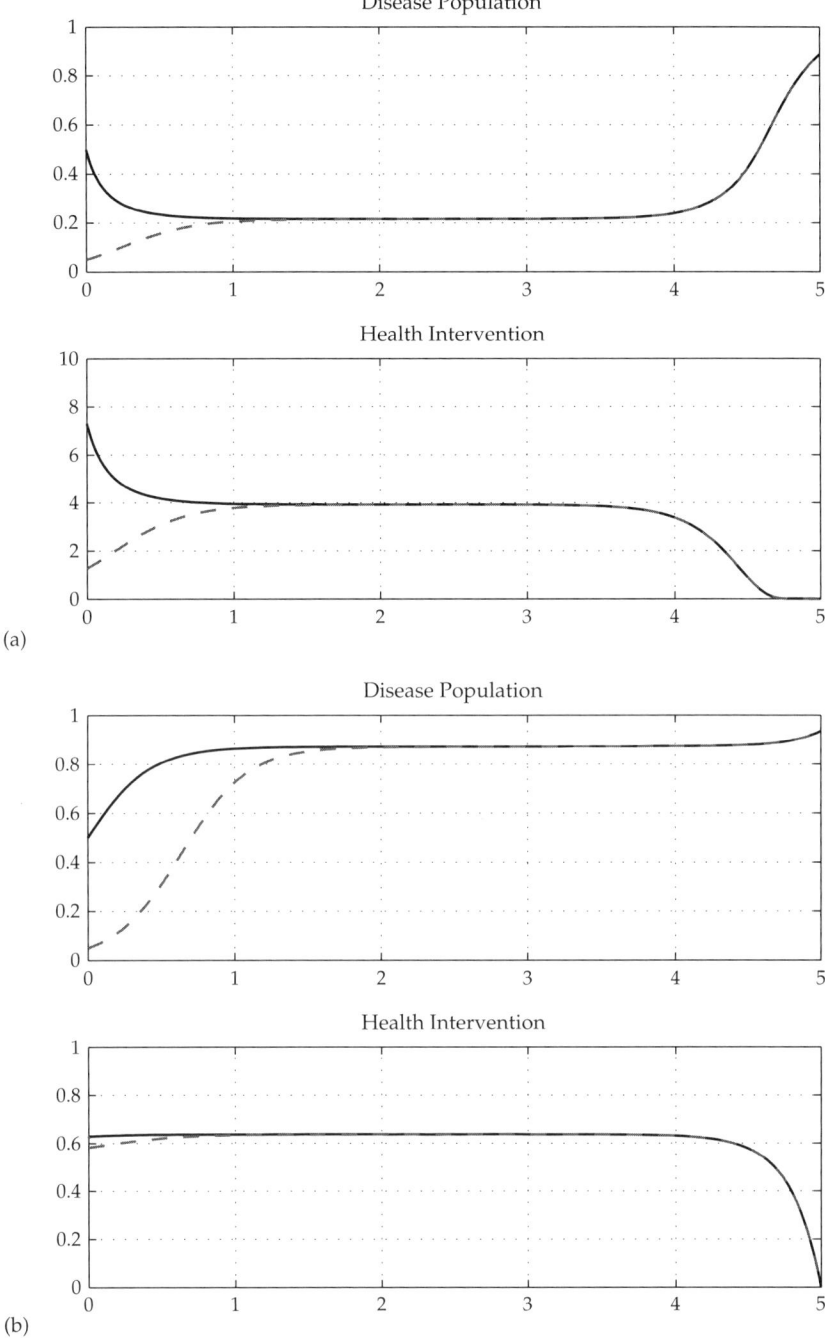

Figure B.13
Optimal disease-population and health-intervention trajectories for $(\alpha, c, r, \bar{u}) = (5, .15, .2, 10)$ and $x_0 \in \{5\%, 50\%\}$ (a) when $\kappa = 1.05$, and (b) when $\kappa = 2$.

Note that the state constraint $x(t) \geq 0$ for $t \in [0, T)$ is automatically satisfied, since otherwise the investor's objective would diverge to minus infinity because the log does not allow a zero consumption over nonzero-measure time intervals. Hence, the state constraint $x(t) \geq 0$ can be written equivalently as the state endpoint inequality constraint $x(T) \geq 0$.

The current-value Hamiltonian associated with the optimal control problem is then given by

$$\hat{H}(t, x, c, \nu) = \ln(c) + \nu(\alpha x - c),$$

where ν is the current-value adjoint variable. Given an optimal admissible state-control trajectory $(x^*(t), c^*(t))$, $t \in [0, T]$, the PMP yields the following necessary optimality conditions:

- Adjoint equation

$$\dot{\nu}(t) = -(\alpha - r)\nu(t), \qquad \forall t \in [0, T].$$

- Transversality

$$\nu(T) \geq 0.$$

- Maximality

$$c^*(t) = \min\left\{\frac{1}{\nu(t)}, \bar{c}\right\} \in \arg\max_{c \in [0, \bar{c}]} \hat{H}(t, x^*(t), c, \nu(t)),$$

for all $t \in [0, T]$.

From the adjoint equation and the transversality condition,

$$\nu(t) = \nu(T)e^{-(\alpha - r)(t - T)} \geq 0, \qquad \forall t \in [0, T].$$

Clearly, the capital stock $x^*(T)$ at the end of the horizon can be positive only if it is not possible to exhaust all resources by consuming at the maximum rate $\bar{c}$ on the entire time interval $[0, T]$, namely,

$$x^*(T) > 0 \Leftrightarrow x_0 > \bar{x}_0 \equiv \frac{\bar{c}}{\alpha}(1 - e^{-\alpha T})$$

$$\Leftrightarrow c^*(t) \equiv \bar{c}.$$

When the initial capital stock x_0 is limited, that is, when it does not exceed $\bar{x}_0 = (\bar{c}/\alpha)(1 - e^{-\alpha T})$, then necessarily $x^*(T) = 0$. By the Cauchy formula it is

Appendix B

$$x^*(t) = x_0 e^{\alpha t} - \int_0^t e^{\alpha(t-s)} c^*(s) \, ds$$

$$= x_0 e^{\alpha t} + \int_0^t e^{\alpha(t-s)} \min\{c_0 e^{(\alpha-r)s}, \bar{c}\} \, ds,$$

where $c_0 = c^*(0) = v(T) e^{(\alpha-r)T}$ is the initial consumption rate, so

$$x^*(T) = x_0 e^{\alpha T} - \int_0^T e^{\alpha(T-s)} \min\{c_0 e^{(\alpha-r)s}, \bar{c}\} \, ds = 0.$$

If the spending limit $\bar{c}$ is large enough, it will never constrain the agent's consumption, so

$$x^*(T) = \left(x_0 - \frac{c_0}{r}(1 - e^{-rT}) \right) e^{\alpha T} = 0,$$

and thus $c_0 = r x_0 / (1 - e^{-rT})$. Hence,[14]

$$x_0 \leq \underline{x}_0 \equiv \frac{\bar{c}}{r}\left(1 - e^{-rT}\right) e^{-(\alpha - r)T}$$

if and only if

$$c^*(t) = \frac{r x_0 e^{(\alpha - r)t}}{1 - e^{-rT}} = \bar{c}\left(\frac{x_0}{\underline{x}_0}\right) e^{-(\alpha-r)(T-t)},$$

for all $t \in [0, T]$. Finally, if the initial capital stock x_0 lies between $\underline{x}_0$ and $\bar{x}_0$, then there exists a switching time $t_s \in (0, T]$ such that $c_0 e^{(\alpha-r)t_s} = \bar{c}$, and

$$x^*(T) = \left(x_0 - \frac{c_0}{r}(1 - e^{-rt_s}) - \frac{\bar{c}}{\alpha}(e^{-\alpha t_s} - e^{-\alpha T}) \right) e^{\alpha T} = 0.$$

Consequently, the initial consumption level c_0 is determined implicitly by

$$x_0 = \frac{c_0}{r}\left(1 - \left(\frac{c_0}{\bar{c}}\right)^{\frac{r}{\alpha-r}}\right) + \frac{\bar{c}}{\alpha}\left(\left(\frac{c_0}{\bar{c}}\right)^{\frac{\alpha}{\alpha-r}} - e^{-\alpha T}\right), \quad (*)$$

or equivalently, by

$$\frac{\alpha \rho}{1-\rho}\left(\frac{x_0}{\bar{c}} + \frac{e^{-\alpha T}}{\alpha}\right) = \frac{1}{1-\rho}\left(\frac{c_0}{\bar{c}}\right) - \left(\frac{c_0}{\bar{c}}\right)^{\frac{1}{1-\rho}},$$

---

14. One can show that $\alpha > r$ implies that $\bar{x}_0 > \underline{x}_0$, for all $T > 0$.

where $\rho = r/\alpha \in (0,1)$. To summarize, the investor's optimal T-horizon consumption plan, as a function of his initial capital x_0, the investment return α, and his opportunity cost of capital r, is given by

$$c_T^*(t) = \begin{cases} \dfrac{rx_0 e^{(\alpha-r)t}}{1-e^{-rT}} & \text{if } x_0 \leq \underline{x}_0, \\ \bar{c} & \text{if } x_0 > \bar{x}_0, \\ \min\{c_0 e^{(\alpha-r)t}, \bar{c}\} & \text{otherwise,} \end{cases} \quad \forall t \in [0,T],$$

where $c_0 = c_0(\alpha, r, x_0)$ is determined by (*).

b. Building on the finite-horizon solution in part a, note first that as $T \to \infty$, the thresholds $\underline{x}_0 \to 0^+$ and $\bar{x}_0 \to \bar{c}/\alpha$. In addition, since $\ln(c(t))$ is bounded from above by $\ln(\bar{c})$, the difference between the finite-horizon objective $J_T(c)$ and the corresponding infinite-horizon objective $J_\infty(c)$ goes to zero as $T \to \infty$ for any feasible consumption plan $c(t) \in [0, \bar{c}]$, $t \geq 0$. One can therefore expect pointwise convergence of the finite-horizon consumption plan $c_T^*(t)$, $t \geq 0$, to the optimal infinite-horizon consumption plan $c_\infty^*(t)$, $t \geq 0$, as $T \to \infty$. The consumption plan

$$c_\infty^*(t) = \lim_{T \to \infty} c_T^*(t) = \begin{cases} \bar{c} & \text{if } x_0 > \bar{c}/\alpha, \\ \min\{c_0 e^{(\alpha-r)t}, \bar{c}\} & \text{otherwise,} \end{cases}$$

for all $t \geq 0$, where c_0 satisfies

$$\frac{rx_0}{\bar{c}} = \left(1 - \left(\frac{c_0}{\bar{c}}\right)^{\frac{r}{\alpha-r}}\right)\left(\frac{c_0}{\bar{c}}\right) + \frac{r}{\alpha}\left(\frac{c_0}{\bar{c}}\right)^{\frac{\alpha}{\alpha-r}},$$

is indeed a solution to the infinite-horizon optimal consumption problem, since there can be no policy that can produce a strictly higher value of the objective function.[15] Note that under the optimal infinite-horizon consumption plan it becomes eventually optimal to consume at the spending limit $\bar{c}$.

Part 2: Myopic Receding-Horizon Policy

c. Periodic updating with limited implementation may be due to uncertainty or limited commitment ability. For example, an elected official may not be able to implement a policy over a horizon that exceeds the length of his mandate. Such receding-horizon decision making may also result from periodic reporting and decision-making schedules in

---

15. If for some $\varepsilon > 0$ there is a policy $\hat{c}$ such that $J_\infty(\hat{c}) \geq J_\infty(c_\infty^*) + \varepsilon$, then that immediately produces a contradiction for large enough horizons, $T > \left|\ln\left(\varepsilon/|\ln(\bar{c})|\right)\right|$, because then the value of the infinite-horizon objective $J_\infty(\hat{c})$ must be less than ε away from the optimal finite-horizon objective $J_T(c_T^*)$.

Appendix B

organizations. Figure B.14 depicts how, given a planning horizon T and an implementation horizon τ, the receding-horizon policy $\hat{c}^*_{T,\tau}$ is obtained as concatenation of partially implemented finite-horizon policies $c^k(t) = c^*_T(t - k\tau)$ for $t \in \mathcal{I}_k = [k\tau, (k+1)\tau], k \in \mathbb{N}$.

d. Yes. When $T \to \infty$, the investor's optimal infinite-horizon consumption policy does depend only on his current capital and not on the time that is left to consume. Indeed, the solution to the infinite-horizon optimal consumption problem does not depend on when it starts in absolute time, but only on the investor's initial capital x_0 (and, of course, on the parameters α, r, which are assumed fixed). By construction, the limiting receding-horizon policy is equal to the infinite-horizon optimal policy,

$$\lim_{T \to \infty} \hat{c}^*_{T,\tau}(t) = c^*_\infty(t), \qquad \forall t \geq 0.$$

e. Yes. Given an initial capital x_0 below the threshold $\bar{x}_0 = \frac{\bar{c}}{\alpha}(1 - e^{-\alpha T})$, and a planning horizon $T > (\frac{1}{r})\ln(\frac{\alpha}{\alpha - r})$, the optimal T-horizon state trajectory is initially upward-sloping, since

$$\frac{\dot{x}^*(t)}{x_0} = \alpha \frac{e^{-rt} - e^{-rT}}{1 - e^{-rT}} e^{\alpha t} - r \frac{e^{(\alpha - r)t}}{1 - e^{-rT}} = \frac{e^{\alpha t}}{1 - e^{-rT}}[(\alpha - r)e^{-rt} - \alpha e^{-rT}]$$

in a right-neighborhood of $t = 0$. If τ is the time such that $x^*(\tau) = x^*(0) = x_0$, then the corresponding receding-horizon state-control path $(\hat{x}^*_{T,\tau}(t), \hat{u}^*_{T,\tau}(t))$ is τ-periodic. Figure B.15b shows numerical examples with and without periodicity, for $(\alpha, r, \bar{c}, x_0) = (.2, .1, 4, 15)$ and $(T, \tau) \in \{(10, 6), (10, 2.7)\}$.

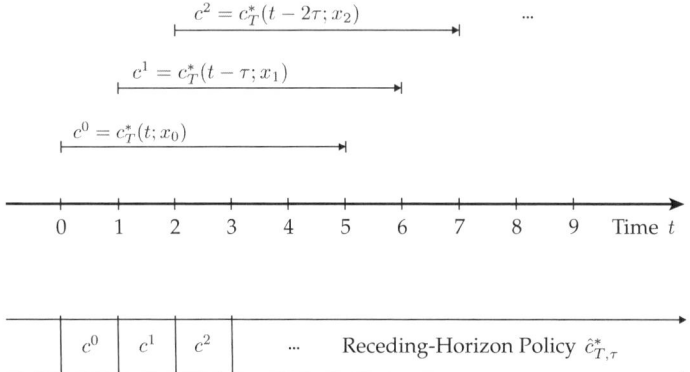

Figure B.14
Receding-horizon decision making with $(T, \tau) = (5, 1)$.

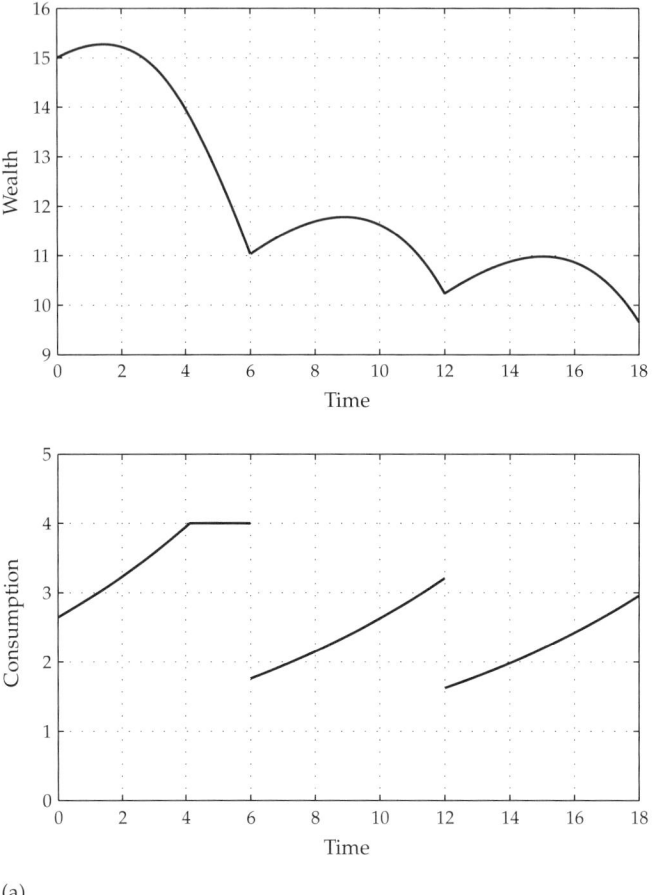

(a)

Figure B.15
(a) Receding-horizon consumption plan with nonmonotonic state-control trajectory and capital exhaustion ($\lim_{t\to\infty} \hat{x}^*_{T,\tau}(t) = 0$). (b) Receding-horizon consumption plan with periodic state-control trajectory.

Part 3: Prescriptive Measures

f. Somewhat surprisingly, a long-run steady state $\bar{x}^*_\infty$ does not always exist. Indeed, if $x_0 > \bar{c}/\alpha$, then the investor's capital continues to grow despite his consuming at the spending limit $\bar{c}$ at all times $t \geq 0$. When $x_0 \leq \bar{c}/\alpha$, the findings in part b imply that $\bar{x}^*_\infty = \bar{c}/\alpha$. Yet, because convergence to the steady state along an optimal state trajectory is possible only from below, it is not stable.

Appendix B

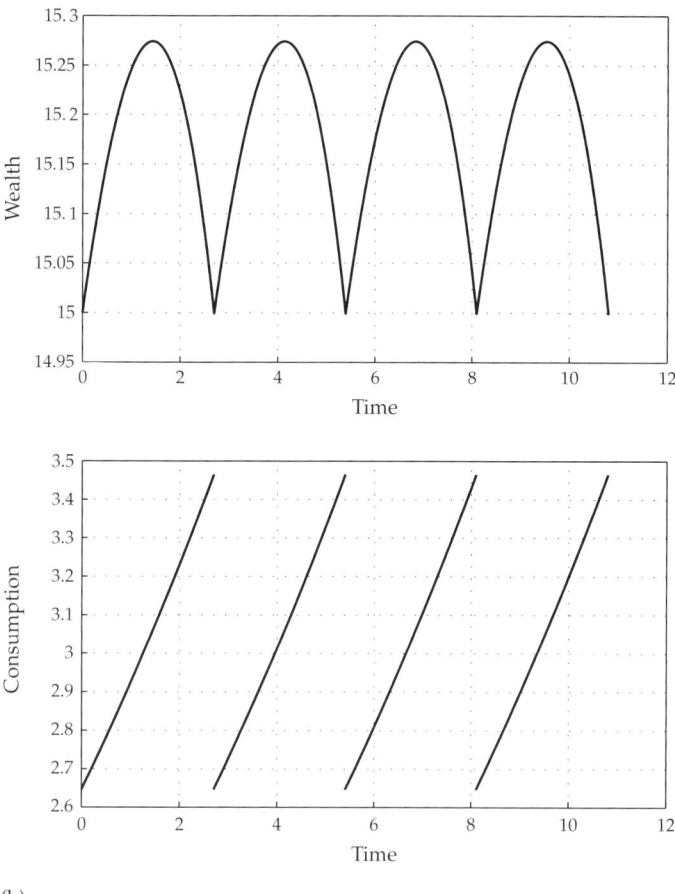

(b)

Figure B.15
(*continued*)

g. The only initial states x_0 for which the modified (T, τ)-receding-horizon consumption plan with the additional endpoint constraint is feasible are such

$$x_0 \leq \bar{x}_\infty^* = \bar{c}/\alpha,$$

implying solutions identical to the corresponding optimal infinite-horizon plans. By setting consumption to zero, the long-run steady state is reached in the fastest possible way, which bounds the planning horizons T that can be used to implement the modified receding-horizon policy from below, requiring that

$$T \geq \underline{T} \equiv \frac{1}{\alpha} \ln\left(\frac{\bar{c}}{\alpha x_0}\right).$$

Note also that under the modified receding-horizon consumption plan it is not possible for the investor to consume at the spending limit toward the end of the planning horizon, since that would not allow the capital to grow to the long-run steady state. Thus, the optimal consumption plan $\tilde{c}_T^*(t)$, $t \in [0, T]$, is interior, in the sense that

$$\tilde{c}_T^*(t) = \tilde{c}_0 e^{(\alpha - r)t} \in (0, \bar{c}), \qquad \forall t \in [0, T].$$

The constant $\tilde{c}_0 = \tilde{c}_T^*(0)$ is determined by the state-endpoint constraint, so (analogous to part a)

$$\tilde{x}^*(T) = \left(x_0 - \frac{\tilde{c}_0}{r}(1 - e^{-rT})\right) e^{\alpha T} = \frac{\bar{c}}{\alpha},$$

whence

$$\tilde{c}_0 = \frac{r}{1 - e^{-rT}}\left(x_0 - \frac{\bar{c}}{\alpha} e^{-\alpha T}\right) < c_0,$$

and

$$\tilde{c}_T^*(t) = \frac{re^{(\alpha - r)t}}{1 - e^{-rT}}\left(x_0 - \frac{\bar{c}}{\alpha} e^{-\alpha T}\right), \qquad \forall t \in [0, T].$$

One can think of the modified finite-horizon consumption plan as a regular finite-horizon consumption plan where initial capital x_0 has been reduced (virtually) by the amount $(\bar{c}/\alpha)e^{-\alpha T}$ ($< x_0$), that is,

$$\tilde{c}_T^*(t; x_0) = c_T^*(t; x_0 - (\bar{c}/\alpha)e^{-\alpha T}), \qquad \forall t \in [0, T].$$

As a consequence, it is not possible that within the planning horizon the trajectory becomes downward-sloping. Hence, under the modified receding-horizon consumption plan $\tilde{c}_{T,\tau}^*(t)$, $t \geq 0$, there cannot be any periodic cyclical consumption behavior. Because of the monotonicity of the state trajectory with the state-endpoint constraint, there is convergence to the dynamically optimal steady state for any implementation horizon $\tau \in (0, T)$, a significant improvement, which comes at the price of a more moderate consumption, effectively excluding consumption levels at the spending limit.

3.6 (Optimal Consumption with Stochastic Lifetime)

Appendix B

a. Given any admissible control trajectory $c(t) > \varepsilon$, $t \geq 0$, for some $\varepsilon \in (0, \min\{\alpha x_0, \bar{c}\})$,[16] the investor's objective functional is (using integration by parts)

$$\bar{J}(c) = \int_0^\infty g(T) \left(\int_0^T e^{-rt} U(c(t))\, dt \right) dT$$

$$= \left[G(T) \int_0^T e^{-rt} U(c(t))\, dt \right]_0^\infty - \int_0^\infty G(T) e^{-rT} U(c(T))\, dT,$$

where $G(T) = \int_0^T g(\tau)\, d\tau = \lambda \int_0^T e^{-\lambda \tau}\, d\tau = 1 - e^{-\lambda T}$ is the cumulative distribution function for the Poisson random variable $\tilde{T}$. Thus,

$$\bar{J}(c) = \int_0^\infty e^{-rt} U(c(t))\, dt - \int_0^\infty \left(1 - e^{-\lambda T}\right) e^{-rT} U(c(T))\, dT$$

$$= \int_0^\infty e^{-(\lambda + r)t} U(c(t))\, dt,$$

since the fact that $U([\varepsilon, \bar{c}])$ is compact implies that the function $U(c(t))$ is bounded (whence all integrals in the last relation converge). One can therefore rewrite the investor's stochastic optimal consumption problem (as formulated in part a) as the following infinite-horizon optimal control problem.

$$\bar{J}(c) = \int_0^\infty e^{-(\lambda + r)t} \ln(c(t))\, dt \longrightarrow \max_{c(\cdot)}$$

s.t. $\dot{x}(t) = \alpha x(t) - c(t)$,

$x(0) = x_0, \quad x(t) \geq 0,$

$c(t) \in [0, \bar{c}]$,

$\forall t \geq 0$.

b. The solution to the optimal control problem in part a can be obtained by approximating the infinite-horizon optimal control problem by a sequence of finite-horizon optimal control problems of the form

---

16. The constant $\varepsilon > 0$ needs to be chosen small enough so that the initial capital x_0 is not exhausted in finite time.

$$J_{T_k}(c) = \int_0^{T_k} e^{-(\lambda+r)t} \ln(c(t))\, dt \longrightarrow \max_{c(\cdot)}$$

s.t. $\dot{x}(t) = \alpha x(t) - c(t)$,

$x(0) = x_0, \qquad x(t) \geq 0$,

$c(t) \in [0, \bar{c}]$,

$\forall t \in [0, T_k]$,

for $k \in \mathbb{N}$, where $0 < T_k < T_{k+1}$ such that $T_k \to \infty$ as $k \to \infty$, and then taking the limit for $k \to \infty$ in the optimal consumption policy $c_{T_k}(t)$, $t \in [0, T_k]$ of the kth finite-horizon problem.

First formulate the deterministic finite-horizon optimal control problem. Let $T_k = T > 0$ and set $\rho = \lambda + r$. Note that, depending on the value of λ, it is possible that $\alpha > \rho$ or $\alpha \leq \rho$. The corresponding current-value Hamiltonian is

$$\hat{H}(x, c, v) = \ln(c) + v(\alpha x - c),$$

and the PMP yields the following necessary optimality conditions for an optimal state-control trajectory $(x_T^*(t), c_T^*(t))$, $t \in [0, T]$:

- Adjoint equation

$$\dot{v}(t) = -(\alpha - \rho)v(t), \qquad \forall t \in [0, T].$$

- Transversality

$$v(T) \geq 0.$$

- Maximality

$$c_T^*(t) = \min\left\{\frac{1}{v(t)}, \bar{c}\right\} \in \arg\max_{c \in [0, \bar{c}]} \hat{H}(x_T^*(t), c, v(t)), \qquad \forall t \in [0, T].$$

The adjoint equation and the transversality condition yield that

$$v(t) = v(T)e^{-(\alpha - \rho)(t - T)} \geq 0, \qquad \forall t \in [0, T].$$

The capital stock $x^*(T)$ at $t = T$ can be positive only if it is not possible to exhaust all resources by consuming at the maximum rate $\bar{c}$ on the entire time interval $[0, T]$, namely,

$$x^*(T) > 0 \Leftrightarrow x_0 > \bar{x}_0 \equiv \frac{\bar{c}}{\alpha}(1 - e^{-\alpha T}) \Leftrightarrow c^*(t) \equiv \bar{c}.$$

Appendix B

When the initial capital stock x_0 is limited, that is, when it does not exceed $\bar{x}_0 = (\bar{c}/\alpha)(1 - e^{-\alpha T})$, then necessarily $x^*(T) = 0$. By the Cauchy formula it is

$$x^*(t) = x_0 e^{\alpha t} - \int_0^t e^{\alpha(t-s)} c^*(s)\, ds$$

$$= x_0 e^{\alpha t} + \int_0^t e^{\alpha(t-s)} \min\{c_0 e^{(\alpha-\rho)s}, \bar{c}\}\, ds,$$

where $c_0 = c^*(0) = e^{(\alpha-\rho)T}/\nu(T)$ is the (possibly fictitious)[17] initial consumption rate. Thus,

$$x^*(T) = x_0 e^{\alpha T} - \int_0^T e^{\alpha(T-s)} \min\{c_0 e^{(\alpha-\rho)s}, \bar{c}\}\, ds = 0.$$

If the spending limit $\bar{c}$ is large enough, it will never constrain the investor's consumption, so

$$x^*(T) = \left(x_0 - \frac{c_0}{\rho}(1 - e^{-\rho T})\right) e^{\alpha T} = 0,$$

and thus $c_0 = \rho x_0/(1 - e^{-\rho T})$. Hence,[18]

$$x_0 \leq \underline{x}_0 \equiv \frac{\bar{c}}{\rho}(1 - e^{-\rho T}) e^{-([\alpha-\rho]_+)T} \Leftrightarrow$$

$$c^*(t) = \frac{\rho x_0 e^{(\alpha-\rho)t}}{1 - e^{-\rho T}} = \bar{c}\left(\frac{x_0}{\underline{x}_0}\right) e^{-([\alpha-\rho]_+)T} e^{(\alpha-\rho)t}, \qquad \forall t \in [0, T].$$

Last, if the initial capital stock x_0 lies between $\underline{x}_0$ and $\bar{x}_0$, then there exist a switching time $t_s \in (0, T]$ and a constant c_0 such that $c_0 e^{(\alpha-\rho)t_s} = \bar{c}$, and

$$x^*(T) = \begin{cases} (x_0 - \frac{c_0}{\rho}(1 - e^{-\rho t_s}) - \frac{\bar{c}}{\alpha}(e^{-\alpha t_s} - e^{-\alpha T}))e^{\alpha T} & \text{if } \alpha > \rho \\ (x_0 - \frac{\bar{c}}{\alpha}(1 - e^{-\alpha t_s}) - \frac{c_0}{\rho}(e^{-\rho t_s} - e^{-\rho T}))e^{\alpha T} & \text{if } \alpha \leq \rho \end{cases}$$

$$= 0.$$

The (possibly fictitious) initial consumption level c_0 is determined implicitly by the equation

---
17. It is possible that $c_0 > \bar{c}$ when $\alpha < \rho$.
18. One can show that $\bar{x}_0 > \underline{x}_0$, for all $T > 0$.

$$x_0 = \begin{cases} \frac{c_0}{\rho}\left(1-\left(\frac{c_0}{\bar{c}}\right)^{\frac{\rho}{\alpha-\rho}}\right) + \frac{\bar{c}}{\alpha}\left(\left(\frac{c_0}{\bar{c}}\right)^{\frac{\alpha}{\alpha-\rho}} - e^{-\alpha T}\right) & \text{if } \alpha > \rho, \\ \frac{\bar{c}}{\alpha}\left(1-\left(\frac{c_0}{\bar{c}}\right)^{\frac{\alpha}{\rho-\alpha}}\right) + \frac{\bar{c}}{\rho}\left(\left(\frac{c_0}{\bar{c}}\right)^{\frac{\rho}{\rho-\alpha}} - e^{-\rho T}\right) & \text{otherwise.} \end{cases} \quad (**)$$

Hence, the investor's optimal T-horizon consumption plan, as a function of his initial capital x_0, the investment return α, and his (effective) opportunity cost of capital ρ, is given by

$$c_T^*(t) = \begin{cases} \dfrac{\rho x_0 e^{(\alpha-\rho)t}}{1-e^{-\rho T}} & \text{if } x_0 \leq \underline{x}_0, \\ \bar{c} & \text{if } x_0 > \bar{x}_0, \qquad \forall t \in [0,T], \\ \min\{c_0 e^{(\alpha-\rho)t}, \bar{c}\} & \text{otherwise,} \end{cases}$$

where $c_0 = c_0(\alpha, \rho, x_0)$ is determined by (**).

Now the deterministic infinite-horizon optimal control problem can be formulated. Take the limit in the foregoing optimal finite-horizon consumption plan.[19] For this, note first that as $T \to \infty$, the thresholds $\underline{x}_0 \to (\bar{c}/\rho)\mathbf{1}_{\{\alpha \leq \rho\}}$ and $\bar{x}_0 \to \bar{c}/\alpha$. The consumption plan

$$c^*(t) = \lim_{T\to\infty} c_T^*(t) = \begin{cases} \rho x_0 e^{(\alpha-\rho)t} & \text{if } x_0 \leq (\bar{c}/\rho)\mathbf{1}_{\{\alpha \leq \rho\}}, \\ \bar{c} & \text{if } x_0 > \bar{c}/\alpha, \\ \min\{c_0 e^{(\alpha-\rho)t}, \bar{c}\} & \text{otherwise,} \end{cases}$$

for all $t \geq 0$, where c_0 satisfies (**), is the solution to the infinite-horizon optimal consumption problem and thus also to the investor's original stochastic optimal consumption problem.

c. When T is perfectly known, it is possible to set $\lambda = 0$ and then obtain from part b the optimal T-horizon consumption plan,

$$c_T^*(t) = \begin{cases} \dfrac{r x_0 e^{(\alpha-r)t}}{1-e^{-rT}} & \text{if } x_0 \leq \frac{\bar{c}}{r}(1-e^{-rT})e^{-([\alpha-r]_+)T}, \\ \bar{c} & \text{if } x_0 > \frac{\bar{c}}{\alpha}(1-e^{-\alpha T}), \\ \min\{c_0 e^{(\alpha-r)t}, \bar{c}\} & \text{otherwise,} \end{cases}$$

for all $t \in [0,T]$, where $c_0 = c_0(\alpha, r, x_0)$ is determined by (**) for $\rho = r$. Unless the initial capital is large, namely, when $x_0 > \bar{x}_0 = \frac{\bar{c}}{\alpha}(1-e^{-\alpha T})$,

---

19. Since $\ln(c(t))$ is bounded from above by $\ln(\bar{c})$, the difference between the finite-horizon objective $J_T(c)$ and the corresponding infinite-horizon objective $J_\infty(c)$ goes to zero as $T \to \infty$ for any feasible consumption plan $c(t) \in [0, \bar{c}]$, $t \geq 0$. One can therefore expect pointwise convergence of the optimal finite-horizon consumption plan $c_T^*(t)$, $t \geq 0$, to the optimal infinite-horizon consumption plan $c^*(t)$, $t \geq 0$, as $T \to \infty$, since there is no policy that can produce a strictly higher value of the objective function. Indeed, if for some $\varepsilon > 0$ there exists a policy $\hat{c}$ such that $J(\hat{c}) \geq J(c_\infty^*) + \varepsilon$, one obtains a contradiction for large enough $T > |\ln(\varepsilon/|\ln(\bar{c})|)|$ because then the value of the infinite-horizon objective $J_\infty(\hat{c})$ must be less than ε away from the optimal finite-horizon objective $J_T(c_T^*)$.

Appendix B

the optimal finite-horizon policy uses up the entire capital within T. The infinite-horizon policy, on the other hand, ensures that the investor's bank balance stays positive.

The expected value of perfect information (EVPI) about the length of the planning horizon is the expected difference of the investor's optimal payoffs when using a policy $c_T^*(t)$, $t \in [0, T]$, that is perfectly adapted to the planning-horizon realization T versus using the general-purpose policy $c^*(t)$, $t \geq 0$, in the case where the horizon is unknown. In other words, EVPI $= E[J_{\tilde{T}}(c_{\tilde{T}}^*)] - \bar{J}(c^*)$, or equivalently,

$$\text{EVPI} = \int_0^\infty g(T) \left(\int_0^T e^{-rt} \left(U(c_T^*(t)) - U(c^*(t)) \right) dt \right) dT.$$

Using similar techniques as in (ii), the expected value of perfect information (which must be nonnegative) becomes

$$\text{EVPI} = \int_0^\infty e^{-rt} \left(\ln \frac{c_\infty^*(t)}{c^*(t)} - (1 - e^{-\lambda t}) \ln \frac{c_t^*(t)}{c^*(t)} \right) dt - \left(T_s - \frac{1 - e^{-\lambda T_s}}{\lambda} \right),$$

where T_s is a switching time such that $c_T^*(t)$ depends nontrivially on the horizon T for $T < T_s$.

d. The optimal (deterministic) infinite-horizon consumption plan $c_\infty^*(t)$, $t \geq 0$, is

$$c_\infty^*(t) = \begin{cases} \bar{c} & \text{if } x_0 > \bar{c}/\alpha, \\ \min\{c_0 e^{(\alpha - r)t}, \bar{c}\} & \text{otherwise,} \end{cases} \quad \forall t \geq 0.$$

Consider first the case where $\rho < \alpha$ and the initial capital is small, so that $x_0 < \bar{c}/\alpha$. From (*), c_0 is implicitly given by

$$x_0 = \left(\frac{\bar{c}}{\alpha} - \frac{\bar{c}}{\rho} \right) \left(\frac{c_0}{\bar{c}} \right)^{\frac{\alpha}{\alpha - \rho}} + \frac{c_0}{\rho}.$$

Calculating the first derivative of c_0 with respect to ρ using this equation shows that c_0 is increasing in ρ. Consequently, the initial optimal consumption rate in the stochastic optimal consumption problem exceeds the optimal deterministic infinite-horizon consumption rate. Note that in this situation both $c^*(t)$ and $c_\infty^*(t)$ will reach $\bar{c}$ at the switching time

$$t_s = \frac{1}{\alpha - \rho} \ln \left(\frac{\bar{c}}{c_0} \right).$$

When $\alpha \leq \rho$, that is, when the expected lifetime is relatively short, the optimal consumption $c^*(t)$ is a nonincreasing function, in contrast to $c^*_\infty(t)$, which is nondecreasing.

e. On the one hand, when the personal lifetime is not known, it is best to spend more conservatively than when the horizon is perfectly known, since in the latter case, nothing has to be left at the end (unless initial capital is so large that it is essentially impossible to consume). Yet, on the other hand, while c^*_∞ is always nondecreasing (as a consequence of the assumption that $\alpha > r$), it is possible (when $\alpha \leq \rho = \lambda + r$) that c^* is decreasing. For a given α, r this happens when λ is large (when the expected lifetime is short). This means that when the remaining lifetime is unknown but likely to be short, it becomes optimal to consume a lot in the present and the near future, and when—against expectations—life turns out to be longer, to reduce consumption later, which is exactly the opposite philosophy of the deterministic infinite-horizon policy when excess return $\alpha - r$ is positive. To interpret both in the same context, one can think of $\alpha - \rho$ as the effective excess return and thus of λ as a price of lifetime risk.

B.4 Game Theory

4.1 (Linear-Quadratic Differential Game)

a. As in the analysis of the linear-quadratic regulator in example 3.3, consider an HJB equation for each player $i \in \{1, \ldots, N\}$,

$$rV^i(t,x) - V^i_t(t,x) =$$

$$\max_{u^i \in \mathcal{U}^i} \left\{ -x' R^i(t) x - \sum_{j=1}^N (u^j)' S^{ij}(t) u^j + \langle V^i_x(t,x), A(t)x + \sum_{j=1}^N B^j(t) u^j \rangle \right\}.$$

Analogous to earlier developments assume that the value function is quadratic, of the form

$$V^i(t,x) = -x' Q^i(t) x,$$

where $Q^i(t)$ a continuously differentiable matrix function with symmetric positive definite values in $\mathbb{R}^{n \times n}$ for all $t \in [0, T]$. Substituting the value function into the HJB equation yields a linear feedback law,

$$\mu^i(t,x) = -(S^{ii}(t))^{-1}(B^i(t))' Q^i(t) x,$$

Appendix B

for all $i \in \{1, \ldots, N\}$. The matrix functions $Q^1, \ldots, Q^N$ satisfy a system of Riccati differential equations,

$$-\dot{R}^i(t) = \dot{Q}^i - rQ^i + Q^i A(t) + A'(t)Q^i$$

$$+ \sum_{j=1}^{N} (2Q^i - Q^j) B^j(t) (S^{jj}(t))^{-1} S^{ij}(t) (S^{jj}(t))^{-1} (B^j(t))' Q^j,$$

for all $t \in [0, T]$, with endpoint conditions

$$Q^i(T) = K^i, \qquad i \in \{1, \ldots, N\}.$$

The equilibrium state trajectory $x^*(t)$, $t \in [0, T]$, is then obtained as solution to the linear IVP

$$\dot{x} = \left[A(t) - \sum_{j=1}^{N} B^j(t) (S^{jj}(t))^{-1} (B^j(t))' Q^j(t) \right] x, \qquad x(0) = x_0.$$

The (unique) open-loop Nash-equilibrium strategy profile is therefore $u^* = (u^{1*}, \ldots, u^{N*})$, with

$$u^{i*}(t) = \mu^i(t, x^*(t)) = -(S^{ii}(t))^{-1} (B^i(t))' Q^i(t) x^*(t), \qquad \forall t \in [0, T],$$

for all $i \in \{1, \ldots, N\}$.

b. Assume that each player $i \in \{1, \ldots, N\}$ uses a linear feedback law of the form

$$\mu^i(t, x) = M^i(t) x,$$

where the matrix function M^i is continuously differentiable with values in $\mathbb{R}^{m \times n}$. Instead of the HJB equation, which is useful for the computation of open-loop Nash equilibria, an application of the PMP (see proposition 3.5), proves more productive. Substituting the other players' feedback laws in player i's objective functional and the state equation, player i's current-value Hamiltonian has the form

$$\hat{H}^i(t, x, u^i, v^i) = -x' \left(R^i + \sum_{j \neq i}^{N} (M^j)' S^{ij} M^j \right) x - (u^i)' S^{ii} u^i$$

$$+ \langle v^i, Ax + \sum_{j \neq i} B^j M^j x + B^i u^i \rangle,$$

where v^i is the current-value adjoint variable. The maximality condition yields that $u^{i*} = (S^{ii})^{-1}(B^i)'(v^i/2)$, so the adjoint equation becomes

$$\dot{v}^i = rv^i + 2\left(R^i + \sum_{j\neq i}^N (M^j)'S^{ij}M^j\right)x - \left(A + \sum_{j\neq i} B^j M^j\right)v^i,$$

with transversality condition

$$v^i(T) = -2K^i x(T).$$

To solve the adjoint equation, use the intuition from the solution for part a, and assume that $v^i = -2Q^i x$, building on the conceptual proximity of the HJB equation and the PMP (see section 3.3). This guess for the relation between v^i is arrived at by first assuming that player i's value function is quadratic, of the form $V^i(t,x) = -x'Q^i(t)x$, as in part a, and then setting $v^i = V^i_x = -2Q^i x$, where $Q^i(t)$ is a continuously differentiable matrix function with symmetric positive definite values in $\mathbb{R}^{n\times n}$. The adjoint equation, together with the transversality condition, is solved by $v^i = -2Q^i x$ (so $M^i = -(S^{ii})^{-1}(B^i)'Q^i$), provided that the functions $Q^1, \ldots, Q^N$ satisfy the system of Riccati differential equations

$$-R^i(t) = \dot{Q}^i - rQ^i + Q^i A(t) + A'(t)Q^i$$
$$+ \sum_{j=1}^N (Q^i - Q^j)B^j(t)(S^{jj}(t))^{-1}S^{ij}(S^{jj}(t))^{-1}(B^j(t))'Q^j,$$

for all $t \in [0,T]$, with endpoint conditions

$$Q^i(T) = K^i, \quad i \in \{1, \ldots, N\}.$$

The corresponding closed-loop equilibrium state-control trajectory $(x^*(t), u^*(t))$, $t \in [0,T]$, is then obtained as for part a.

c. The fundamental difference between the open-loop equilibrium in part a and the closed-loop equilibrium in part b is that in the open-loop formulation each player i considers the other players' strategies only as a function of time, whereas in the closed-loop formulation the control law for the other players appears in player i's objective and in his considerations about the evolution of the state (see remark 4.6). Consequently, the Riccati differential equations in parts a and b differ by the term $Q^i B^j(t)(S^{jj}(t))^{-1}S^{ij}(S^{jj}(t))^{-1}(B^j(t))'Q^j$, although the structural properties (such as linearity) of the equilibria under the two solution concepts are essentially the same.

Appendix B

4.2 (Cournot Oligopoly)

a. The differential game $\Gamma(x_0)$ consists of a set of players $\mathcal{N} = \{1,\ldots,N\}$, a set of objective functionals, $\{J^i(u^i)\}_{i=1}^N$, where $J^i(u^i)$ (for $i \in \mathcal{N}$) is specified in the problem, a description of the evolution of the state $p(t)$ from its known initial value p_0, given by the IVP

$$\dot{p} = f(p, u^1, \ldots, u^N), \qquad p(0) = p_0,$$

and the control constraints $u^i \in \mathcal{U} = [0, \bar{u}]$, where $\bar{u} > 0$ is given.

b. In the stationary Cournot game, in which the price p_0 and all output strategies $u_0^i \in \mathcal{U} = [0, \bar{u}]$, $i \in \mathcal{N}$, stay constant, the equilibrium price p_0 is determined by

$$p_0 = a - \sum_{i=1}^N u_0^i.$$

Given the other firms' stationary strategy profile $u_0^{-i} \in [0, \bar{u}]^{N-1}$, each firm $i \in \mathcal{N}$ determines its best-response correspondence

$$\begin{aligned}\mathrm{BR}^i(u_0^{-i}) &= \arg\max_{u^i \in [0,\bar{u}]} \left\{\left(a - u^i - \sum_{j \neq i} u_0^j\right) u^i - C(u^i)\right\} \\ &= \left\{\min\left\{\bar{u}, \frac{1}{3}\left[a - c - \sum_{j \neq i} u_0^j\right]_+\right\}\right\}.\end{aligned}$$

At the stationary Nash-equilibrium strategy profile $u_0 = (u_0^1, \ldots, u_0^N)$ one obtains

$$u_0^i \in \mathrm{BR}^i(u_0^{-i}), \qquad \forall\, i \in \mathcal{N}.$$

Using the symmetry yields

$$u_0^i = \min\left\{\bar{u}, \frac{a-c}{N+2}\right\} \in [0, \bar{u}], \qquad \forall\, i \in \mathcal{N}.$$

The corresponding stationary equilibrium price is

$$p_0 = \frac{c + (2a/N)}{1 + (2/N)}.$$

Note that $p_0 \to c$ as $N \to \infty$.

c. Given the other players' strategy profile $u^{-i}(t)$, $t \geq 0$, player $i \in \mathcal{N}$ solves the optimal control problem

$$J^i(u^i) = \int_0^\infty e^{-rt}(p(t)u^i(t) - C(u^i(t)))\,dt \longrightarrow \max_{u^i(\cdot)}$$

s.t. $\dot{p} = \alpha\left(a - p - u^i - \sum_{j \neq i} u^j\right),$

$p(0) = p_0,$

$u^i(t) \in [0, \bar{u}], \qquad \forall\, t \geq 0.$

The current-value Hamiltonian of player i's optimal control problem is given by

$$\hat{H}^i(t, p, u^i, v^i) = pu^i - cu^i - \frac{(u^i)^2}{2} + \alpha\left(a - p - u^i - \sum_{j \neq i} u^j(t)\right)v^i,$$

where v^i denotes the current-value adjoint variable. Applying the PMP yields the following necessary optimality conditions for any optimal state-control trajectory $(p^*(t), u^{i*}(t))$, $t \geq 0$:

- Adjoint equation

$\dot{v}^i(t) = (\alpha + r)\, v^i(t) - u^{i*}(t), \qquad \forall\, t \geq 0.$

- Transversality

$e^{-rt} v^i(t) \to 0, \quad \text{as } t \to \infty.$

- Maximality

$u^{i*}(t) = p^*(t) - \alpha v^i(t) - c \in \arg\max_{\hat{u}^i \in [0, \bar{u}]}\, \hat{H}^i(t, p^*(t), \hat{u}^i, v^i(t)),$

for all $t \geq 0$.

From the maximality condition one finds that $v^i = \left(p^* - u^{i*} - c\right)/\alpha$, and by differentiating with respect to time, $\dot{v}^i = (\dot{p}^* - \dot{u}^{i*})/\alpha$. Using the adjoint equation and the state equation, one can eliminate the adjoint variable and obtain that in equilibrium

$$\dot{u}^{i*}(t) = \alpha\left(a - p^*(t) - \sum_{j \neq i} u^{j*}(t)\right) - (\alpha + r)\left(p^*(t) - c - u^{i*}(t)\right),$$

for all $t \geq 0$ and all $i \in \mathcal{N}$. Together with the state equation, these ODEs determine the evolution of the optimal state-control trajectory.

Appendix B

Restricting attention to a symmetric solution (which by uniqueness of solutions to ODEs is also the only solution), one obtains a system of two linear ODEs (with constant coefficients),

$$\dot{p}(t) = \alpha(a - p(t) - Ny(t)),$$

$$\dot{y}(t) = \alpha a + (\alpha + r)c - (2\alpha + r)p(t) - ((N-2)\alpha - r)y(t),$$

where $u^{i*} = y$ for all $i \in \mathcal{N}$; any remaining superscripts have been dropped for convenience. Note first that the system described by these ODEs has the unique equilibrium

$$(\bar{p}, \bar{y}) = \left(\frac{c + \frac{2\alpha+r}{\alpha+r}\left(\frac{a}{N}\right)}{1 + \frac{2\alpha+r}{\alpha+r}\left(\frac{1}{N}\right)}, \frac{a-c}{N+2-\frac{r}{\alpha+r}} \right),$$

and can therefore be written in the form

$$\begin{bmatrix} \dot{p} \\ \dot{y} \end{bmatrix} = A \begin{bmatrix} p - \bar{p} \\ y - \bar{y} \end{bmatrix},$$

with the system matrix

$$A = -\begin{bmatrix} \alpha & \alpha N \\ 2\alpha + r & \alpha(N-2) - r \end{bmatrix}.$$

Thus, given an initial value (p_0, y_0), the Cauchy formula yields that

$$\begin{bmatrix} p(t) \\ y(t) \end{bmatrix} = \begin{bmatrix} \bar{p} \\ \bar{y} \end{bmatrix} + \begin{bmatrix} p_0 - \bar{p} \\ y_0 - \bar{y} \end{bmatrix} e^{At}, \quad \forall t \geq 0.$$

There are two remaining problems. First, the determinant of A is negative,

$$\det(A) = \lambda_1 \lambda_2 = -(N+1)\alpha r - (N+2)\alpha^2 < 0,$$

so the (real) eigenvalues $\lambda_1 < 0 < \lambda_2$ of A,

$$\lambda_{1,2} = \frac{r - (N-1)\alpha}{2} \pm \frac{1}{2}\sqrt{((N-1)\alpha + r)^2 + 8\alpha r + 4(N+2)\alpha^2},$$

must have different signs. This means that the system is unstable and cannot be expected to converge to the steady state $(\bar{p}, \bar{y})$ computed earlier, as long as $(p_0, y_0) \neq (\bar{p}, \bar{y})$. Second, the initial value y_0 is not known and has to be determined from the transversality condition. The first difficulty can be addressed by realizing that the equilibrium $(\bar{p}, \bar{y})$, though unstable, is in fact a saddle point, so there are trajectories (in the direction of eigenvectors associated with the negative eigenvalue λ_1) that converge to it. The second difficulty is resolved by choosing y_0 such that $(p_0 - \bar{p}, y_0 - \bar{y})$ becomes an eigenvector associated with the negative eigenvalue λ_1, namely,[20]

$$\begin{bmatrix} p_0 - \bar{p} \\ y_0 - \bar{y} \end{bmatrix} \perp A - \lambda_1 I,$$

where I is the 2×2 identity matrix. The last relation is equivalent with

$$(A - \lambda_1 I) \begin{bmatrix} p_0 - \bar{p} \\ y_0 - \bar{y} \end{bmatrix} = 0,$$

which in turn implies that

$$y_0 = \bar{y} + \frac{p_0 - \bar{p}}{2\alpha N}((N-3)\alpha - r$$

$$+ \sqrt{((N-1)\alpha + r)^2 + 8\alpha r + 4(N+2)\alpha^2}).$$

Note that the convergence of $(p(t), y(t))$ to the steady state $(\bar{p}, \bar{y})$ is compatible with the transversality condition, which can be used as an alternative justification for the choice of the initial value y_0. This yields

$$p(t) = \bar{p} + (p_0 - \bar{p})e^{\lambda_1 t},$$

and

$$y(t) = \bar{y} + (y_0 - \bar{y}) e^{\lambda_1 t}.$$

Noting that $p^*(t) = p(t)$ and $u^{i*}(t) = y(t)$, the equilibrium turnpike $(\bar{p}^*, \bar{u}^*)$ can be directly compared to the static solution in part a,

$$(\bar{p}^*, \bar{u}^{i*}) = \left(\frac{c + \frac{2\alpha + r}{\alpha + r}\left(\frac{a}{N}\right)}{1 + \frac{2\alpha + r}{\alpha + r}\left(\frac{1}{N}\right)}, \frac{a - c}{N + 2 - \frac{r}{\alpha + r}} \right)$$

$$= \left(\frac{p_0 - \frac{r}{\alpha + r}\left(\frac{1}{N+2}\right)}{1 - \frac{r}{\alpha + r}\left(\frac{1}{N+2}\right)}, \frac{u_0}{1 - \frac{r}{\alpha + r}\left(\frac{1}{N}\right)} \right),$$

for all $i \in \mathcal{N}$, provided that the control bound $\bar{u}$ is large enough (and thus nonbinding). While the dynamic steady-state production is always

---

20. $\perp$ denotes "is orthogonal to."

Appendix B

larger than for the static solution, the same holds true for price only if the static solution p_0 is large (greater than 1).

The equilibrium price monotonically decreases if $p_0 > \bar{p}$ and increases otherwise, converging to the steady state $\bar{p}$. Firm i's equilibrium production output $u^{i*}(t)$ either decreases or increases, depending on the sign of $y_0 - \bar{u}^{i*}$, converging to the steady state $\bar{u}^{i*}$ as $t \to \infty$.

Remark B.5 Instead of solving the preceding system of first-order ODEs, it is possible instead to transform this system into a single second-order ODE for price,

$$\ddot{p}(t) + \sigma \dot{p}(t) + \rho p(t) = \theta,$$

where $\sigma = (N-1)\alpha - r$, $\rho = -\alpha((N+2)\alpha + (N+1)r)$, $\theta = -a\alpha^2 - \alpha(a + cN)(\alpha + r)$. Using the substitution $\xi = Bp - R$, this inhomogeneous second-order linear differential equation with constant coefficients can be transformed to the corresponding homogeneous equation,

$$\ddot{\xi}(t) + \sigma \dot{\xi}(t) + \rho \xi(t) = 0,$$

with solution

$$\xi(t) = C_1 e^{\lambda_1 t} + C_2 e^{\lambda_2 t},$$

where $\lambda_{1,2}$ are the roots of the characteristic equation

$$\lambda^2 + \sigma \lambda + \rho = 0 \Leftrightarrow \lambda_{1,2} = \frac{-\sigma \pm \sqrt{\sigma^2 - 4\rho}}{2},$$

which are identical to the eigenvalues of A determined earlier. Note that since $\rho < 0$, the roots λ_1, λ_2 are always real and distinct. Then,

$$p(t) = \frac{\theta}{\rho} + \bar{C}_1 e^{\lambda_1 t} + \bar{C}_2 e^{\lambda_2 t},$$

where $\bar{C}_1 = C_1/\rho$ and $\bar{C}_2 = C_2/\rho$. The control becomes

$$u^{i*}(t) = \frac{a\rho - \theta}{\rho N} - \frac{\bar{C}_1}{N}\left(1 + \frac{\lambda_1}{N}\right)e^{\lambda_1 t} - \frac{\bar{C}_2}{N}\left(1 + \frac{\lambda_2}{N}\right)e^{\lambda_2 t},$$

and the adjoint variable is

$$v^i(t) = \frac{(N+1)\theta - a\rho}{N\alpha\rho} - c + \frac{\bar{C}_1}{N\alpha}\left(N + 1 + \frac{\lambda_1}{N}\right)e^{\lambda_1 t} + \frac{\bar{C}_2}{N\alpha}\left(N + 1 + \frac{\lambda_2}{N}\right)e^{\lambda_2 t}.$$

The transversality condition implies that

$$e^{-rt}\left(\frac{(N+1)\theta - a\rho}{N a \rho} - c\right) \frac{\bar{C}_1}{N\alpha}\left(N+1+\frac{\lambda_1}{n}\right) e^{(\lambda_1 - r)t}$$

$$+ \frac{\bar{C}_2}{N\alpha}\left(N+1+\frac{\lambda_2}{N}\right) e^{(\lambda_2 - r)t} \to 0, \quad \text{as } t \to \infty.$$

Note that $\lambda_1 < 0$ (λ_1 is the smallest root of the characteristic equation). Hence, $e^{(\lambda_1 - r)t} \to 0$ as $t \to \infty$. The sign of $\lambda_2 - r$ is analyzed as

$$\lambda_2 - r = \frac{-(\alpha(N-1)+r) + \sqrt{(\alpha(N-1)+r)^2 + 4\alpha^2(N+2) + 8\alpha r}}{2} > 0.$$

Therefore, the transversality condition holds if and only if $\bar{C}_2 = 0$. The initial condition $p(0) = p_0$ yields

$$\bar{C}_1 = p_0 - \frac{\theta}{\rho},$$

where θ/ρ is equal to the steady state $\bar{p}^*$. The resulting equilibrium price,

$$p(t) = \frac{\theta}{\rho} + \left(p_0 - \frac{\theta}{\rho}\right) e^{\lambda_1 t} = \bar{p}^* + (p_0 - \bar{p}^*) e^{\lambda_1 t},$$

corresponds to the earlier solution. The equilibrium production output becomes

$$u^{i*}(t) = \frac{1}{N}\left(a - \frac{\theta}{\rho}\right) - \frac{1}{N}\left(p_0 - \frac{\theta}{\rho}\right)\left(1 + \frac{\lambda_1}{N}\right) e^{\lambda_1 t}$$

$$= \frac{a - \bar{p}^*}{N} + \left(\frac{a - p_0}{N} - \frac{a - \bar{p}^*}{N}\right)\left(1 + \frac{\lambda_1}{N}\right) e^{\lambda_1 t}$$

$$= \bar{u}^{i*} + \left(\frac{a - p_0}{N} - \bar{u}^{i*}\right)\left(1 + \frac{\lambda_1}{N}\right) e^{\lambda_1 t}$$

$$= \bar{u}^{i*} + (y_0 - \bar{u}^{i*}) e^{\lambda_1 t},$$

for all $i \in \mathcal{N}$, where

$$y_0 = \frac{a - p_0}{N} + \frac{\lambda_1}{N}\left(\frac{a - p_0}{N} - \bar{u}^{i*}\right)$$

is firm i's initial production as previously determined. □

d. First compute the feedback law corresponding to the open-loop Nash equilibrium determined for part c. By eliminating time from the relations

Appendix B

for $p^*(t)$ and $u^{i*}(t)$, it is

$$\frac{u^i - \bar{u}^{i*}}{y_0 - \bar{u}^{i*}} = \frac{p - \bar{p}^*}{p_0 - \bar{p}^*}$$

for any (p, u^i) on the equilibrium path. The corresponding feedback law,

$$\mu^{i*}(t, p) = \bar{u}^{i*} + (y_0 - \bar{u}^{i*})\frac{p - \bar{p}^*}{p_0 - \bar{p}^*}, \qquad \forall i \in \mathcal{N}$$

is affine in the price p and independent of time. To determine firm i's best response, assume that all other firms have affine production strategies of the form

$$\mu(p) = \gamma_1 + \gamma_2 p,$$

where γ_1, γ_2 are appropriate constants. In that case, player $i \in \mathcal{N}$ solves the optimal control problem

$$J^i(u^i) = \int_0^\infty e^{-rt}(p(t)u^i(t) - C(u^i(t)))\, dt \longrightarrow \max_{u^i(\cdot)}$$

s.t. $\dot{p} = \alpha(a - p - u^i - (N-1)(\gamma_1 + \gamma_2 p))$,

$p(0) = p_0$,

$u^i(t) \in [0, \bar{u}], \qquad \forall t \geq 0$.

The corresponding current-value Hamiltonian is

$$\hat{H}^i(t, p, u^i, v^i) = pu^i - cu^i - \frac{(u^i)^2}{2}$$
$$+ \alpha(a - p - u^i - (N-1)(\gamma_1 + \gamma_2 p))v^i,$$

where v^i denotes the current-value adjoint variable. Applying the PMP yields the following necessary optimality conditions for any optimal state-control trajectory $(p^*(t), u^{i*}(t)), t \geq 0$:

- Adjoint equation

$$\dot{v}^i(t) = (\alpha + \alpha(N-1)\gamma_2 + r)v^i(t) - u^{i*}(t), \qquad \forall t \geq 0.$$

- Transversality

$$e^{-rt}v^i(t) \to 0, \qquad \text{as } t \to \infty.$$

- Maximality

$$u^{i*}(t) = p^*(t) - \alpha v^i(t) - c \in \arg\max_{\hat{u}^i \in [0, \bar{u}]} \hat{H}^i(t, p^*(t), \hat{u}^i, v^i(t)),$$

for all $t \geq 0$.

Substituting the firms' symmetric closed-loop strategy profile in the state equation, together with the initial condition $p(0) = p_0$, yields

$$p^*(t) = \frac{a - N\gamma_1}{1 + N\gamma_2} + \left(p_0 - \frac{a - N\gamma_1}{1 + N\gamma_2}\right) e^{-\alpha(1 + N\gamma_2)t}, \quad \forall t \geq 0.$$

The maximality condition implies that

$$v^i = \frac{p^* - u^{i*} - c}{\alpha} = \frac{(1 - \gamma_2)p^* - \gamma_1 - c}{\alpha}.$$

Differentiating this relation with respect to time and using the state equation as well as the adjoint equation one obtains

$$\frac{1 - \gamma_2}{\alpha} \dot{p} = (1 - \gamma_2)(a - p^* - N(\gamma_1 + \gamma_2 p^*))$$

$$= (\alpha + \alpha(N-1)\gamma_2 + r) \frac{(1 - \gamma_2)p^* - \gamma_1 - c}{\alpha} - \gamma_1 - \gamma_2 p^*.$$

The last equation holds on the entire (nonconstant) price path $p^*(t)$, $t \geq 0$, if and only if

$$\frac{\gamma_2}{1 - \gamma_2} - (2N - 1)\gamma_2 - 2 = \frac{r}{\alpha},$$

and

$$(1 - \gamma_2)(a - N\gamma_1) + \frac{c + \gamma_1}{\alpha}(\alpha + r + \alpha(N-1)\gamma_2) + \gamma_1 = 0.$$

Thus,

$$\gamma_2 = \frac{2N\alpha - r - 4\alpha \pm \sqrt{(2N\alpha + r)^2 + 4\alpha(\alpha + r)}}{2\alpha(2N - 1)},$$

and

$$\gamma_1 = \frac{-\alpha(a + c) - rc + \alpha(a - Nc + c)\gamma_2}{-\alpha(N - 2) + r + \alpha(2N - 1)\gamma_2}.$$

Note that because of the transversality condition and the explicit solution for $p^*(t)$, one must have that $1 + N\gamma_2 \geq 0$. Consequently, the negative solution for γ_2 will not be acceptable, since

Appendix B

$$1 + N\gamma_2 = 1 + N \frac{2N\alpha - r - 4\alpha - \sqrt{(2N\alpha + r)^2 + 4\alpha(\alpha + r)}}{2\alpha(2N-1)}$$

$$< 1 - N \frac{2\alpha + r}{(2N-1)\alpha} < 0.$$

Using the explicit solution for $p^*(t)$ and $u^*(t) = \gamma_1 + \gamma_2 p^*(t)$, the long-run closed-loop state-control tuple $(\bar{p}_c^*, \bar{u}_c^*)$ is given by

$$\bar{p}_c^* = \frac{a - N\gamma_1}{1 + N\gamma_2},$$

and

$$\bar{u}_c^* = \gamma_1 + \gamma_2 \frac{a - N\gamma_1}{1 + N\gamma_2}.$$

e. When the market becomes competitive, that is, as $N \to \infty$, in all the solutions (static, open-loop, and closed-loop) the price converges to c and aggregate production approaches $a - c$.

4.3 (Duopoly Pricing Game)

a. Given a strategy $p^j(t)$, $t \geq 0$, for firm $j \in \{1, 2\}$, firm $i \neq j$ solves the optimal control problem

$$J^i(p^i|p^j) = \int_0^\infty e^{-rt} p^i(t)[x_i(t)(1 - x_i(t) - x_j(t))$$

$$\times (\alpha(x_i(t) - x_j(t)) - (p^i(t) - p^j(t)))] dt \longrightarrow \max_{p^i(\cdot)}$$

s.t. $\dot{x}^i = x_i(1 - x_i - x_j)[\alpha(x_i - x_j) - (p^i(t) - p^j(t))],$

$x_i(0) = x_{i0},$

$p^i(t) \in [0, P], \quad \forall t \geq 0,$

where the constant $P > 0$ is a (sufficiently large) maximum willingness to pay.

b. From the state equations for the evolution of $x(t) = (x_i(t), x_j(t))$, $t \geq 0$, it is possible to eliminate time, since

$$\frac{dx_2}{dx_1} = \frac{x_2(1 - x_1 - x_2)[\alpha(x_1 - x_2) - (p^1 - p^2)]}{x_1(1 - x_1 - x_2)[\alpha(x_2 - x_1) - (p^2 - p^1)]} = -\frac{x_2}{x_1},$$

so that by direct integration

$$\ln\left(\frac{x_2}{x_{20}}\right) = \int_{x_{20}}^{x_2} \frac{d\xi_2}{\xi_2} = -\int_{x_{10}}^{x_1} \frac{d\xi_1}{\xi_1} = -\ln\left(\frac{x_1}{x_{10}}\right),$$

which implies that

$$x_1 x_2 = x_{10} x_{20}$$

along any admissible state trajectory. Thus, any admissible state trajectory $x(t)$ of the game $\Gamma(x_0)$ moves along the curve $\mathcal{C}(x_0) = \{(x_1, x_2) \in [0,1]^2 : x_1 x_2 = x_{10} x_{20}\}$.

c. The current-value Hamiltonian for firm i's optimal control problem is

$$\hat{H}^i(t, x, p, v^i) = (1 - x_i - x_j)(\alpha(x_i - x_j) - (p^i - p^j))$$
$$\times (x_i(p^i + v_i^i) - x_j v_j^i),$$

where $v^i = (v_i^i, v_j^i)$ denotes the current-value adjoint variable associated with the state $x = (x_i, x_j)$. The PMP yields the following necessary optimality conditions for any equilibrium state-control trajectory $(x^*(t), p^*(t))$, $t \geq 0$:

• Adjoint equation

$$\dot{v}_i^i = r v_i^i - (p^{i*} + v_i^i)[(1 - 2x_i^* - x_j^*)(\alpha(x_i^* - x_j^*)$$
$$- (p^{i*} - p^{j*})) + \alpha x_i^*(1 - x_i^* - x_j^*)]$$
$$- (p^{j*} + v_j^i)x_j^*[(p^j - p^i) - \alpha(1 - 2x_i^*)],$$

$$\dot{v}_j^i = r v_j^i - (p^i + v_j^i)x_i^*[(p^i - p^j) - \alpha(1 - 2x_j^*)]$$
$$- (p^j + v_j^i)[(1 - x_i^* - 2x_j^*)(\alpha(x_j^* - x_i^*))$$
$$- (p^j - p^i) + \alpha x_j^*(1 - x_i^* - x_j^*)].$$

• Transversality

$$e^{-rt} v^i(t) \to 0, \quad \text{as } t \to \infty.$$

• Maximality

$$p^{i*} = \frac{1}{2}[x_i^*(p^{j*} - v_i^i + \alpha(x_i^* - x_j^*)) - x_j^*(p^{j*} + v_j^i)].$$

To examine the qualitative behavior of the optimal state-control trajectories it is useful to first examine any turnpikes that the model may have. Note first that

Appendix B

$$\dot{x}_i = 0 \Rightarrow \begin{cases} \text{either} & x_i = 0, \\ \text{or} & x_i + x_j = 1, \\ \text{or} & \alpha(x_i - x_j) = p^i - p^j. \end{cases}$$

Since $x_0 > 0$ by assumption, the first point, $x_i = 0$, does not belong to the curve $C(x_0)$ determined for part b and therefore cannot be a steady state. The second option implies two equilibria where the line $x_i + x_j = 1$ intersects the curve $C(x_0)$. The third option,

$$\alpha(x_i - x_j) = p^i - p^j, \tag{B.10}$$

implies, together with the maximality condition, that

$$x_i(p^i + v_i^i) = -x_j(p^j + v_j^i). \tag{B.11}$$

In addition, $\dot{v}_i^i = \dot{v}_j^i = 0$ implies that

$$(p^i + v_i^i)[(1 - 2x_i - x_j)(\alpha(x_i - x_j) - (p^i - p^j) + \alpha x_i(1 - x_i - x_j))]$$
$$= rv_i^i - (p^j + v_j^i)x_j[(p^j - p^i) - \alpha(1 - 2x_i)], \tag{B.12}$$

and

$$(p^j + v_j^i)[(1 - x_i - 2x_j)(\alpha(x_j - x_i)) - (p^j - p^i) + \alpha x_j(1 - x_i - x_j)]$$
$$= rv_j^i - (p^i + v_i^i)x_i[(p^i - p^j) - \alpha(1 - 2x_j)], \tag{B.13}$$

so that combining (B.10)–(B.13) yields $\bar{v}_i^i = \bar{v}_j^i = 0$, and

$$\bar{p}^i = \frac{\alpha(x_i - x_j)x_j}{x_j + x_i}, \quad \bar{p}^j = \frac{\alpha(x_j - x_i)x_j}{x_j + x_i},$$

for all $i, j \in \{1, 2\}$ with $i \neq j$. Since the same holds for firm j, one can conclude that $\bar{p}^i = \bar{p}^j = 0$, and $\bar{x}_i = \bar{x}_j$. Taking into account the fact that the steady state lies on the curve $C(x_0)$, one obtains that

$$\bar{x} = (\bar{x}_i, \bar{x}_j) = \left(\frac{1 - \sqrt{1 - 4x_{i0}x_{j0}}}{2}, \frac{1 + \sqrt{1 - 4x_{i0}x_{j0}}}{2} \right).$$

Thus, the model has three stationary points: two of them are given by the intersections of $C(x_0)$ with the line $x_i + x_j = 1$, and the third stationary point lies at the intersection of $C(x_0)$ with the line $x_i = x_j$ (figure B.16). The dynamics are such that optimal state trajectory lies on the curve $C(x_0)$.

For $x_{10} = x_{20}$, the initial state coincides with a stationary point, so the system will remain at this state forever.

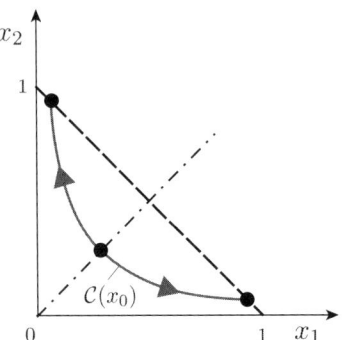

Figure B.16
Equilibrium phase trajectories.

Consider a small neighborhood of the stationary point $\bar{x}$ with $\bar{x}_j > \bar{x}_i$, and assume that the system starts in this neighborhood (at a point $x_0 \neq \bar{x}$ on $\mathcal{C}(x_0)$). Then $\dot{x}^i < 0$ and $\dot{x}^j > 0$, so the system moves along the curve from the third stationary point toward this first stationary point (see figure B.16).

d. For a typical Nash-equilibrium state trajectory $x^*(t)$, $t \geq 0$, see figure B.16.

e. When firm i starts out with a higher installed base, then it is best for firm j to charge nothing for its product, since it cannot prevent the erosion of its customer base. Charging zero is the best that firm i can do in order to have positive sales in the future (which, at least in the standard open-loop equilibrium, will not happen). On the other hand, firm i is able to charge a positive price and still gain new customers, slowly spending its brand premium $x^i - x^j$ (i.e., the difference between its installed base and firm j's installed base).

Remark B.6 Using trigger-strategy equilibria it is possible to obtain periodic equilibria in which the firms "pump" customers back and forth along $\mathcal{C}(x_0)$, alternating between approaching neighborhoods of stationary points. □

4.4 (Industrial Pollution)
It is convenient to consider throughout the general case where

$$(\gamma^1, \gamma^2) \in \{0, 1\}^2.$$

a. In the static version of the duopoly game the state of the system is not evolving, so necessarily

Appendix B

$$q_0^1 + q_0^2 = \beta x_0$$

and

$$u_0^i = \delta y_{i0}$$

for all $i \in \{1,2\}$, where (x_0, y_{i0}) and (q_0^i, u_0^i) correspond to the stationary values of the state and control variables. Given firm j's stationary strategy (q_0^j, u_0^j), firm $i \in \{1,2\} \setminus \{j\}$ chooses a (set of) best response(s) that maximizes its total payoff,

$$BR^i(q_0^j) = \arg\max_{q_0^i \in [0, \bar{u}/\delta]} \left\{ (1 - q_0^i - q_0^j)q_0^i - c^i q_0^i - \gamma^i \frac{(q_0^i + q_0^j)^2}{2\beta^2} - \kappa \frac{\delta^2 (q_0^i)^2}{2} \right\},$$

where the fact that $q_0^i = y_{i0}$ is used because any overcapacity is not in firm i's best interest. Carrying out the straightforward maximization yields

$$BR^i(q_0^j) = \left\{ \min \left\{ \frac{\left[1 - c^i - \left(1 + \frac{\gamma^i}{\beta^2}\right) q_0^j \right]_+}{2 + \frac{\gamma^i}{\beta^2} + \kappa \delta^2}, \frac{\bar{u}}{\delta} \right\} \right\}.$$

At a Nash equilibrium (q_0^{i*}, q_0^{j*}) of the static duopoly game, $q_0^{i*} \in BR^i(q_0^{j*})$, or equivalently,

$$q_0^{i*} = \min \left\{ \left[\frac{(1 - c^i)\left(2 + \frac{\gamma^j}{\beta^2} + \kappa\delta^2\right) - (1 - c^j)\left(1 + \frac{\gamma^i}{\beta^2}\right)}{\left(2 + \frac{\gamma^i}{\beta^2} + \kappa\delta^2\right)\left(2 + \frac{\gamma^j}{\beta^2} + \kappa\delta^2\right) - \left(1 + \frac{\gamma^i}{\beta^2}\right)\left(1 + \frac{\gamma^j}{\beta^2}\right)} \right]_+, \frac{1 - c^i}{2 + \frac{\gamma^i}{\beta^2} + \kappa\delta^2} \right\},$$

provided that the (by assumption large) upper bound $\bar{u}$ on the capacity-expansion rate (here only used for maintenance) is not binding. Firm i's equilibrium expansion rate is $u_0^{i*} = \delta q_0^{i*}$. In addition, at an interior Nash equilibrium the pollution level is

$$x_0^* = \frac{q_0^{i*} + q_0^{j*}}{\beta}$$

$$= \frac{1}{\beta} \frac{(1 + \kappa\delta^2)(2 - c^i - c^j)}{\left(2 + \frac{\gamma^i}{\beta^2} + \kappa\delta^2\right)\left(2 + \frac{\gamma^j}{\beta^2} + \kappa\delta^2\right) - \left(1 + \frac{\gamma^i}{\beta^2}\right)\left(1 + \frac{\gamma^j}{\beta^2}\right)},$$

and firm i's production capacity is

$y_{i0}^* = q_0^{i*}.$

b. Given an admissible initial state (x_0, y_0), the differential game $\Gamma(x_0, y_0)$ is defined by player i's optimal control problem, given player j's strategy. To keep the solution simple, assume that y_0 is sufficiently small so that $q^i = y_i$, that is, the firm produces at its full capacity.

$$J(u^i|u^j) \longrightarrow \max_{u^i(\cdot)}$$

s.t. $\dot{x}(t) = y_i(t) + y_j(t) - \beta x(t), \quad x(0) = x_0,$
$\dot{y}_i(t) = u^i(t) - \delta y_i(t), \quad y_i(0) = y_0^i,$
$\dot{y}_j(t) = u^j(t) - \delta y_j(t), \quad y_j(0) = y_0^j,$

$u^i(t) \in [0, \bar{u}], \quad \forall t \geq 0,$

where

$$J(u^i|u^j) = \int_0^\infty e^{-rt}\left((1 - y_i(t) - y_j(t) - c^i)y_i(t) - \frac{\kappa(u^i(t))^2 + \gamma^i x^2(t)}{2}\right)dt.$$

The current-value Hamiltonian corresponding to player i's optimal control problem is

$$\hat{H}^i(x, y, u, v^i) = (1 - y_i - y_j - c^i)y_i - \frac{\kappa(u^i)^2 + \gamma^i x^2(t)}{2}$$
$$+ v_x^i(y_i + y_j - \beta x) + v_i^i(u^i - \delta y_i) + v_j^i(u^j - \delta y_j),$$

where $v^i = (v_x^i, v_i^i, v_j^i)$ is the current-value adjoint variable associated with the state (x, y_i, y_j). The PMP yields the following necessary optimality conditions for an open-loop Nash-equilibrium state-control trajectory $(x^*(t), y^*(t), u^*(t))$, $t \geq 0$:

• Adjoint equation

$\dot{v}_x^i(t) = (r + \beta)v_x^i(t) + \gamma^i x^*(t),$

$\dot{v}_i^i(t) = (r + \delta)v_i^i(t) - v_x^i(t) + 2y_i^*(t) + y_j^*(t) - (1 - c^i),$

$\dot{v}_j^i(t) = (r + \delta)v_j^i(t) - v_x^i(t) + y_i^*(t),$

for all $t \geq 0$.

• Transversality

$$\lim_{t \to \infty} e^{-rt}(v_x^i(t), v_i^i(t), v_j^i(t)) = 0.$$

Appendix B   319

- Maximality

$$u^{i*}(t) \in \arg\max_{u^i \in [0,\bar{u}]} \hat{H}^i(x^*(t), y^*(t), (u^i, u^{j*}(t)), v^i(t)),$$

for all $t \geq 0$, so

$$u^{i*}(t) = \min\left\{\frac{v_i^i(t)}{\kappa}, \bar{u}\right\}, \quad \forall t \geq 0.$$

The resulting combined linear Hamiltonian system for both players (which includes the maximality condition) is

$$\dot{z} = Az - b,$$

where

$$z = \begin{bmatrix} x^* \\ y_i^* \\ y_j^* \\ v_x^i \\ v_i^i \\ v_j^i \\ v_x^j \\ v_i^j \\ v_j^j \end{bmatrix}, \quad b = \begin{bmatrix} 0 \\ 0 \\ 0 \\ 0 \\ 1-c^i \\ 0 \\ 0 \\ 1-c^j \\ 0 \end{bmatrix},$$

and

$$A = \begin{bmatrix} -\beta & 1 & 1 & 0 & 0 & 0 & 0 & 0 & 0 \\ 0 & -\delta & 0 & 0 & 1/\kappa & 0 & 0 & 0 & 0 \\ 0 & 0 & -\delta & 0 & 0 & 0 & 0 & 0 & 1/\kappa \\ \gamma^i & 0 & 0 & r+\beta & 0 & 0 & 0 & 0 & 0 \\ 0 & 2 & 1 & 0 & r+\delta & 0 & 0 & 0 & 0 \\ 0 & 1 & 0 & -1 & 0 & r+\delta & 0 & 0 & 0 \\ \gamma^j & 0 & 0 & 0 & 0 & 0 & r+\beta & 0 & 0 \\ 0 & 0 & 1 & 0 & 0 & 0 & -1 & r+\delta & 0 \\ 0 & 1 & 2 & 0 & 0 & 0 & 0 & 0 & r+\delta \end{bmatrix}.$$

The equilibrium turnpike $(\bar{x}^*, \bar{y}^*, \bar{u}^*, \bar{v}^i, \bar{v}^j)' = \bar{z}$ of the Hamiltonian system is determined by

$$\bar{z} = A^{-1}b.$$

Omitting the expressions for the turnpike values of the adjoint variables one obtains

$$\bar{x}^* = \frac{2 - c^i - c^j}{(3 + \delta \kappa r + \delta^2 \kappa)\beta},$$

$$\bar{y}_i^* = \frac{1 - 2c^i + c^j - \delta(\delta + r)(1 - c^i)\kappa}{3 + 4\delta\kappa r + \delta^2(\kappa^2 r^2 + 4\kappa) + 2\delta^3 \kappa^2 r + \delta^4 \kappa^2},$$

$$\bar{y}_j^* = \frac{1 - 2c^j + c^i - \delta(\delta + r)(1 - c^j)\kappa}{3 + 4\delta\kappa r + \delta^2(\kappa^2 r^2 + 4\kappa) + 2\delta^3 \kappa^2 r + \delta^4 \kappa^2},$$

and thus $\bar{u}^{i*} = \delta \bar{y}_i^*$ and $\bar{u}^{j*} = \delta \bar{y}_j^*$. The eigenvalues $\lambda_1, \ldots, \lambda_9$ of the system matrix A are all real and such that

$$\lambda_i \in \left\{ -\beta, r+\beta, r+\delta, \frac{\kappa r \pm \sqrt{\kappa^2 r^2 + 4\kappa^2 r \delta + 4\kappa + 4\delta^2 \kappa^2}}{2\kappa}, \right.$$

$$\left. \frac{\kappa r \pm \sqrt{\kappa^2 r^2 + 4\kappa^2 r \delta + 12\kappa + 4\delta^2 \kappa^2}}{2\kappa} \right\};$$

except for $r+\beta$ and $r+\delta$ the multiplicity of all eigenvalues is 1. Since some of the eigenvalues are positive and some are negative, the turnpike $\bar{z}$, which is an equilibrium of the linear system

$$\dot{z} = Az - b = A(z - \bar{z}),$$

is a saddle point. This insight is consistent with the transversality condition and provides conditions for obtaining the initial values for the adjoint variables. Indeed, the initial values (including the six missing ones) have to be chosen such that the starting vector z_0 is orthogonal to all eigenvectors associated with the six positive eigenvalues. The corresponding six conditions provide the six missing components of the initial value $z(0) = z_0$ (those for the adjoint variable), and then

$$z(t) = \bar{z} + (z_0 - \bar{z})e^{At}, \tag{B.14}$$

which converges to $\bar{z}$ as $t \to \infty$. Explicitly, using the eigenvalue decomposition of A, equation (B.14) for $[x^*(t), y_i^*(t), y_j^*(t)]'$ becomes

$$\begin{bmatrix} x^*(t) - \bar{x}^* \\ y_i^*(t) - \bar{y}_i^* \\ y_j^*(t) - \bar{y}_j^* \end{bmatrix} = \begin{bmatrix} 0 & -\frac{(2\beta+r)(r+\delta+\beta)}{\gamma^j} & \frac{2}{\kappa(\lambda_3+\delta)(\lambda_3+\beta)} \\ \frac{-1}{\kappa(\lambda_2+\delta)} & 0 & \frac{1}{\kappa(\lambda_3+\delta)} \\ \frac{1}{\kappa(\lambda_2+\delta)} & 0 & \frac{1}{\kappa(\lambda_3+\delta)} \end{bmatrix} \begin{bmatrix} \rho_1 e^{\lambda_1 t} \\ \rho_2 e^{\lambda_2 t} \\ \rho_3 e^{\lambda_3 t} \end{bmatrix},$$

Appendix B

where

$$\rho_1 = \frac{1}{2}\kappa(\lambda_2 + \delta)(y_{j0} - \bar{y}_j^* - (y_{i0} - \bar{y}_i^*)),$$

$$\rho_2 = \frac{\gamma_2(y_{i0} - \bar{y}_i^*) - \lambda_3(x_0 - \bar{x}^*) + (y_{j0} - \bar{y}_j^*) - \beta(x_0 - \bar{x}^*)}{(\lambda_3 + \beta)(r\delta + r^2 + 2\beta\delta + 2\beta^2) + 3\beta r(\lambda_3 + \beta^2)},$$

$$\rho_3 = \frac{1}{2}\kappa(\lambda_3 + \delta)\lambda_3(y_{i0} - \bar{y}_i^* + y_{j0} - \bar{y}_j^*),$$

and

$$\lambda_1 = \frac{\kappa r - \sqrt{\kappa^2 r^2 + 4\kappa^2 r\delta + 4\kappa + 4\delta^2 \kappa^2}}{2\kappa},$$

$$\lambda_2 = -\beta,$$

$$\lambda_3 = \frac{\kappa r - \sqrt{\kappa^2 r^2 + 4\kappa^2 r\delta + 12\kappa + 4\delta^2 \kappa^2}}{2\kappa}.$$

c. It is now possible to derive a closed-loop Nash equilibrium of $\Gamma(x_0, y_0)$ in strategies that are affine in the players' capacities, such that

$$u^{i*}(t) = \mu^i(x^*(t), y^*(t)) = \alpha_0^i - \alpha_x^i x(t) - \alpha_i^i y_i^*(t) - \alpha_j^i y_j^*(t),$$

where $\alpha_0^i, \alpha_x^i, \alpha_i^i, \alpha_j^i$ are appropriate coefficients. Firm i's corresponding current-value Hamiltonian becomes

$$\hat{H}^i(x, y, u^i, v^i) = (1 - y_i - y_j - c^i)y_i - \frac{\kappa(u^i)^2 + \gamma^i x^2(t)}{2}$$

$$+ v_x^i(y_i + y_j - \beta x) + v_i^i(u^i - \delta y_i)$$

$$+ v_j^i(\mu^j(x, y) - \delta y_j),$$

where, as in part b, $v^i = (v_x^i, v_i^i, v_j^i)$ is the current-value adjoint variable associated with the state (x, y_i, y_j). The PMP yields the following necessary optimality conditions for a closed-loop Nash-equilibrium state-control trajectory $(x^*(t), y^*(t), u^*(t)), t \geq 0$:

- Adjoint equation

$$\dot{v}_x^i(t) = (r + \beta)v_x^i(t) + \gamma^i x^*(t) + \alpha_x^j v_j^i(t),$$

$$\dot{v}_i^i(t) = (r + \delta)v_i^i(t) - v_x^i(t) + 2y_i^*(t) + y_j^*(t) + \alpha_i^j v_j^i(t) - (1 - c^i),$$

$$\dot{v}_j^i(t) = (r + \delta + \alpha_j^j)v_j^i(t) - v_x^i(t) + y_i^*(t),$$

for all $t \geq 0$.

- Transversality

$$\lim_{t \to \infty} e^{-rt}(v_x^i(t), v_i^i(t), v_j^i(t)) = 0.$$

- Maximality

$$u^{i*}(t) \in \arg\max_{u^i \in [0, \bar{u}]} \hat{H}^i(x^*(t), y^*(t), u^i, v^i(t)),$$

for all $t \geq 0$, so that

$$u^{i*}(t) = \min\left\{\frac{v_i^i(t)}{\kappa}, \bar{u}\right\}, \quad \forall t \geq 0.$$

The combined linear Hamiltonian system for both players (including the maximality condition) becomes

$$\dot{z} = Az - b,$$

where z and b are as in part b and

$$A = \begin{bmatrix} -\beta & 1 & 1 & 0 & 0 & 0 & 0 & 0 & 0 \\ 0 & -\delta & 0 & 0 & 1/\kappa & 0 & 0 & 0 & 0 \\ 0 & 0 & -\delta & 0 & 0 & 0 & 0 & 0 & 1/\kappa \\ \gamma^i & 0 & 0 & r+\beta & 0 & \alpha_x^j & 0 & 0 & 0 \\ 0 & 2 & 1 & 0 & r+\delta & \alpha_i^j & 0 & 0 & 0 \\ 0 & 1 & 0 & -1 & 0 & r+\delta+\alpha_j^j & 0 & 0 & 0 \\ \gamma^j & 0 & 0 & 0 & 0 & 0 & r+\beta & \alpha_x^i & 0 \\ 0 & 0 & 1 & 0 & 0 & 0 & -1 & r+\delta+\alpha_i^i & 0 \\ 0 & 1 & 2 & 0 & 0 & 0 & 0 & \alpha_j^i & r+\delta \end{bmatrix}.$$

The equilibrium turnpike $(\bar{x}^*, \bar{y}^*, \bar{u}^*, \bar{v}^i, \bar{v}^j)' = \bar{z}$ of the Hamiltonian system is determined by

$$\bar{z} = A^{-1}b.$$

The detailed analysis of the closed-loop equilibrium turns out to be complicated but can be sketched as follows. It is convenient to use the Laplace transform because instead of a linear system of ODEs it allows one to consider an equivalent algebraic system. Taking the Laplace transform of the initial value problem $\dot{z} = Az - b, z(0) = z_0$ yields

$$sZ - z_0 = AZ - b,$$

Appendix B

where $Z(s) = \mathcal{L}[z](s) = \int_0^\infty e^{-st}z(t)\,dt$ for $s \in \mathbb{C}$ is the Laplace transform of $z(\cdot)$. Hence,

$$Z = (A - sI)^{-1}(b - z_0),$$

which yields that

$$N_i^i(s) \equiv \mathcal{L}[v_i^i](s) = [0,0,0,0,1,0,0,0,0](A - sI)^{-1}(b - z_0)$$
$$= \mathcal{L}[\kappa u^{i*}](s) = \kappa \mathcal{L}[\alpha_0^i + \alpha_x^i x + \alpha_i^i y_i + \alpha_j^i y_j](s),$$

or equivalently,

$$\frac{\alpha_0^i}{s} + [\alpha_x^i, \alpha_i^i, \alpha_j^i, 0, -1/\kappa, 0, 0, 0, 0](A - sI)^{-1}(b - z_0) = 0,$$

for all $s \neq 0$ (as long as convergent). An analogous relation is obtained for $N_j^j(s) = \kappa \mathcal{L}[\alpha_0^j + \alpha_x^j x + \alpha_i^j y_i + \alpha_j^j y_j](s)$:

$$\frac{\alpha_0^j}{s} + [\alpha_x^j, \alpha_i^j, \alpha_j^j, 0, 0, 0, 0, 0, -1/\kappa](A - sI)^{-1}(b - z_0) = 0,$$

for all $s \neq 0$ (as long as convergent). If one can determine α (possibly together with the missing initial conditions in z_0), then a linear closed-loop equilibrium exists. Yet, the corresponding algebraic calculations turn out to be complicated and are therefore omitted.

d. To illustrate the differences between the static and open-loop Nash equilibria, compare the respective steady states in the symmetric case with $\delta = 0$, as discussed in part c. In that case the static solution becomes

$$x_0^* = \frac{2(1-c)}{3\beta + (1/\beta)}, \qquad y_{i0}^* = \frac{1-c}{3 + (1/\beta^2)},$$

and the open-loop solution is given by

$$\bar{x}^*\big|_{\text{open-loop}} = \frac{2(1-c)}{3\beta}, \qquad \bar{y}_i^*\big|_{\text{open-loop}} = \frac{1-c}{3}.$$

Because of the time consistency of the open-loop equilibrium compared to the stationary solution, which is not a Nash equilibrium, output is larger and thus pollution levels higher in the long run.

e. The static and open-loop equilibria were derived for the general case with $(\gamma^1, \gamma^2) \in \{0,1\}^2$ in parts a and b, respectively. It is important to note that the long-run steady state in the open-loop equilibrium is in fact independent of γ^1, γ^2.

f. When firms lack the ability to commit to their production paths, the aggregate production increases. While this is generally good for markets, as prices become more competitive, decreasing a deadweight loss in the economy, the drawback is that pollution levels are also increasing. In order to reduce publicly harmful stock pollution levels, a regulator needs to find a way to force firms to internalize at least a portion of the social costs of pollution. This could be achieved by a tax on the firms' output or a cap-and-trade market, for instance.

B.5 Mechanism Design

5.1 (Screening)

a. Assume without any loss of generality that the total mass of consumers is 1, divided into masses μ of type-H consumers and $1-\mu$ of type-L consumers. In order to generate a positive utility for consumers, $(1-x)^2$ must be less than 1, which implies that $x \in [0, 2]$. Further observe that since customers care only about the quality through $(1-x)^2$ and that a higher quality is more costly to the producer than a lower one, the only economically meaningful qualities are $x \in [0, 1]$. If a single camera model is produced, there are two cases: either all consumers buy it, or only type-H consumers buy it. In the former case, for a given quality x^m, set the price so that type-L consumers get zero utility, that is, $t^m = \theta_L(1-(1-x^m)^2)/2$. Since the consumers' mass sums to 1, the profit becomes

$$\theta_L \frac{(1-(1-x^m)^2)}{2} - cx^m;$$

it is maximized at

$$x^m = 1 - c/\theta_L.$$

The corresponding price and profit are

$$t^m = \frac{\theta_L}{2}\left[1 - \frac{c^2}{\theta_L^2}\right]$$

and

$$\Pi^m = \frac{(\theta_L - c)^2}{2\theta_L}.$$

In the second case, to sell only to type-H consumers, raise the price to $t^m = \theta_H(1-(1-x^m)^2)/2$, setting the utility of the high-type consumers to zero. The profit is then (since there is a mass μ of type-H consumers)

Appendix B

$$\mu\left[\theta_H \frac{(1-(1-x^m)^2)}{2} - cx^m\right],$$

which is maximized at

$$x^m = 1 - c/\theta_H.$$

The corresponding price and profit are

$$t^m = \frac{\theta_H}{2}\left[1 - \frac{c^2}{\theta_H^2}\right]$$

and

$$\Pi^m = \mu \frac{(\theta_H - c)^2}{2\theta_H}.$$

In particular, shutdown is avoided as long as

$$\mu \leq \frac{\theta_H}{\theta_L}\left(\frac{\theta_L - c}{\theta_H - c}\right)^2.$$

b. If one can perfectly distinguish consumers types, set their respective utilities to zero by letting $t_i = \theta_i(1-(1-x_i)^2)/2$, and optimize with respect to x_i. In part a, this corresponds to

$$\hat{x}_L = 1 - c/\theta_L, \qquad \hat{t}_L = \theta_L\left[1 - \frac{c^2}{\theta_L^2}\right],$$

$$\hat{x}_H = 1 - c/\theta_H, \qquad \hat{t}_H = \theta_H\left[1 - \frac{c^2}{\theta_H^2}\right].$$

The resulting profit is

$$\hat{\Pi} = \mu \frac{(\theta_H - c)^2}{2\theta_H} + (1-\mu)\frac{(\theta_L - c)^2}{2\theta_L},$$

clearly higher than the profit derived earlier in part a.

c. Observe that the utility function $u(t, x, \theta) = \theta(1-(1-x)^2)/2 - t$ satisfies the Spence-Mirrlees condition (5.6), since $\frac{\partial^2 u}{\partial x \partial \theta} = \theta(1-x) \geq 0$. Therefore (see section 5.3), at the optimal contract,

- the individual-rationality constraint of type L is binding,

$$t_L^* = \theta_L \frac{(1-(1-x_L^*)^2)}{2}; \tag{IR-L}$$

- the incentive-compatibility constraint of type H is binding,

$$\theta_H \frac{(1-(1-x_H^*)^2)}{2} - t_H^* = \theta_H \frac{(1-(1-x_L^*)^2)}{2} - t_L^*; \quad \text{(IC-H)}$$

- $x_H^* = x_H$ (where x_H was determined in part b);
- the individual-rationality constraint of type H and the incentive-compatibility constraint of type-L consumers can be ignored.

Moreover,

$$x_L^* \in \arg\max_{x_L \in [0,1]} \left\{ u(x_L, \theta_L) - cx_L - \frac{\mu}{1-\mu} \left(u(x_L, \theta_H) - u(x_L, \theta_L) \right) \right\},$$

which in the current setting yields

$$x_L^* = 1 - \frac{c}{\theta_L - \frac{\mu}{1-\mu}(\theta_H - \theta_L)}.$$

Observe that the no-shutdown condition $\mu \leq \frac{\theta_L - c}{\theta_H - c}$ is equivalent to $\frac{\mu}{1-\mu}(\theta_H - \theta_L) < \theta_L - c$, which has two consequences. First, the denominator, $\theta_L - \frac{\mu}{1-\mu}(\theta_H - \theta_L)$, in the expression for x_L^* is nonnegative. This, together with the fact that $\frac{\mu}{1-\mu}(\theta_H - \theta_L)$ is nonnegative, implies that $x_L^* \leq x_L$, since $x_L = 1 - \frac{c}{\theta_L}$. This result was expected, given the results in section 5.3. Second, the denominator is in fact greater than c, which implies that x_L^* is nonnegative, so that under the no-shutdown condition of part a, screening implies that both types of consumers will be able to buy cameras. From x_L^*, the prices can be easily computed. t_L^* is given by (IR-L):

$$t_L^* = \frac{\theta_L}{2}\left(1 - \frac{c^2}{[\theta_L - \frac{\mu}{1-\mu}(\theta_H - \theta_L)]^2}\right).$$

Similarly, t_H^* is given by (IC-H):

$$t_H^* = \theta_H \frac{1-(1-x_H)^2}{2} - (\theta_H - \theta_L)(1 - (1-x_L^*)^2)$$

$$= t_H - (\theta_H - \theta_L)(1 - (1-x_L^*)^2) < t_H.$$

Therefore, type-H consumers receive an information rent, since they are charged a lower price. The second-best profit is

$$\Pi^* = \mu(t_H^* - cx_H^*) + (1-\mu)(t_L^* - cx_L^*),$$

which, after some computation, can be rewritten as

Appendix B

$$\Pi^* = \mu\{(t_H - cx_H) - [(\theta_H - \theta_L)(1 - (1-x_L^*)^2)]\} \tag{B.15}$$
$$+ (1-\mu)\{(t_L - cx_L) - [(x_L - x_L^*)(c + \theta_L(x_L - x_L^*))]\},$$

from which it is clear that $\Pi^* \leq \hat{\Pi}$. It can also verified that $\Pi^* \geq \Pi^m = (\theta_L - c)$, the last equality holding when both consumer types can buy cameras. From (B.15),

$$\Pi^* - (\theta_L - c) = \mu(t_H^* - cx_H^*) - \mu(\theta_L - c)$$
$$- (1-\mu)((x_L - x_L^*)(c + \theta_L(x_L - x_L^*))) \tag{B.16}$$
$$= (1-\mu)\left[\beta(1-x_L)^2 - \frac{c}{\theta_L - \beta}(c + \theta_L(x_L - x_L^*))\right],$$

where $\beta = \frac{\mu}{1-\mu}(\theta_H - \theta_L)$. Using the fact that $1 - x_L^* = \frac{c}{\theta_L} + \frac{c\beta}{\theta_L - \beta}$ and $x_L - x_L^* = \frac{c\beta}{\theta_L - \beta}$, expression (B.16) is nonnegative if and only if

$$\left(\frac{c}{\theta_L} + \frac{c\beta}{\theta_L - \beta}\right)^2 - \frac{c}{\theta_L - \beta}\left(c + \theta_L \frac{c\beta}{\theta_L - \beta}\right) \geq 0,$$

or equivalently, after simplifications,

$$(1 - \theta_L)(\theta_L - \beta) \geq 0,$$

which is true, since $\beta \leq \theta_L - c$. Thus, it has been verified that $\Pi^* \geq \Pi^m$. When $\mu > \frac{\theta_L - c}{\theta_H - c}$, the market for low-type consumers shuts down, and the profits Π^* and Π^m are equal, with value $\mu(\theta_H - c)$.

d. Referring to section 5.3,

$$\Phi(x, \theta) = S(x, \theta) - u_\theta(x, \theta)(1 - F(\theta))/f(\theta),$$

where $f(\theta) \equiv 1$, $F(\theta) \equiv \theta$,

$$S(x, \theta) = \theta \frac{1 - (1-x)^2}{2} - cx,$$

and

$$u_\theta(x, \theta) = \frac{1 - (1-x)^2}{2}.$$

Therefore, $x^*(\theta)$ maximizes

$$(2\theta - 1)\frac{1 - (1-x^2)}{2} - cx,$$

which yields

$$x^*(\theta) = \begin{cases} 1 - c/(2\theta - 1) & \text{if } \theta \geq (1+c)/2, \\ 0 & \text{otherwise.} \end{cases}$$

With this, the optimal nonlinear pricing schedule becomes

$$t^*(\theta) = \int_{(1+c)/2}^{\theta} \left(u_x(x^*(\vartheta), \vartheta) \frac{dx^*(\vartheta)}{d\vartheta} \right) d\vartheta$$

$$= \begin{cases} \frac{1}{4} + \frac{c}{2} - \left(\frac{4\theta - 1}{4(2\theta - 1)^2} \right) c^2 & \text{if } \theta \geq (1+c)/2, \\ 0 & \text{otherwise.} \end{cases}$$

Note that for $c \geq 1$ a complete shutdown would occur, with the firm not providing any product to any type.

5.2 (Nonlinear Pricing with Congestion Externalities)

a. By the revelation principle, the principal can restrict attention to direct revelation mechanisms $\mathcal{M} = (\Theta, \rho)$ with message space Θ (equal to the type space) and allocation function $\rho = (t, x) : \Theta \to \mathbb{R} \times [0,1]$, which satisfy the incentive-compatibility constraint

$$\theta \in \arg\max_{\hat{\theta} \in \Theta} U(t(\hat{\theta}), x(\hat{\theta}), y, \theta) = \arg\max_{\hat{\theta} \in \Theta} \{\theta x(\hat{\theta})(1 - \alpha y) - t(\hat{\theta})\}, \quad \text{(B.17)}$$

for all $\theta \in \Theta$, where

$$y = \int_0^1 x(\theta) \, dF(\theta).$$

The first-order condition corresponding to agent θ's problem of sending his optimal message in (B.17) is

$$U_t(t(\theta), x(\theta), y, \theta)\dot{t}(\theta) + U_x(t(\theta), x(\theta), y, \theta)\dot{x}(\theta) = -\dot{t}(\theta) + \theta(1 - \alpha y)\dot{x}(\theta)$$

$$= 0,$$

for all $\theta \in \Theta$. As shown in Weber (2005b), this first-order condition and the single-crossing condition,

$$(\hat{\theta} - \theta)(U_\theta(t(\hat{\theta}), x(\hat{\theta}), y, \theta) - U_\theta(t(\theta), x(\theta), y, \theta))$$

$$= (\hat{\theta} - \theta)(x(\hat{\theta}) - x(\theta))(1 - \alpha y) \geq 0,$$

for all $\hat{\theta}, \theta \in \Theta$, together are equivalent to the incentive-compatibility constraint (B.17). The single-crossing condition can be simplified to $(\hat{\theta} - \theta)(x(\hat{\theta}) - x(\theta)) \geq 0$ for all $\hat{\theta}, \theta \in \Theta$ (since in general $\alpha y < 1$), which is equivalent to $x(\theta)$ being nondecreasing on the type space Θ. Thus, the

Appendix B

principal's mechanism design problem can be formulated equivalently in terms of the following optimal control problem:

$$J(u) = \int_0^1 \left(t(\theta) - \frac{c(x(\theta))^2}{2} \right) f(\theta)\, d\theta \longrightarrow \max_{u(\cdot)}$$

s.t. $\dot{t} = \theta(1-\alpha y)u,\qquad t(0) = 0,$

$\dot{x} = u,\qquad x(0) = 0,$

$\dot{z} = xf(\theta),\qquad (z(0), z(1)) = (0, y),$

$u(\theta) \in [0, \bar{u}],$

$\forall\, \theta \in [0, 1],$

where $\bar{u} > 0$ is a given (large) control bound.

b. The Hamiltonian for the optimal control problem formulated in part a. is

$$H(\theta, t, x, z, u, \psi) = \left(t - \frac{cx^2}{2} \right) f(\theta) + \psi_t \theta(1-\alpha y)u + \psi_x u + \psi_z x f(\theta),$$

where $\psi = (\psi_t, \psi_x, \psi_z)$ is the adjoint variable associated with the state (t, x, z) of the system. The PMP provides the following necessary optimality conditions, satisfied by an optimal state-control trajectory $(t^*(\theta), x^*(\theta), z^*(\theta))$, $\theta \in [0, 1]$:

- Adjoint equation

$-\dot{\psi}_t(\theta) = f(\theta),$

$-\dot{\psi}_x(\theta) = \big(\psi_z(\theta) - cx^*(\theta)\big)f(\theta),$

$-\dot{\psi}_z(\theta) = 0,$

for all $\theta \in [0, 1]$.

- Transversality

$(\psi_t(1), \psi_x(1), \psi_z(1)) = 0.$

- Maximality

$$u^*(\theta) \in \arg\max_{u \in [0, \bar{u}]} \{(\psi_t(\theta)\theta(1-\alpha y) + \psi_x(\theta))u\},$$

for (almost) all $\theta \in [0, 1]$.

From the adjoint equation and transversality condition it follows immediately that

$$\psi_t(\theta) = 1 - F(\theta), \qquad \psi_x(\theta) = -c \int_\theta^1 x^*(\vartheta)\, dF(\vartheta), \qquad \psi_z(\theta) = 0,$$

for all $\theta \in [0,1]$. The maximality condition yields that

$$u^*(\theta) \in \begin{cases} \{\bar{u}\} & \text{if } (1-\alpha y)\theta\, \psi_t(\theta) + \psi_x(\theta) > 0, \\ [0, \bar{u}] & \text{if } (1-\alpha y)\theta\, \psi_t(\theta) + \psi_x(\theta) = 0, \\ \{0\} & \text{otherwise,} \end{cases}$$

for all $\theta \in [0,1]$. Because of the linearity of the Hamiltonian in the control u, first examine the possibility of a singular control, that is, the situation where $(1-\alpha y)\theta\, \psi_t(\theta) + \psi_x(\theta) \equiv 0$ on an interval $[\theta_0, 1]$, for some $\theta_0 \in [0,1)$ that represents the lowest participating type. This is equivalent to

$$(1-\alpha y)\theta(1 - F(\theta)) = c \int_\theta^1 x^*(\vartheta)\, dF(\vartheta), \qquad \forall\, \theta \in [\theta_0, 1].$$

Differentiating with respect to θ yields that

$$x^*(\theta) = \frac{1-\alpha y}{c}\left(\theta - \frac{1-F(\theta)}{f(\theta)}\right), \qquad \forall\, \theta \in [\theta_0, 1].$$

The latter satisfies the control constraint $\dot{x}^*(\theta) = u^*(\theta) \geq 0$ as long as the function

$$\theta - \frac{1-F(\theta)}{f(\theta)}$$

is nondecreasing, which is a standard distributional assumption in mechanism design. For simplicity, this is assumed here.[21] In addition, $x^*(\theta) \geq 0$ is needed, so (omitting some of the details),

$$\begin{aligned} x^*(\theta) &= \frac{1-\alpha y}{c}\left[\theta - \frac{1-F(\theta)}{f(\theta)}\right]_+ \\ &= \begin{cases} \frac{1-\alpha y}{c}\left(\theta - \frac{1-F(\theta)}{f(\theta)}\right) & \text{if } \theta \in [\theta_0, 1], \\ 0 & \text{otherwise,} \end{cases} \end{aligned}$$

---

21. For example, this assumption is satisfied by any distribution F that has a nondecreasing hazard rate, such as the normal, exponential, or uniform distributions. When this assumption is not satisfied, then based on the maximality condition, it becomes necessary to set $u(\theta)$ on one (or several) type intervals, which is usually referred to as ironing in the mechanism design literature. Ironing leads to bunching of types, i.e., different types are pooled together, and all obtain the same bundle for the same price. For bunched types, the principal does not obtain full revelation; the types are not separated by a revenue-maximizing screening mechanism.

Appendix B

for all $\theta \in [0, 1]$. The missing constants θ_0 and y are such that

$$\theta_0 - \frac{1 - F(\theta_0)}{f(\theta_0)} = 0,$$

and (because $y = \int_0^1 x^*(\vartheta)\, dF(\vartheta)$)

$$y = \frac{\int_{\theta_0}^1 \left(\theta - \frac{1-F(\theta)}{f(\theta)}\right) dF(\theta)}{c + \alpha \int_{\theta_0}^1 \left(\theta - \frac{1-F(\theta)}{f(\theta)}\right) dF(\theta)}.$$

In addition,

$$t^*(\theta) = (1 - \alpha y) \int_{\theta_0}^\theta \vartheta \dot{x}^*(\vartheta)\, d\vartheta$$

$$= \frac{(1-\alpha y)^2}{c} \int_{\theta_0}^\theta \vartheta \left(1 - \frac{d}{d\vartheta} \frac{1-F(\vartheta)}{f(\vartheta)}\right) d\vartheta.$$

Remark B.7 If the agent types are distributed uniformly on $\Theta = [0, 1]$, so that $F(\theta) \equiv \theta$, then $\theta_0 = 1/2$, and

$$y = \frac{\int_{1/2}^1 (2\theta - 1)\, d\theta}{c + \alpha \int_{1/2}^1 (2\theta - 1)\, d\theta} = \frac{1}{\alpha + 4c},$$

whence

$$x^*(\theta) = \frac{[2\theta - 1]_+}{c + (\alpha/4)} \quad \text{and} \quad t^*(\theta) = \frac{c\left[\theta^2 - (1/4)\right]_+}{(c + (\alpha/4))^2},$$

for all $\theta \in [0, 1]$. □

c. The taxation principle states that given an optimal screening contract $(t^*(\theta), x^*(\theta))$, $\theta \in \Theta$, the optimal price-quantity schedule $\tau^*(x)$, $x \in [0, 1]$, can be found as follows:

$$\tau^*(x) = \begin{cases} t^*(\theta) & \text{if } \exists \theta \in \Theta \text{ s.t. } x^*(\theta) = x, \\ \infty & \text{otherwise.} \end{cases}$$

Given that $x^*(\theta)$ is nondecreasing on Θ, let $\varphi(x)$ be its (set-valued) inverse such that

$$\theta \in \varphi(x^*(\theta)), \qquad \forall \theta \in \Theta.$$

By incentive compatibility,

$$\forall x \in x^*(\Theta): \quad \hat{\theta}, \theta \in \varphi(x) \Rightarrow t^*(\hat{\theta}) = t^*(\theta).$$

Hence, the optimal price-quantity schedule $\tau^*(x)$ is single-valued, and
$$\tau^*(x) \in t^*(\varphi(x)), \qquad \forall x \in x^*(\Theta),$$
whereas $\tau^*(x) = \infty$ for all $x \notin x^*(\Theta)$.

Remark B.8 A uniform type distribution (see remark B.7) results in $x^*(\Theta) = [0, 1/(c + \alpha/4)]$ and
$$\varphi(x) = \begin{cases} \{\frac{1}{2}\left(1 + \left(c + \frac{\alpha}{4}\right)x\right)\} & \text{if } 0 < x \leq \frac{1}{c+(\alpha/4)}, \\ [0, \frac{1}{2}] & \text{if } x = 0. \end{cases}$$

This implies that
$$\tau^*(x) = \begin{cases} \frac{c/4}{(c+(\alpha/4))^2}\left(\left(1 + \left(c + \frac{\alpha}{4}\right)x\right)^2 - 1\right) & \text{if } 0 < x \leq \frac{1}{c+(\alpha/4)}, \\ \infty & \text{otherwise}. \end{cases}$$

□

d. Clearly, an increase in all agents' sensitivity α to the congestion externality decreases any type θ's consumption $x^*(\theta; \alpha)$ as well as the total transfer $t^*(\theta; \alpha)$ that this type pays to the principal. Somewhat less obvious is the fact that for a given bandwidth, the optimal price-quantity schedule $\tau^*(x; \alpha)$ is also decreasing in α (which can be shown by straightforward differentiation, at least for the simple example discussed in the preceding remarks).

Appendix C: Intellectual Heritage

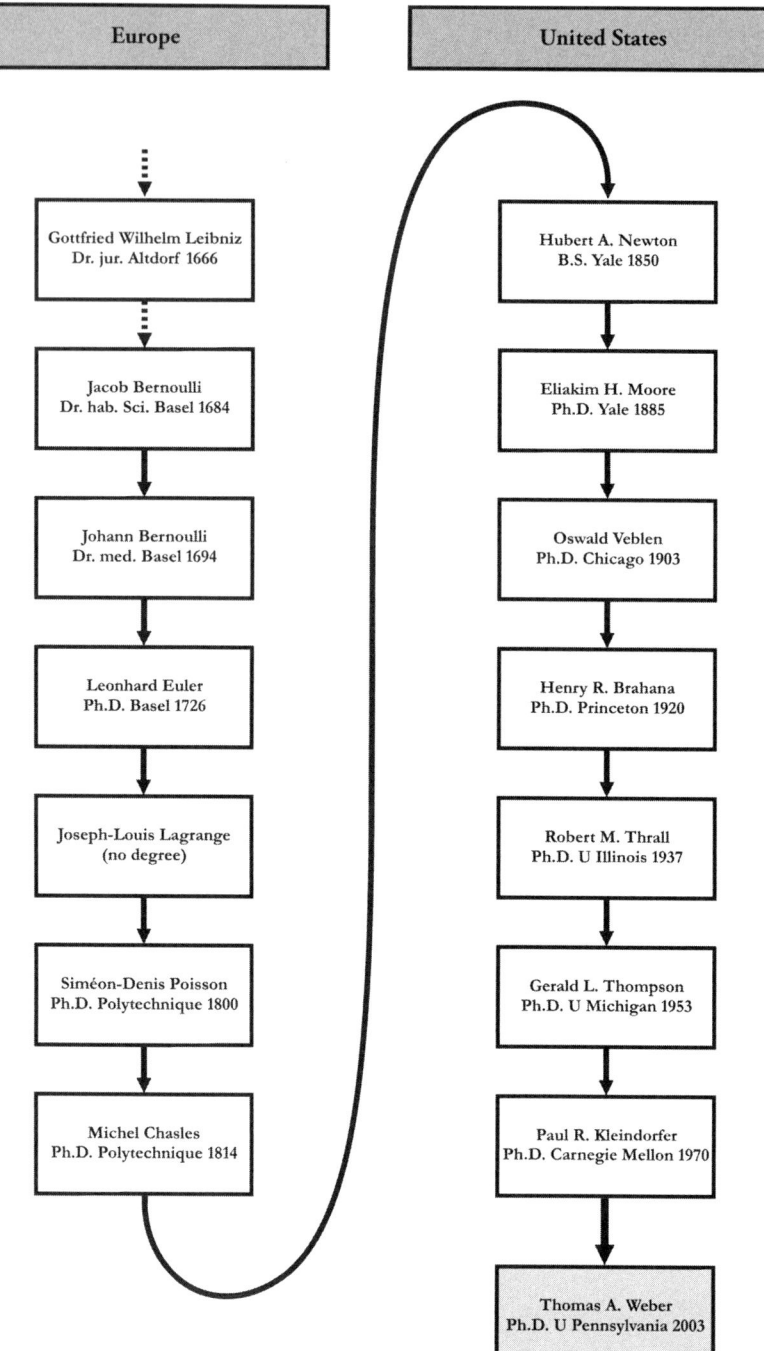

Figure C.1
Source: Mathematics Genealogy Project, http://genealogy.math.ndsu.nodak.edu.

References

Abreu, D., P. K. Dutta, and L. Smith. 1994. The Folk Theorem for Repeated Games: A NEU Condition. *Econometrica* 62 (2): 939–948.

Acemoglu, D. 2009. *Introduction to Modern Economic Growth.* Princeton, N.J.: Princeton University Press.

Afanas'ev, A. P., V. V. Dikusar, A. A. Milyutin, and S. V. Chukanov. 1990. *Necessary Conditions in Optimal Control.* Moscow: Nauka. In Russian.

Aghion, P., and P. Howitt. 2009. *The Economics of Growth.* Cambridge, Mass.: MIT Press.

Akerlof, G. A. 1970. The Market for "Lemons": Quality Uncertainty and the Market Mechanism. *Quarterly Journal of Economics* 84 (3): 488–500.

Aleksandrov, A. D., A. N. Komogorov, and M. A., Lavrentyev. 1969. *Mathematics: Its Content, Methods, and Meaning.* Cambridge, Mass.: MIT Press. Reprinted: Dover Publications, Mineola, N.Y., 1999.

Allee, W. C. 1931. *Animal Aggregations: A Study in General Sociology.* Chicago: University of Chicago Press.

———. 1938. *The Social Life of Animals.* London: Heinemann.

Ampère, A.-M. 1843. *Essai sur la Philosophie des Sciences. Seconde Partie.* Paris: Bachelier.

Anderson, B.D.O., and J. B. Moore. 1971. *Linear Optimal Control.* Englewood Cliffs, N.J.: Prentice-Hall.

Aoki, M. 2001. *Toward a Comparative Institutional Analysis.* Cambridge, Mass.: MIT Press.

Apostol, T. M. 1974. *Mathematical Analysis.* 2d ed. Reading, Mass.: Addison Wesley.

Armstrong, M. 1996. Multiproduct Nonlinear Pricing. *Econometrica* 64 (1): 51–75.

Armstrong, M. A. 1983. *Basic Topology.* New York: Springer.

Arnold, V. I. 1973. *Ordinary Differential Equations.* Cambridge, Mass.: MIT Press.

———. 1988. *Geometrical Methods in the Theory of Ordinary Differential Equations.* 2d ed. New York: Springer.

———. 1989. *Mathematical Methods of Classical Mechanics.* New York: Springer.

———. 1992. *Catastrophe Theory.* 3d ed. New York: Springer.

———. 2000. *"Zhestkie" i "Miagkie" Matematicheskie Modeli ("Rigid" and "Flexible" Models in Mathematics)*. Moscow: Mcnmo.

Arnold, V. I., V. S. Afrajmovich, Y. S. Il'yashenko, and L. P. Shil'nikov. 1999. *Bifurcation Theory and Catastrophe Theory*. New York: Springer.

Arnold, V. I., and Y. S. Il'yashenko. 1988. Ordinary Differential Equations. In *Encyclopedia of Mathematical Sciences*, Vol. 1, ed. D. V. Anosov and V. I. Arnold, 1–148.

Arrow, K. J. 1968. Applications of Control Theory to Economic Growth. In *Mathematics of the Decision Sciences. Part 2*, ed. G. B. Dantzig and A. F. Veinott, 85–119. Lectures in Applied Mathematics, Vol. 12. Providence, R.I.: American Mathematical Society.

Arrow, K. J., and M. Kurz. 1970a. Optimal Growth with Irreversible Investment in a Ramsey Model. *Econometrica* 38 (2): 331–344.

———. 1970b. *Public Investment, the Rate of Return, and Optimal Fiscal Policy*. Baltimore, Md.: Johns Hopkins Press.

Arutyunov, A. V. 1999. Pontryagin's Maximum Principle in Optimal Control Theory. *Journal of Mathematical Sciences* 94 (3): 1311–1365.

———. 2000. *Optimality Conditions: Abnormal and Degenerate Problems*. Boston: Kluwer.

Aseev, S. M. 1999. Methods of Regularization in Nonsmooth Problems of Dynamic Optimization. *Journal of Mathematical Sciences* 94 (3): 1366–1393.

———. 2009. *Infinite-Horizon Optimal Control with Applications in Growth Theory*. Lecture Notes. Moscow: Moscow State University; MAKS Press.

Aseev, S. M., and A. V. Kryazhimskii. 2007. The Pontryagin Maximum Principle and Optimal Economic Growth Problems. *Proceedings of the Steklov Institute of Mathematics* 257 (1): 1–255.

Aubin, J.-P. 1998. *Optima and Equilibria*. 2d ed. New York: Springer.

Aumann, R. J. 1976. Agreeing to Disagree. *Annals of Statistics* 4 (6): 1236–1239.

Axelrod, R. 1984. *The Evolution of Cooperation*. New York: Basic Books.

Balder, E. J. 1988. Generalized Equilibrium Results for Games with Incomplete Information. *Mathematics of Operations Research* 13 (2): 265–276.

Banach, S. 1922. Sur les Opérations dans les Ensembles Abstraits et leur Application aux Équations Intégrales. *Fundamenta Mathematicae* 3: 133–181.

Barbashin, E. A., and N. N. Krasovskii. 1952. On the Stability of Motion as a Whole. *Doklady Akademii Nauk SSSR* 86 (3): 453–456. In Russian.

Barrow-Green, J. 1997. *Poincaré and the Three-Body Problem*. History of Mathematics, Vol. 11. Providence, R.I.: American Mathematical Society.

Başar, T., and G. J. Olsder. 1995. *Dynamic Noncooperative Game Theory*. 2d ed. New York: Academic Press.

Bass, F. M. 1969. A New Product Growth Model for Consumer Durables. *Management Science* 15 (5): 215–227.

References

Bass, F. M., T. V. Krishnan, and D. C. Jain. 1994. Why the Bass Model Fits without Decision Variables. *Marketing Science* 13 (3): 203–223.

Bellman, R. E. 1953. *Stability Theory of Differential Equations.* New York: McGraw-Hill.

———. 1957. *Dynamic Programming.* Princeton, N.J.: Princeton University Press.

Bendixson, I. O. 1901. Sur les Courbes Définies par des Équations Différentielles. *Acta Mathematica* 24 (1): 1–88.

Berge, C. 1959. *Espaces Topologiques et Fonctions Multivoques.* Paris: Dunod. English translation: *Topological Spaces,* trans. E. M. Patterson. Edinburgh: Oliver and Boyd, 1963. Reprinted: Dover Publications, Mineola, N.Y., 1997.

Bertsekas, D. P. 1995. *Nonlinear Programming.* Belmont, Mass.: Athena Scientific.

———. 1996. *Neuro-Dynamic Programming.* Belmont, Mass.: Athena Scientific.

———. 2007. *Dynamic Programming and Optimal Control.* 2 vols. Belmont, Mass.: Athena Scientific.

Besicovitch, A. S. 1928. On Kakeya's Problem and a Similar One. *Mathematische Zeitschrift* 27 (1): 312–320.

Blåsjö, V. 2005. The Isoperimetric Problem. *American Mathematical Monthly* 112 (6): 526–566.

Bluman, G. W., and S. Kumei. 1989. *Symmetries and Differential Equations.* New York: Springer.

Boltyanskii, V. G. 1958. The Maximum Principle in the Theory of Optimal Processes. *Doklady Akademii Nauk SSSR* 119 (6): 1070–1073. In Russian.

———. 1994. The Maximum Principle: How It Came to Be. Report No. 526, Technical University Munich, Germany. Also in Boltyanski[i], Martini, and Soltan 1998, 204–230.

Boltyanski[i], V. G., H. Martini, and P. S. Soltan. 1998. *Geometric Methods and Optimization Problems.* New York: Springer.

Border, K. C. 1985. *Fixed Point Theorems with Applications to Economics and Game Theory.* Cambridge: Cambridge University Press.

Boyd, S., and L. Vandenberghe. 2004. *Convex Optimization.* Cambridge: Cambridge University Press.

Brinkhuis, J., and V. Tikhomirov. 2005. *Optimization: Insights and Applications.* Princeton, N.J.: Princeton University Press.

Bronshtein, I. N., K. A. Semendyayev, G. Musiol, and H. Muehlig. 2004. *Handbook of Mathematics.* 4th ed. New York: Springer.

Buckingham, E. 1914. On Physically Similar Systems: Illustrations of the Use of Dimensional Equations. *Physical Review* 4 (4): 345–376.

Bulow, J. 1982. Durable Goods Monopolists. *Journal of Political Economy* 90 (2): 314–332.

Byrne, O. 1847. *The First Six Books of the Elements of Euclid.* London: Pickering.

Cajori, F. 1919. *A History of the Conceptions of Limits and Fluxions in Great Britain.* London: Open Court Publishing.

Campbell, D. M., and J. C. Higgins, eds. 1984. *Mathematics: People, Problems, Results*. 3 vols. Belmont, Calif.: Wadsworth.

Cantor, M. B. 1907. *Vorlesungen über Geschichte der Mathematik. Erster Band.* 3d ed. Leipzig: B.G. Teubner.

Carlson, D. A., A. B. Haurie, and A. Leizarowitz. 1991. *Infinite Horizon Optimal Control.* 2d ed. New York: Springer.

Cass, D. 1965. Optimum Growth in an Aggregative Model of Capital Accumulation. *Review of Economic Studies* 32 (3): 233–240.

Cauchy, A.-L. 1824/1913. *Oeuvres Complètes. Deuxième Série. Tome XI.* Académie des Sciences, Ministère de l'Instruction Publique. Paris: Gauthiers-Villars.

Cesari, L. 1973. Closure Theorems for Orientor Fields. *Bulletin of the American Mathematical Society* 79 (4): 684–689.

———. 1983. *Optimization Theory and Applications: Problems with Ordinary Differential Equations.* New York: Springer.

Četaev, N. G. 1934. Un Théoreme sur l'Instabilité. *Doklady Akademii Nauk SSSR* 2: 529–534.

Cho, I.-K., and D. M. Kreps. 1987. Signaling Games and Stable Equilibria. *Quarterly Journal of Economics* 102 (2): 179–222.

Clarke, F. H. 1983. *Optimization and Nonsmooth Analysis.* New York: Wiley Interscience.

Clarke, F. H., Y. S. Ledyaev, R. J. Stern, and P. R. Wolenski. 1998. *Nonsmooth Analysis and Control Theory.* New York: Springer.

Coase, R. H. 1972. Durability and Monopoly. *Journal of Law and Economics* 15 (1): 143–149.

Coddington, E. A., and N. Levinson. 1955. *Theory of Ordinary Differential Equations.* New York: McGraw-Hill.

Dikusar, V. V., and A. A. Milyutin. 1989. *Qualitative and Numerical Methods in Maximum Principle.* Moscow: Nauka. In Russian.

Dixit, A. K., and B. J. Nalebuff. 1991. *Thinking Strategically.* New York: Norton.

Dmitruk, A. V. 1993. Maximum Principle for a General Optimal Control Problem with State and Regular Mixed Constraints. *Computational Mathematics and Modeling* 4 (4): 364–377.

Dockner, E., S. Jørgensen, N. Van Long, and G. Sorger. 2000. *Differential Games in Economics and Management Science.* Cambridge: Cambridge University Press.

Dorfman, R. 1969. An Economic Interpretation of Optimal Control Theory. *American Economic Review* 59 (5): 817–831.

Douglas, J. 1931. Solution of the Problem of Plateau. *Transactions of the American Mathematical Society* 33 (1): 263–321.

Dubovitskii, A.Y., and A. A. Milyutin. 1965. Extremum Problems in the Presence of Restrictions. *Zhurnal Vychislitel'noi Matematiki i Matematicheskoi Fiziki* 5 (3): 395–453. In Russian. English translation: *U.S.S.R. Computational Mathematics and Mathematical Physics* 5 (3): 1–80.

References

———. 1981. Theory of the Maximum Principle. In *Methods of the Theory of Extremal Problems in Economics*, ed. V. L. Levin, 138–177. Moscow: Nauka. In Russian.

Dunford, N., and J. T. Schwartz. 1958. *Linear Operators. Part I: General Theory*. New York: Wiley Interscience.

Euler, L. 1744. *Methodus Inveniendi Lineas Curvas Maximi Minimive Proprietate Gaudentes, Sive Solutio Problematis Isoperimetrici Latissimo Sensu Accepti*. Geneva: M.M. Bousquet.

Feinstein, C. D., and D. G. Luenberger. 1981. Analysis of the Asymptotic Behavior of Optimal Control Trajectories: The Implicit Programming Problem. *SIAM Journal on Control and Optimization* 19 (5): 561–585.

Filippov, A. F. 1962. On Certain Questions in the Theory of Optimal Control. *SIAM Journal on Control* Ser. A. 1 (1): 76–84.

———. 1988. *Differential Equations with Discontinuous Righthand Sides*. Dordrecht, Netherlands: Kluwer.

Fomenko, A. T. 1990. *The Plateau Problem: Historical Survey*. New York: Gordon and Breach.

Fraser-Andrews, G. 1996. Shooting Method for the Numerical Solution of Optimal Control Problems with Bounded State Variables. *Journal of Optimization Theory and Applications* 89 (2): 351–372.

Friedman, J. 1971. A Non-Cooperative Equilibrium for Supergames. *Review of Economic Studies* 28 (1): 1–12.

Fudenberg, D., and E. Maskin. 1986. The Folk Theorem in Repeated Games with Discounting or with Incomplete Information. *Econometrica* 54 (3): 533–554.

Fudenberg, D., and J. Tirole. 1991. *Game Theory*. Cambridge, Mass.: MIT Press.

Galilei, G. 1638. *Discorsi e Dimonstrazioni Matematiche Intorno à Due Nuoue Scienze*. Leiden, Netherlands: Elsevirii.

Gamkrelidze, R. V. 1959. Time-Optimal Processes with Bounded State Coordinates. *Doklady Akademii Nauk SSSR* 125: 475–478.

———. 1978. *Principles of Optimal Control Theory*. New York: Plenum Press.

———. 1999. Discovery of the Maximum Principle. *Journal of Dynamical and Control Systems* 5 (4): 437–451.

———. 2008. The "Pontryagin Derivative" in Optimal Control. *Doklady Mathematics* 77 (3): 329–331.

Geanakoplos, J. 1992. Common Knowledge. *Journal of Economic Perspectives* 6 (4): 53–82.

Gelfand, I. M., and S. V. Fomin. 1963. *Calculus of Variations*. Englewood Cliffs, N.J.: Prentice-Hall. Reprinted: Dover Publications, Mineola, N.Y., 2000.

Giaquinta, M., and S. Hildebrandt. 1996. *Calculus of Variations*. 2 vols. New York: Springer.

Gibbard, A. 1973. Manipulation of Voting Schemes: A General Result. *Econometrica* 41 (4): 587–601.

Gibbons, R. 1992. *Game Theory for Applied Economists*. Princeton, N.J.: Princeton University Press.

Godunov, S. K. 1994. *Ordinary Differential Equations with Constant Coefficient*. Providence, R.I.: American Mathematical Society.

Goldstine, H. H. 1980. *A History of the Calculus of Variations from the Seventeenth through the Nineteenth Century*. New York: Springer.

Granas, A., and J. Dugundji. 2003. *Fixed Point Theory*. New York: Springer.

Green, J., and J.-J. Laffont. 1979. On Coalition Incentive Compatibility. *Review of Economic Studies* 46 (2): 243–254.

Gronwall, T. H. 1919. Note on the Derivatives with Respect to a Parameter of the Solutions of a System of Differential Equations. *Annals of Mathematics* 20 (3): 292–296.

Guesnerie, R., and J.-J. Laffont. 1984. A Complete Solution to a Class of Principal-Agent Problems with an Application to the Control of a Self-Managed Firm. *Journal of Public Economics* 25 (3): 329–369.

Guillemin, V., and A. Pollack. 1974. *Differential Topology*. Englewood Cliffs, N.J.: Prentice-Hall.

Gül, F., H. Sonnenschein, and R. B. Wilson. 1986. Foundations of Dynamic Monopoly and the Coase Conjecture. *Journal of Economic Theory* 39 (1): 155–190.

Hadamard, J. 1902. Sur les Problèmes aux Dérivées Partielles et Leur Signification Physique. *Princeton University Bulletin*, no. 23, 49–52.

Halkin, H. 1974. Necessary Conditions for Optimal Control Problems with Infinite Horizons. *Econometrica* 42 (2): 267–272.

Hamilton, W. R. 1834. On a General Method in Dynamics. *Philosophical Transactions of the Royal Society of London* 124: 247–308.

Harsanyi, J. C. 1967. Games with Incomplete Information Played by "Bayesian" Players. Part I: The Basic Model. *Management Science* 14 (3): 159–182.

Hartl, R. F., S. P. Sethi, and R. G. Vickson. 1995. A Survey of the Maximum Principles for Optimal Control Problems with State Constraints. *SIAM Review* 37 (2): 181–218.

Hartman, P. 1964. *Ordinary Differential Equations*. New York: Wiley.

Hildebrandt, S. 1989. The Calculus of Variations Today. *Mathematical Intelligencer* 11 (4): 50–60.

Hildebrandt, S., and A. Tromba. 1985. *Mathematics and Optimal Form*. New York: Freeman.

Hotelling, H. 1931. The Economics of Exhaustible Resources. *Journal of Political Economy* 39 (2): 137–175.

Hungerford, T. W. 1974. *Algebra*. New York: Springer.

Hurwicz, L., and S. Reiter. 2006. *Designing Economic Mechanisms*. New York: Cambridge University Press.

Hurwitz, A. 1895. Über die Bedingungen, unter welchen eine Gleichung nur Wurzeln mit negativen reellen Theilen besitzt. *Mathematische Annalen* 46 (2): 273–284.

Huygens, C. 1673. *Horologium Oscillatorium*. Paris: F. Muguet.

References

Ioffe, A. D., and V. M. Tikhomirov. 1979. *Theory of Extremal Problems*. Amsterdam: North Holland.

Isidori, A. 1995. *Nonlinear Control Systems*. New York: Springer.

Jacobi, C.G.J. 1884. Vorlesungen über Dynamik. In *Gesammelte Werke: Supplementband*, ed. E. Lottner. Berlin: G. Reimer.

Jordan, C. 1909. *Cours d'Analyse de l'École Polytechnique*. 3 vols. Paris: Gauthier-Villars.

Jowett, B. 1881.*The Republic of Plato*. 3d ed. Oxford: Clarendon Press.

Kakeya, S. 1917. Some Problems on Maximum and Minimum Regarding Ovals. *Tôhoku Science Reports* 6: 71–88.

Kakutani, S. 1941. A Generalization of Brouwer's Fixed Point Theorem. *Duke Mathematical Journal* 8 (3): 457–459.

Kalman, R. E., and R. S. Bucy. 1961. New Results in Linear Filtering and Prediction Theory. *Transactions of the ASME Ser. D: Journal of Basic Engineering* 83 (3): 95–108.

Kamien, M. I., and N. L. Schwartz. 1991. *Dynamic Optimization: The Calculus of Variations and Optimal Control in Economics and Management*. 2d ed. Amsterdam: Elsevier.

Kant, I. 1781. *Kritik der Reinen Vernunft*. Riga: J.F. Hartknoch.

Karp, L., and D. M. Newbery. 1993. Intertemporal Consistency Issues in Depletable Resources. In *Handbook of Natural Resource and Energy Economics*, vol. 2, ed. A. V. Kneese and J. L. Sweeney, 881–931. Amsterdam: Elsevier.

Keerthi, S. S., and E. G. Gilbert. 1988. Optimal Infinite-Horizon Feedback Laws for a General Class of Constrained Discrete-Time Systems: Stability and Moving-Horizon Approximation. *Journal of Optimization Theory and Applications* 57 (2): 265–293.

Khalil, H. K. 1992. *Nonlinear Systems*. New York: Macmillan.

Kihlstrom, R. E., and M. H. Riordan. 1984. Advertising as a Signal. *Journal of Political Economy* 92 (3): 427–450.

Kirillov, A. A., and A. D. Gvishiani. 1982. *Theorems and Problems in Functional Analysis*. New York: Springer.

Kolmogorov, A. N., and S. V. Fomin. 1957. *Elements of the Theory of Functions and Functional Analysis*. 2 vols. Rochester, N.Y.: Graylock Press. Reprinted: Dover Publications, Mineola, N.Y., 1999.

Koopmans, T. C. 1965. On the Concept of Optimal Economic Growth. *The Econometric Approach to Development Planning. Pontificiae Academiae Scientiarum Scripta Varia* 28: 225–300. Reissued: North-Holland, Amsterdam, 1966.

Kreps, D. M. 1990. *Game Theory and Economic Modelling*. Oxford: Clarendon Press.

Kress, R. 1998. *Numerical Analysis*. New York: Springer.

Kurian, G. T., and B. A. Chernov. 2007. *Datapedia of the United States*. Lanham, Md.: Bernan Press.

Kydland, F. E., and E. C. Prescott. 1977. Rules Rather Than Discretion: The Inconsistency of Optimal Plans. *Journal of Political Economy* 85 (3): 473–491.

Laffont, J.-J. 1989. *The Economics of Uncertainty and Information.* Cambridge, Mass.: MIT Press.

Laffont, J.-J., E. Maskin, and J.-C. Rochet. 1987. Optimal Nonlinear Pricing with Two-Dimensional Characteristics. In *Information, Incentives, and Economic Mechanisms: Essays in Honor of Leonid Hurwicz*, ed. T. Groves, R. Radner, and S. Steiger, 255–266. Minneapolis: University of Minnesota Press.

Lagrange, J.-L. 1760. Essai d'une Nouvelle Méthode pour Déterminer les Maxima et les Minima des Formules Intégrales Indéfinies. Reprinted in *Oeuvres*, vol. 1, ed. J.-A. Serret, 335–362. Hildesheim, Germany: G. Olms, 1973.

———. 1788/1811. *Mécanique Analytique, Nouvelle Édition.* Vol. 1. Paris: Mme Ve Courcier. Reprinted: Cambridge University Press, New York, 2009.

Lancaster, K. J. 1966. A New Approach to Consumer Theory. *Journal of Political Economy* 74 (2): 132–157.

Lang, S. 1993. *Real and Functional Analysis.* 3d ed. New York: Springer.

LaSalle, J. P. 1968. Stability Theory for Ordinary Differential Equations. *Journal of Differential Equations* 4 (1): 57–65.

Lee, E. B., and L. Markus. 1967. *Foundations of Optimal Control Theory.* New York: Wiley.

Lee, J. M. 2003. *Introduction to Smooth Manifolds.* New York: Springer.

Leibniz, G. W. 1684. Nova Methodus pro Maximis et Minimis, Itemque Tangentibus, quaenec Fractas, nec Irrationales Quantitates Moratur, et Singularis pro Illis Calculi Genus. *Acta Eruditorum.* Leipzig: J. Grossium and J. F. Gletitschium.

Lorenz, E. N. 1963. Deterministic Nonperiodic Flow. *Journal of the Atmospheric Sciences* 20 (2): 130–141.

Lotka, A. J. 1920. Undamped Oscillations Derived from the Law of Mass Action. *Journal of the American Chemical Society* 42 (8): 1595–1599.

Luenberger, D. G. 1969. *Optimization by Vector Space Methods.* New York: Wiley.

Lyapunov, A. M. 1892. *The General Problem of the Stability of Motion.* Kharkov, Ukraine: Kharkov Mathematical Society. French translation: Problème Général de la Stabilité du Mouvement, trans. A. Liapounoff. *Annales de la Faculté des Sciences de Toulouse, Deuxième Série* 9 (1907): 203–474. English translation: *The General Problem of the Stability of Motion*, trans. A. T. Fuller. London: Taylor and Francis, 1992.

Magaril-Il'yaev, G. G., and V. M. Tikhomirov. 2003. *Convex Analysis: Theory and Applications.* Providence, R.I.: American Mathematical Society.

Mailath, G. J., and L. Samuelson. 2006. *Repeated Games and Reputations: Long-Run Relationships.* Oxford: Oxford University Press.

Mangasarian, O. L. 1966. Sufficient Conditions for the Optimal Control of Nonlinear Systems. *SIAM Journal on Control* 4 (1): 139–152.

Matthews, S. A., and J. H. Moore. 1987. Monopoly Provision of Quality and Warranties: An Exploration in the Theory of Multidimensional Screening. *Econometrica* 55 (2): 441–467.

References

Maupertuis, P.-L. 1744. Accord de Différentes Loix de la Nature qui Avoient Jusqu'ici Paru Incompatibles. *Histoire de l'Académie Royale des Sciences de Paris*, 417–426.

Maxwell, J. C. 1868. On Governors. *Proceedings of the Royal Society* 16 (10): 270–283.

Mayne, D. Q., and H. Michalska. 1990. Receding Horizon Control of Nonlinear Systems. *IEEE Transactions on Automatic Control* 35 (7): 814–824.

Mayr, O. 1970. *The Origins of Feedback Control.* Cambridge, Mass.: MIT Press.

Megginson, R. E. 1998. *An Introduction to Banach Space Theory.* New York: Springer.

Milgrom, P. R., and I. Segal. 2002. Envelope Theorems for Arbitrary Choice Sets. *Econometrica* 70 (2): 583–601.

Milgrom, P. R., and R. J. Weber. 1985. Distributional Strategies for Games with Incomplete Information. *Mathematics of Operations Research* 10: 619–632.

Milyutin, A. A., and N. P. Osmolovskii. 1998. *Calculus of Variations and Optimal Control.* Providence, R.I.: American Mathematical Society.

Mirrlees, J. A. 1971. An Exploration in the Theory of Optimal Income Taxation. *Review of Economic Studies* 38 (2): 175–208.

Moler, C. B., and C. F. Van Loan. 2003. Nineteen Dubious Ways to Compute the Exponential of a Matrix: Twenty-Five Years Later. *SIAM Review* 45 (1): 3–49.

Mordukhovich, B. S. 2006. *Variational Analysis and Generalized Differentiation.* 2 vols. New York: Springer.

Murray, J. D. 2007. *Mathematical Biology.* 3d ed. 2 vols. New York: Springer.

Mussa, M., and S. Rosen. 1978. Monopoly and Product Quality. *Journal of Economic Theory* 18 (2): 301–317.

Myerson, R. B. 1979. Incentive Compatibility and the Bargaining Problem. *Econometrica* 47 (1): 61–74.

Nash, J. F. 1950. Equilibrium Points in *n*-Person Games. *Proceedings of the National Academy of Sciences* 36 (1): 48–49.

Nerlove, M. L., and K. J. Arrow. 1962. Optimal Advertising Policy under Dynamic Conditions. *Economica* 29 (144): 129–142.

Palfrey, T. R., and S. Srivastava. 1993. *Bayesian Implementation.* Chur, Switzerland: Harwood Academic Publishers.

Peano, G. 1890. Démonstration de l'Integrabilité des Équations Différentielles Ordinaires. *Mathematische Annalen* 37 (2): 182–238.

Pearl, J. 2000. *Causality: Models, Reasoning, and Inference.* Cambridge: Cambridge University Press.

Perron, O. 1913. Zur Existenzfrage eines Maximums oder Minimums. *Jahres-bericht der Deutschen Mathematiker-Vereinigung* 22 (5/6): 140–144.

Petrovski[i], I. G. 1966. *Ordinary Differential Equations.* Englewood Cliffs, N.J.: Prentice-Hall. Reprinted: Dover Publications, Mineola, N.Y., 1973.

Poincaré, H. 1892. *Les Méthodes Nouvelles de la Mécanique Céleste.* Vol. I. Paris: Gauthier-Villars.

———. 1928. *Oeuvres de Henri Poincaré.* Vol. 1. Paris: Gauthier-Villars.

Pontryagin, L. S. 1962. *Ordinary Differential Equations.* Reading, Mass.: Addison Wesley.

Pontryagin, L. S., V. G. Boltyanskii, R. V. Gamkrelidze, and E. F. Mishchenko. 1962. *The Mathematical Theory of Optimal Processes.* New York: Wiley Interscience.

Powell, W. B. 2007. *Approximate Dynamic Programming.* New York: Wiley Interscience.

Primbs, J. 2007. Portfolio Optimization Applications of Stochastic Receding Horizon Control. In *Proceedings of the 26th American Control Conference*, New York, 1811–1816.

Radner, R., and R. W. Rosenthal. 1982. Private Information and Pure-Strategy Equilibria. *Mathematics of Operations Research* 7: 401–409.

Radó, T. 1930. On Plateau's Problem. *Annals of Mathematics* 31 (2): 457–469.

Ramsey, F. P. 1928. A Mathematical Theory of Saving. *Economic Journal* 38 (152): 543–559.

Roberts, S. M., and J. S. Shipman. 1972. *Two-Point Boundary Value Problems: Shooting Methods.* New York: Elsevier.

Robertson, R. M. 1949. Mathematical Economics before Cournot. *Journal of Political Economy* 57 (6): 523–536.

Rochet, J.-C. 1985. The Taxation Principle and Multitime Hamilton-Jacobi Equations. *Journal of Mathematical Economics* 14 (2): 113–128.

Rochet, J.-C., and P. Choné. 1998. Ironing, Sweeping and Multidimensional Screening. *Econometrica* 66 (4): 783–826.

Rochet, J.-C., and L. A. Stole. 2003. The Economics of Multidimensional Screening. In *Advances in Economics and Econometrics: Theory and Applications,* vol. 1, ed. M. Dewatripont, L.-P. Hansen, and S. J. Turnovsky, 150–197. New York: Cambridge University Press.

Rockafellar, R. T., and R.J.B. Wets. 2004. *Variational Analysis.* New York: Springer.

Ross, I. M., and F. Fahroo. 2003. Legendre Pseudospectral Approximations of Optimal Control Problems. In *New Trends in Nonlinear Dynamics and Control, and Their Applications,* ed. W. Kang, M. Xiao, and C. Borges, 327–341. Lecture Notes in Control and Information Sciences. Vol. 295. New York: Springer.

Ross, W. D. 1951. *Plato's Theory of Ideas.* Oxford: Clarendon Press.

Routh, E. J. 1877. *A Treatise on the Stability of a Given State of Motion.* London: Macmillan.

Rudin, W. 1976. *Principles of Mathematical Analysis.* 3d ed. New York: McGraw-Hill.

Russell, B. 1961. *A History of Western Philosophy.* 2d ed. New York: Simon and Schuster. Reprinted: Folio Society, London, 2004.

Rutquist, P. E., and M. M. Edvall. 2009. *PROPT: MATLAB Optimal Control Software.* Technical Report. Pullman, Wash.: TOMLAB Optimization.

Sachdev, P. L. 1997. *A Compendium on Nonlinear Ordinary Differential Equations.* New York: Wiley Interscience.

References

Samuelson, P. A. 1970. What Makes for a Beautiful Problem in Science? *Journal of Political Economy* 78 (6): 1372–1377.

Schechter, M. 2004. *An Introduction to Nonlinear Analysis.* Cambridge: Cambridge University Press.

Schwartz, A. L. 1996. Theory and Implementation of Numerical Methods Based on Runge-Kutta Integration for Solving Optimal Control Problems. Ph.D. diss., University of California, Berkeley.

Schwartz, A. L., E. Polak, and Y. Chen. 1997. RIOTS: A MATLAB Toolbox for Solving Optimal Control Problems. Technical Report. Department of Electrical Engineering and Computer Sciences, University of California, Berkeley.

Seierstad, A., and K. Sydsæter. 1987. *Optimal Control Theory with Economic Applications.* Amsterdam: North-Holland.

Sethi, S. P. 1977. Optimal Advertising for the Nerlove-Arrow Model under a Budget Constraint. *Operational Research Quarterly* 28 (3): 683–693.

Sethi, S. P., and G. L. Thompson. 1974. Quantitative Guidelines for a Communicable Disease: A Complete Synthesis. *Biometrics* 30 (4): 681–691.

———. 2000. *Optimal Control Theory: Applications to Management Science and Economics.* 2d ed. Boston: Kluwer.

Sim, Y. C., S. B. Leng, and V. Subramaniam. 2000. A Combined Genetic Algorithms–Shooting Method Approach to Solving Optimal Control Problems. *International Journal of Systems Science* 31 (1): 83–89.

Smirnov, G. V. 2002. *Introduction to the Theory of Differential Inclusions.* Providence, R.I.: American Mathematical Society.

Sontag, E. D. 1998. *Mathematical Control Theory: Deterministic Finite-Dimensional Systems.* 2d ed. New York: Springer.

Spence, A. M. 1973. Job Market Signaling. *Quarterly Journal of Economics* 87 (3): 355–374.

Steiner, J. 1842. Sur le Maximum et le Minimum des Figures dans le Plan, sur la Sphère et dans l'Espace en Général. *Journal für die Reine und Angewandte Mathematik* 1842 (24): 93–162, 189–250.

Stephens, P. A., W. J. Sutherland, and R. P. Freckleton. 1999. What Is the Allee Effect? *Oikos* 87 (1): 185–190.

Stiglitz, J. E. 1975. The Theory of "Screening," Education, and the Distribution of Income. *American Economic Review* 65 (3): 283–300.

Stokey, N. L. 1981. Rational Expectations and Durable Goods Pricing. *Bell Journal of Economics* 12 (1): 112–128.

Strang, G. 2009. *Introduction to Linear Algebra.* 4th ed. Wellesley, Mass.: Wellesley-Cambridge Press.

Struwe, M. 1989. *Plateau's Problem and the Calculus of Variations.* Princeton, N.J.: Princeton University Press.

Sutton, R. S., and A. G. Barto. 1998. *Reinforcement Learning: An Introduction.* Cambridge, Mass.: MIT Press.

Taylor, A. E. 1965. *General Theory of Functions and Integration.* New York: Blaisdell Publishing. Reprinted: Dover Publications, Mineola, N.Y., 1985.

Thom, R. F. 1972. *Stabilité Structurelle et Morphogénèse: Essai d'Une Théorie Générale des Modèles.* Reading, Mass.: W.A. Benjamin.

Thomas, I. 1941. *Greek Mathematical Works.* 2 vols. Cambridge, Mass.: Harvard University Press.

Tikhomirov, V. M. 1986. *Fundamental Principles of the Theory of Extremal Problems.* New York: Wiley Interscience.

Tikhonov, A. N. 1963. Solution of Incorrectly Formulated Problems and the Regularization Method. *Doklady Akademii Nauk SSSR* 151 (4): 501–504. In Russian.

Verri, P. 1771. *Meditazioni Sulla Economia Politica.* Livorno: Stamperia dell'Enciclopedia.

Vidale, M. L., and H. B. Wolfe. 1957. An Operations Research Study of Sales Response to Advertising. *Operations Research* 5 (3): 370–381.

Vinter, R. B. 1988. New Results on the Relationship between Dynamic Programming and the Maximum Principle. *Mathematics of Control, Signals, and Systems* 1 (1): 97–105.

———. 2000. *Optimal Control.* Boston: Birkhäuser.

Volterra, V. 1926. Variazione Fluttuazioni del Numero d'Individui in Specie Animali Conviventi. *Memorie dela R. Academia Nationale dei Lincei.* Ser. 6, vol. 2, 31–113. English translation: Variations and Fluctuations of a Number of Individuals in Animal Species Living Together. In *Animal Ecology*, by R. N. Chapman, 409–448. New York: McGraw-Hill, 1931.

von Neumann, J. 1928. Zur Theorie der Gesellschaftsspiele. *Mathematische Annalen* 100: 295–320.

von Neumann, J., and O. Morgenstern. 1944. *Theory of Games and Economic Behavior.* Princeton, N.J.: Princeton University Press.

Walter, W. 1998. *Ordinary Differential Equations.* New York: Springer.

Warga, J. 1972. *Optimal Control of Differential and Functional Equations.* New York: Academic Press.

Weber, T. A. 1997. *Constrained Predictive Control for Corporate Policy.* Technical Report LIDS-TH 2398. Laboratory for Information and Decision Systems, Massachusetts Institute of Technology, Cambridge, Mass.

———. 2005a. Infinite-Horizon Optimal Advertising in a Market for Durable Goods. *Optimal Control Applications and Methods* 26 (6): 307–336.

———. 2005b. *Screening with Externalities.* Technical Report 2005-05-17. Department of Management Science and Engineering, Stanford University, Stanford, Calif.

———. 2006. An Infinite-Horizon Maximum Principle with Bounds on the Adjoint Variable. *Journal of Economic Dynamics and Control* 30 (2): 229–241.

References

Weierstrass, K. 1879/1927. *Mathematische Werke.* Vol. 7: *Vorlesungen über Variationsrechnung.* Leipzig: Akademische Verlagsgesellschaft.

Weitzman, M. L. 2003. *Income, Wealth, and the Maximum Principle.* Cambridge, Mass.: Harvard University Press.

Wiener, N. 1948. *Cybernetics, or, Control and Communications in the Animal and the Machine.* New York: Wiley.

———. 1950. *The Human Use of Human Beings: Cybernetics and Society.* Boston: Houghton Mifflin.

Wiggins, S. 2003. *Introduction to Applied Nonlinear Dynamic Systems and Chaos.* New York: Springer.

Willems, J. C. 1996. 1696: The Birth of Optimal Control. In *Proceedings of the 35th Conference on Decision and Control,* Kobe, Japan.

Williams, S. R. 2008. *Communication in Mechanism Design: A Differential Approach.* New York: Cambridge University Press.

Wilson, R. B. 1971. Computing Equilibria of n-Person Games. *SIAM Journal on Applied Mathematics* 21 (1): 80–87.

———. 1993. *Nonlinear Pricing.* Oxford: Oxford University Press.

Zermelo, E. 1913. Über eine Anwendung der Mengenlehre auf die Theorie des Schachspiels. In *Proceedings of the Fifth Congress of Mathematicians,* Cambridge, 501–504.

Zorich, V. A. 2004. *Mathematical Analysis.* 2 vols. New York: Springer.

Index

$\{\exists, \forall, \dot{\forall}, \gg, \ldots\}$, 231

a.a. (almost all), 103
Abreu, Dilip, 179
Absolute continuity, 243
Absolute risk aversion, 102
Accumulation point, 236
Acemoglu, Darron (1967–), 11
Acta Eruditorum, 7
Action set, 155
Adjoint equation, 107, 110
Adjoint variable, 82, 96
　current-value, 110
Admissible control, 90, 104
Adverse selection, 182, 207
Advertising, 25, 47, 76, 117, 141
a.e. (almost everywhere), 124
Afanas'ev, Alexander Petrovich (1945–), 141
Afrajmovich, Valentin Senderovich (1945–), 75
Aghion, Philippe (1956–), 11
Akerlof, George Arthur (1940–), 181, 182
Alaoglu, Leonidas (1914–1981), 126
Aleksandrov, Aleksandr Danilovich (1912–1999), 15
Alembert, Jean le Rond d' (1717–1783), 9
Algebraic group, 40
Allee, Warder Clyde (1885–1955), 78
Allee effect, 78, 79
Allegory of the Cave, 6
Allocation function, 216
Ampère, André-Marie (1775–1836), 12
Anderson, Brian David Outram, 140
Anti-derivative, 243
Anti-razor, 11
Aoki, Masahiko (1938–), 179

Apostol, Tom M. (1923–), 252
Approximate dynamic programming, 258
Aristotle (384 B.C.–322 B.C.), 9
Armstrong, Christopher Mark (1964–), 227
Armstrong, Mark Anthony, 241
Arnold, Vladimir Igorevich (1937–2010), 10, 13, 64, 75
Arrow, Kenneth Joseph (1921–), 7, 102
Arrow-Pratt coefficient, 102
Arutyunov, Aram V., 106, 120, 133, 141
Arzelà, Cesare (1847–1912), 241
Arzelà-Ascoli theorem, 241
Ascoli, Giulio (1843–1896), 241
Aseev, Sergey M., 115, 140
Asymptotic stability, 58
Aubin, Jean-Pierre, 251, 252
Auction
　first-price, 163
　second-price, 152
Augmented history, 175
Aumann, Robert John (1930–), 162
Autonomous system. *See* Ordinary differential equation
Axelrod, Robert (1943–), 177

Backward induction, 92, 151
Baker, Henry Frederick (1866–1956), 71
Balder, Erik J., 166, 203
Banach, Stefan (1892–1945), 31, 126, 237, 238
Banach-Alaoglu theorem, 126
Banach fixed-point theorem, 238
Banach space, 31
Banach space (complete normed vector space), 237
Barbashin, Evgenii Alekseevich (1918–1969), 58

Barrow-Green, June (1953–), 13
Barto, Andrew G., 258
Başar, Tamer (1949–), 203
Bass, Frank M. (1926–2006), 18, 76
Bass diffusion model, 18, 19, 76, 259
Battle of the sexes, 157, 171
Bayes, Thomas (1702–1761), 152
Bayesian perfection, 181
Bayes-Nash equilibrium (BNE), 152, 163, 184
  existence, 165
Behavioral strategy, 166, 184
Bellman, Richard Ernest (1920–1984), 14, 92, 118, 140, 245
Bellman equation, 118, 119
Bendixson, Ivar Otto (1861–1935), 13, 63
Berge, Claude (1926–2002), 251
Berge maximum theorem, 251
Bernoulli, Daniel (1700–1782), 24
Bernoulli, Jakob (1654–1705), 27
Bernoulli, Johann (1667–1748), 7, 8
Bertsekas, Dimitri P., 140, 151, 246, 252, 258
Besicovitch, Abram Samoilovitch (1891–1970), 7
Besicovitch set. See Kakeya set
Best-response correspondence, 157
Bidding function, 152, 164
Blåsjö, Viktor, 15
Bluman, George W. (1943–), 44
Boltyanskii, Vladimir Grigorevich (1925–), 14, 15, 106, 114, 140, 141
Bolza, Oskar (1857–1942), 105
Bolza problem, 105
Bolzano, Bernhard Placidus Johann Nepomuk (1781–1848), 236, 237
Bolzano-Weierstrass theorem, 236
Border, Kim C., 252
Boundary value problem (BVP), 254
Boundedness condition (condition B), 106
Boyd, Stephen P., 252
Brachistochrone, 7, 8
Branding, 277
Brinkhuis, Jan (1952–), 252
Bronshtein, Ilja N., 74
Buckingham, Edgar (1867–1940), 44
Bucy, Richard Snowden (1935–), 13
Bulow, Jeremy, 168
Bunching, 220, 224, 330
Byrne, Oliver (1810–1880), 6

Cajori, Florian (1859–1930), 7
Calculus of variations, 7–9, 11, 14, 100, 140
Campbell, Douglas M., 15
Cantor, Moritz Benedikt (1829–1920), 6
Carathéodory, Constantin (1873–1950), 140
Carathéodory's theorem, 140
Carlson, Dean A. (1955–), 140
Cass, David (1937–2008), 11
Cauchy, Augustin-Louis (1789–1857), 3, 20, 25
Cauchy formula, 3, 24
  generalized, 72
Cauchy sequence, 237
Cayley, Arthur (1821–1895), 85
Cayley-Hamilton theorem, 85
Center, 48
Centipede game, 172
Cesari, Lamberto (1910–1990), 137, 141
Četaev, Nikolaj Gurjevič, 49
Characteristic equation, 233
Chen, YanQuan, 257
Chernov, Barbara Ann, 260
Cho, In-Koo, 188
Choné, Philippe, 227
Chukanov, S. V., 141
Clairaut, Alexis Claude de (1713–1765), 242
Clairaut's theorem, 242
Clarke, Francis H. (1948–), 15, 140
Closed-loop control, 12
Closed trajectory, 63
Closure (of a set), 236
Coase, Ronald Harry (1910–), 168
Coase conjecture, 168
Coase problem, 168
Coddington, Earl A. (1920–1991), 75
Co-domain (of a function), 241
Commitment, 169, 215, 217
Common knowledge, 155
Compactness condition (condition C), 106
Competitive exclusion, 77, 260
Complementary slackness condition, 249
Complete contingent plan. See Strategy
Complete normed vector space. See Banach space
Complete shutdown, 328
Congestion externality, 228
Conjugate (Young-Fenchel transform), 101
Consumption smoothing, 278
Contractability (observability plus verifiability), 209

Index

Contraction mapping, 32
Contraction mapping principle. *See* Banach fixed-point theorem
Control (variable), 4, 83
  admissible, 90, 104
Control constraint, 89
Control constraint set. *See* Control set
Controllability, 4, 81, 84, 88
Control set, 84, 87
Control system, 11, 83
  linear, 84
Convergence (of a sequence), 236
Convergence of trajectory to set, 62
Convex hull, 127
Coordination game, 157
Co-state. *See* Adjoint variable
Cournot, Antoine Augustin (1801–1877), 197
Cournot oligopoly, 160, 203
  with cost uncertainty, 164
  repeated, 179
Cournot-Stackelberg duopoly, 197
Cumulative distribution function (cdf), 162
Current-value formulation, 110
Cybernetics, 12
Cycloid, 8

De-dimensionalize (an ODE), 44, 77
δ-calculus, 8
Dido (queen of Carthage) (ca. 800 B.C.), 6
Dido's problem. *See* Isoperimetric problem
Differentiable function, 242
Differential game, 188
Dikusar, Vasily Vasilievich (1937–), 141
Dirac, Paul Adrien Maurice (1902–1984), 75
Dirac distribution, 75
Discount factor, 110, 173, 174
Discount rate, 92
Dixit, Avinash Kamalakar (1944–), 177
Dmitruk, Andrei, 141
Dockner, Engelbert, 203
Domain, 20
  contractible, 41
Domain (of a function), 241
Dorfman, Robert (1916–2002), 141
Douglas, Jesse (1897–1965), 10
DuBois-Reymond, Paul David Gustav (1831–1889), 100
DuBois-Reymond equation, 100

Dubovitskii, Abraham Yakovlevich (1923–), 103, 106, 141
Dugundji, James (1920–1985), 252
Duhamel, Jean-Marie (1797–1872), 25
Duhamel principle, 25
Dunford, Nelson (1906–1986), 126, 252
Dutta, Prajit K., 179

Eberlein, William F. (1917–1986), 126
Eberlein-Šmulian theorem, 126
Edvall, Markus M., 257
Egorov, Dmitri Fyodorovich (1869–1931), 130
Eigenvalue, 233
Einstein, Albert (1879–1955), 11, 81
Ekeland, Ivar (1944–), 252
Endpoint constraint, 89
  componentwise, 242
  regularity, 104
Endpoint optimality, 108, 110
Entry game, 166, 167, 169, 170
Envelope condition, 97, 108, 110
Envelope theorem, 250
Equicontinuity, 242
Equilibrium, 3
  Bayes-Nash, 152, 163, 184
  Markovian Nash, 189
  Markov-perfect, 154
  mixed-strategy Nash, 158
  multiplicity, 179
  Nash, 1, 4, 149, 156, 174
  non-Markovian, 190
  odd number of Nash, 161
  perfect Bayesian, 181, 185
  pooling (*see* Signaling)
  refinements, 4, 151
  separating, 154
  subgame-perfect Nash, 169, 172, 191
  trigger strategy, 199
Equilibrium (point), 45
  asymptotically stable, 47
  classification, 48
  exponentially stable, 47
  region of attraction, 56
  stable, 19, 47
  unstable, 47
Equilibrium path, 151
  on vs. off, 153–155, 181, 185–187, 191, 202
Essential supremum, 125
Euclid (of Alexandria) (ca. 300 B.C.), 5, 6
Euler, Leonhard Paul (1707–1783), 8, 24, 42

Euler equation, 8, 10, 100
Euler-Lagrange equation. *See*
  Euler equation
Euler multiplier, 42
Euler's identity, 231
Event, 88
  controllability, 88
Existence
  Bayes-Nash equilibrium, 165
  fixed point: Banach, 238
  fixed point: Kakutani, 251
  global solution to IVP, 35
  local solution to IVP, 31
  mixed-strategy NE, 160
  Nash equilibrium (NE), 159
  remark, 6
  solution to IVP, 28
  solution to OCP (*see* Filippov existence theorem)
  subgame-perfect NE, 172
Expected value of perfect information (EVPI), 301
Exponential growth, 21
Exponential stability, 60
Extensive form, 166, 169, 170
Extremal principles, 9

Fahroo, Fariba, 257
Feedback control, 11
Feedback law, 11, 90, 94, 119, 193, 194
Feinstein, Charles D., 115
Fenchel, Werner (1905–1988), 101
Fermat, Pierre de (1601–1665), 9, 247
Fermat's lemma, 120, 247
Fermat's principle. *See*
  Principle, of least time
Filippov, Aleksei Fedorovich (1923–), 75, 135, 136, 138
Filippov existence theorem, 135, 136
Finite escape time, 35
First-best solution. *See* Mechanism, design problem
First-price auction, 163
Flow
  of autonomous system, 40
  of vector field, 40
Focus
  stable, 48
  unstable, 48

Folk theorem
  Nash reversion, 178
  subgame-perfect, 178
Fomenko, Anatoly Timofeevich (1945–), 10
Fomin, Sergei Vasilovich (1917–1975), 126, 252
Fraser-Andrews, G., 255
Freckleton, Robert P., 79
Friedman, James W., 178
Frisi, Paolo (1728–1784), 7
Fromovitz, S., 105, 249
Fudenberg, Drew, 178, 202
Function, 240
  coercive, 58
  continuous, 241
  essentially bounded, 241
  homogeneous (of degree k), 22
  Lipschitz, 29
  locally Lipschitz, 30
  measurable, 241
  separable, 21
  set-valued, 251
  upper semicontinuous, 251
Function family
  equicontinuous, 242
  uniformly bounded, 242
Fundamental matrix, 69
  properties, 69
Fundamental theorem of algebra, 233
Fundamental theorem of calculus, 243

Galerkin, Boris Grigoryevich (1871–1945), 257
Galerkin methods, 257
Galilei, Galileo (1564–1642), 8
Game, 149
  Bayesian, 162
  differential, 188
  dynamic, 149
    complete-information, 166
    hierarchical, 198
    imperfect-information, 171
    incomplete-information, 180
    perfect-information, 172
  extensive-form, 169, 170
  nodes (terminal/nonterminal), 169
  repeated, 173
  signaling, 181
  static, 149
    complete-information, 155
    incomplete-information, 161

Index

Game theory, 4, 149
Gamkrelidze, Revaz Valerianovich (1927–), 14, 15, 101, 106, 114, 140
Geanakoplos, John (1955–), 155
Gelfand, Israel Moiseevich (1913–2009), 126
Giaquinta, Mariano (1947–), 141
Gibbard, Allan (1942–), 216
Gibbons, Robert S., 203
Gilbert, Elmer G., 14
Global asymptotic stability, 58, 60
Godunov, Sergei Konstantinovich (1929–), 75
Going concern, 83, 113
Golden rule, 115, 117, 142
Goldstine, Herman Heine (1913–2004), 8, 9, 15
Gompertz, Benjamin (1779–1865), 76
Gompertz growth, 76
Gradient method (steepest ascent), 255
Granas, Andrzej, 252
Green, George (1793–1841), 67
Green, Jerry R., 216
Gronwall, Thomas Hakon (1877–1932), 245
Gronwall-Bellman inequality, 245
  simplified, 245
Group, 40
Growth
  exponential, 21
  Gompertz, 76
  logistic, 21, 76
Guesnerie, Roger, 226
Guillemin, Victor (1937–), 241
Gül, Faruk R., 168
Gvishiani, Alexey Dzhermenovich (1948–), 130, 158, 241, 243

Hadamard, Jacques Salomon (1865–1963), 28
Halkin, Hubert, 114
Hamilton, William Rowan (1805–1865), 9, 85
Hamiltonian, 9, 96, 99, 107
  current-value, 111
  geometric interpretation, 100, 101
  maximized, 97
Hamiltonian system, 9
Hamilton-Jacobi-Bellman (HJB) equation, 4, 13–15, 17, 82, 90, 95, 140, 151
  with discounting and salvage value, 92, 93
  principle of optimality, 92
  uniqueness of solution, 91
Hamilton-Jacobi-Bellman inequality, 90
Hamilton-Jacobi equation, 10, 13, 14
Hamilton-Pontryagin function.
  See Hamiltonian
Harmonic oscillator, 87
Harsanyi (Harsányi), John Charles (János Károly) (1920–2000), 180
Hartl, Richard F., 106
Hartman, Philip (1915–), 75
Haurie, Alain (1940–), 140
Hazard rate, 18, 224
Helly, Eduard (1884–1943), 131
Heron of Alexandria (ca. 10 B.C.–A.D. 70), 9
Hesse, Ludwig Otto (1811–1874), 42, 242
Hessian, 242
Higgins, John C., 15
Hildebrandt, Stefan (1936–), 9, 15, 141
Hoëne-Wroński, Józef Maria (1778–1853), 69
Homotopy, 41
Hotelling, Harold (1895–1973), 143
Hotelling rule, 143, 278
Howitt, Peter Wilkinson (1946–), 11
Hungerford, Thomas W., 85, 251
Hurwicz, Leonid (1917–2008), 227
Hurwitz, Adolf (1859–1919), 12, 52, 53
Hurwitz property, 53
Huygens, Christiaan (1629–1695), 8

Ill-posed problems, 28, 75
Il'yashenko, Yuliy S., 64, 75
Implementation horizon, 146
Implementation theorem, 218
Implicit differentiation, 244
Implicit function theorem, 244
Implicit programming problem, 115, 116
Inada, Ken-Ichi (1925–2002), 161
Inada conditions, 161
Individual-rationality constraint.
  See Participation constraint
Information
  hidden, 207
  private, 151
  rent, 208, 212
Information set, 169

Information structure, 174, 196
  causal (nonanticipatory), 196
  delayed-control, 197
  delayed-state, 197
  Markovian, 197
  regular, 196
  sampled-observation, 197
Initial-value problem (IVP), 20
  initial condition, 20
  maximal solution, 29
  nominal, 36
  parameterized, 36
  representation in integral form, 31
Institutional design, 179
Instrument, 209
Integral curve of ODE, 20
Integrating factor, 42
Invariance, 61
Invariance principle, 62
Invariant set, 56, 59
Inverse function theorem, 244
Ioffe, Aleksandr Davidovich (1938–), 140
Ironing, 224, 330
Isidori, Alberto, 88
Isoperimetric constraint, 7
Isoperimetric problem, 6
IVP. *See* Initial-value problem

Jacobi, Carl Gustav Jacob (1804–1851), 10, 61, 242
Jacobian, 242
Jain, Dipak C., 76
Job market signaling, 153
Jordan, Marie Ennemond Camille (1838–1922), 51, 241
Jordan canonical form, 51
Jordan curve theorem, 241
Jørgensen, Steffen, 203
Jowett, Benjamin (1817–1893), 6

Kakeya, Soichi (1886–1947), 7
Kakeya set, 7
Kakutani, Shizuo (1911–2004), 251
Kakutani fixed-point theorem, 251
Kalman (Kálmán), Rudolf Emil (1930–), 13
Kalman filter, 13
Kamien, Morton I., 140
Kant, Immanuel (1724–1804), 11
Karp, Larry S., 203
Karush, William (1917–1997), 249
Karush-Kuhn-Tucker conditions, 249

Keerthi, S. Sathiya, 14
Khalil, Hassan K. (1950–), 75
Kihlstrom, Richard E., 203
Kirillov, Alexandre Aleksandrovich (1936–), 130, 158, 241, 243
Kolmogorov, Andrey Nikolaevich (1903–1987), 15, 252
Koopmans, Tjalling Charles (1910–1985), 11
Krasovskii, Nikolay Nikolayevich (1924–), 58
Kreps, David Marc, 179, 188
Kress, Rainer (1941–), 28
Krishnan, Trichy V., 76
Kryazhimskii, Arkady V., 115, 140
Ktesibios of Alexandria (ca. 285 B.C.–222 B.C.), 11
Kuhn, Harold William (1925–), 249
Kumei, Sukeyuki, 44
Kurian, George Thomas, 260
Kurz, Mordecai, 15
Kydland, Finn Erling (1943–), 203

Laffont, Jean-Jacques Marcel (1947–2004), 216, 226, 227
Lagrange, Joseph-Louis (1736–1813), 8–10, 24, 25, 101, 105, 248
Lagrangian, 249
  small, 107
Lagrange multiplier, 249
Lagrange problem, 105
Lagrange's theorem, 248
Lancaster, Kelvin John (1924–1999), 226
Landau, Edmund Georg Hermann (Yehezkel) (1877–1938), 39
Landau notation, 39
Lang, Serge (1927–2005), 243
Laplace, Pierre-Simon (1749–1827), 73
Laplace transform, 19, 73
  common transform pairs, 74
  inverse, 74
  properties, 74
LaSalle, Joseph Pierre (1916–1983), 62
LaSalle invariance principle, 62
Lavrentyev, Mikhail Alekseevich (1900–1980), 15
Law of parsimony, 11
Lebesgue integration, 241
Lebesgue point, 122
Ledyaev, Yuri S., 15, 140
Lee, E. Bruce, 256
Lee, John M. (1950–), 67

Index

Left-sided limit, 60
Leibniz, Gottfried Wilhelm (1646–1716), 7–9, 11, 17
Leizarowitz, Arie, 140
Leng, S. B., 255
Leonardo da Vinci (1452–1519), 11
Levinson, Norman (1912–1975), 75
L'Hôpital, Guillaume de (1661–1704), 8, 258
L'Hôpital's rule, 258
Limit (of a sequence), 236
Limit cycle, 61, 63
Lindelöf, Ernst Leonard (1870–1946), 34
Linear dependence, 231
Linearization criterion, 54
 generalized, 61
 generic failure, 56
Linear-quadratic differential game, 194, 203, 302
Linear-quadratic regulator, 28, 93
 infinite-horizon, 119
Linear space, 234. *See also* Vector space
Linear system
 nilpotent, 73
Liouville, Joseph (1809–1882), 69
Liouville formula, 69
Lipschitz, Rudolf Otto Sigismund (1832–1903), 29
Lipschitz constant, 32
Lipschitz property, 29
Locally Lipschitz, 30
Logistic growth, 21
Lorenz, Edward Norton (1917–2008), 63
Lorenz oscillator, 63
Lotka, Alfred James (1880–1949), 13, 44
Lower contour set, 149
Luenberger, David G. (1937–), 101, 115, 238, 252
Lyapunov, Aleksandr Mikhailovich (1857–1918), 3, 12, 47
Lyapunov differential equation, 61
Lyapunov equation, 53
Lyapunov function, 3, 4, 19, 47, 49, 53, 93

Magaril-Il'yaev, Georgii Georgievich (1944–), 101
Mailath, Georg Joseph, 203
Mangasarian, Olvi L. (1934–), 105, 111, 112, 249
Mangasarian-Fromovitz conditions (constraint qualification), 105, 249

Mangasarian sufficiency theorem, 112, 113, 115
Market for lemons, 182
Markov, Andrey Andreyevich (1856–1922), 190
Markovian Nash equilibrium, 189
Markov-perfect equilibrium, 154
Markov property, 154, 190
Markus, Lawrence, 256
Martini, Horst, 141
Maskin, Eric Stark (1950–), 178, 227
Mathematical notation, 231
Matrix, 232
 determinant, 232
 eigenvalue, 233
 eigenvector, 233
 identity, 232
 Laplace expansion rule, 232
 nonsingular, 232
 positive/negative (semi)definite, 233
 rank, 232
 symmetric, 232
 transpose, 232
Matrix exponential, 73
Matthews, Steve A. (1955–), 227
Maupertuis, Pierre-Louis Moreau de (1698–1759), 9
Maupertuis's principles. *See* Principle of least action
Maximality condition, 96, 107, 110
Maximizing sequence, 126
Maxwell, James Clerk (1831–1879), 12
Maxwell equation, 66
Mayer, Christian Gustav Adolph (1839–1907), 105
Mayer problem, 105
Mayne, David Quinn (1930–), 14
Mayr, Otto, 11, 15
Mazur, Barry (1937–), 127
Mean-value theorem, 248
Measurable selector, 104
Mechanism
 allocation function, 216
 definition, 216
 design problem, 210
 direct (revelation), 216, 217
 first-best, 211
 instrument, 209
 message space, 215, 216
 second-best, 211
 type space, 209

Mechanism design, 2, 5, 181, 207
Megginson, Robert E., 126
Menu of contracts, 209
Message space, 215, 216
Michalska, Hannah, 14
Milgrom, Paul Robert (1948–), 203, 250
Milyutin, Alexey Alekseevich (1925–2001), 103, 106, 140, 141
Minmax payoff, 177
Mirrlees, James Alexander (1936–), 218, 226, 227
Mishchenko, Evgenii Frolovich (1922–), 14, 15, 106, 114, 140
Mixed strategy, 157
Moler, Cleve B., 73
Moore, John B., 140
Moore, John Hardman, 227
Mordukhovich, Boris Sholimovich, 141
Morgenstern, Oskar (1902–1977), 14, 202
Muehlig, Heiner, 74
Murray, James Dickson (1931–), 76
Musiol, Gerhard, 74
Mussa, Michael L. (1944–), 224, 227
Myerson, Roger Bruce (1951–), 216

Nalebuff, Barry (1958–), 177
Nash, John Forbes (1928–), 1, 149, 152, 156, 203
Nash equilibrium (NE), 1, 4, 149, 156
 closed-loop, 190
 existence, 159
 Markov-perfect, 191
 mixed-strategy, 158
 open-loop, 190
 subgame-perfect, 4, 169, 172, 191
 supergame, 174
Nash reversion folk theorem, 178
Nature (as player), 162, 180
Nerlove, Mark Leon (1933–), 7
Neumann, John von (1903–1957), 14, 202
Newbery, David M., 203
Newton, Isaac (1643–1727), 7, 8
Nilpotent system, 73
Node
 nonterminal, 169
 stable, 48
 terminal, 166, 169, 172
 unstable, 48

Noncredible threat, 168, 169
Nonlinear pricing, 224
 with congestion externality, 228
Nonsmooth analysis, 15, 140
Nontriviality condition, 108, 133
Norm
 equivalence, 236
 properties, 235
Nullcline, 46, 79
Numerical methods, 253
 direct, 256
 indirect, 253
 overview, 254

Objective functional, 89
Observability, 84. *See* Contractability
Observation/sample space, 174
Occam, William of (ca. 1288–1348), 11
Occam's razor, 11
OCP. *See* Optimal control problem
Odd number of Nash equilibria, 161
ODE. *See* Ordinary differential equation
Olsder, Geert Jan (1944–), 203
Olympiodorus the Younger (ca. 495–570), 9
One-parameter group action, 40
One-shot deviation principle, 172
 infinite-horizon, 173
One-sided limit, 60
Open-loop control, 12
Open set, 236
Optimal consumption, 97, 101, 147, 296
Optimal control problem (OCP), 4, 89, 103, 189, 201, 221
 infinite-horizon, 113
 simplified, 108
 time-optimal, 141
Optimal screening contract, 221
Ordinary differential equation (ODE), 2, 17, 20
 autonomous, 44, 47
 completely integrable, 42
 de-dimensionalization, 44, 77
 dependent variable, 17, 20
 exact, 41
 first integral, 42
 homogeneous, 22
 independent variable, 17, 20
 linear, 24
  homogeneous, 3, 24
  particular solution, 3, 24

Index

representation
 explicit, 20, 68
 implicit, 20, 41
 separable, 21
 solution methods (overview), 26
Osmolovskii, Nikolaj Pavlovich (1948–), 140
Output (variable), 83
Output controllability, 88
Output controllability matrix, 88
Output function, 84

p-norm, 235
Parameterized optimization problem, 249
Pareto optimality, 179
Pareto perfection, 179
Partial differential equation (PDE), 17
Partial fraction expansion, 259
Participation constraint, 209, 220
Pascal, Blaise (1623–1662), 1
Payoff function, 149, 155, 169. *See also* Utility function
 average, 174
 quasilinear, 222
Peano, Giuseppe (1858–1932), 28, 71
Peano-Baker formula, 71
Pearl, Judea (1936–), 11
Penetration pricing, 277
Perfect Bayesian equilibrium (PBE), 181, 185
Periodic orbit/trajectory. *See* Limit cycle
Perron, Oskar (1880–1975), 6
Petrovskii, Ivan Georgievich (1901–1973), 75
Phase diagram, 44
Picard, Charles Émile (1856–1941), 34
Picard-Lindelöf error estimate, 34
Planning horizon, 146
Plateau, Joseph Antoine Ferdinand (1801–1883), 10
Plateau's problem, 10
Plato (ca. 427–347 B.C.), 6
Player function, 169
Player set, 155
Poincaré, Jules Henri (1854–1912), 13, 41
Poincaré-Bendixson theorem, 63
Polak, Elijah, 257
Pollack, Alan, 241
Pontryagin, Lev Semenovich (1908–1988), 14, 15, 75, 82, 97, 106, 114, 140

Pontryagin maximum principle (PMP), 4, 15, 83, 97, 107, 221
 proof, 119
 simplified, 109
Positive limit set, 62
Potential (function), 40
 computation, 42
 of linear first-order ODE, 43
Powell, Warren B., 258
Prandtl, Ludwig (1875–1953), 63
Prandtl number, 63
Pratt, John W., 102
Predator-prey system, 44–46, 61, 77, 78
Predecessor function, 169
Preference relation, 149
 quasilinear, 218
Preference representation. *See* Utility function
Preimage, 241
Preorder, 149
Prescott, Edward Christian (1940–), 203
Primbs, James Alan, 14
Principle
 of least action, 9
 of least time, 9
 maximum (*see* Pontryagin maximum principle)
 of optimality, 92, 255
 of plenitude, 11
 of stationary action, 9
Prisoner's dilemma, 155, 156
 differential game, 199
 finitely repeated, 175
 infinitely repeated, 176
Product diffusion, 18
PROPT (MATLAB toolbox), 257
Pseudospectral methods, 257

Rademacher, Hans Adolph (1892–1969), 246
Rademacher theorem, 246
Radner, Roy (1927–), 203
Radó, Tibor (1895–1965), 10
Ramsey, Frank Plumpton (1903–1930), 11
Rapoport, Anatol Borisovich (1911–2007), 177
Rationality, 155
Rayleigh (Third Baron). *See* Strutt, John William
Rayleigh number, 63
Reachability, 84

Rectifiability theorem (for vector fields), 64
Recursive Integration Optimal Trajectory Solver (RIOTS), 257
Region of attraction, 56
  of an equilibrium, 56
Reiter, Stanley (1925–), 227
Renegotiation, 217
Revelation principle, 216, 217
Riccati, Jacopo Francesco (1676–1754), 27, 68, 94
Riccati differential equation, 94, 303, 304
Riccati equation, 27, 68, 119
Right-sided limit, 60
Riordan, Michael H., 203
Risk aversion, 102
  constant absolute (CARA), 102
  constant relative (CRRA), 102, 192
Roberts, Sanford M., 255
Robertson, Ross M., 7
Rochet, Jean-Charles, 227
Rockafellar, R. Tyrrell, 141
Rolle, Michel (1652–1719), 247
Rolle's theorem, 247
Root of a polynomial, 233
Rosen, Sherwin (1938–2001), 224, 227
Rosenthal, Robert W. (1945–2002), 203
Ross, I. Michael, 257
Ross, William David (1877–1971), 6
Routh, Edward John (1831–1907), 12
Rudin, Walter (1921–2010), 252
Russell, Bertrand Arthur William (1872–1970), 11
Rutquist, Per E., 257

Sachdev, P. L. (1943–2009), 75
Saddle, 48
Salvage/terminal value, 92
Sample/observation space, 174
Samuelson, Larry (1953–), 203
Samuelson, Paul Anthony (1915–2009), 11
Scalar product, 231
Schechter, Martin, 252
Schwartz, Adam Lowell, 257
Schwartz, Jacob Theodore (1930–2009), 126, 252
Schwartz, Nancy L., 140
Schwarz, Hermann Amandus (1864–1951), 242
Schwarz's theorem, 242
Screening, 215

Second-best solution. *See* Mechanism, design problem
Second-price auction, 152
Segal, Ilya R., 250
Seierstad, Atle, 140
Semendyayev, Konstantin A., 74
Sensitivity analysis, 39
Sensitivity matrix, 39, 121
Separatrix, 263
Set
  bounded, 236
  closed, 236
  compact, 236
  dense, 104
  open, 236
  open cover, 236
Sethi, Suresh P., 7, 106, 140
Set-valued mapping
  measurable, 104
Shakespeare, William (1564–1616), 207
Shil'nikov, Leonid Pavlovich (1934–), 75
Shipman, Jerome S., 255
Shooting algorithm, 254
Shutdown
  complete, 328
  no-shutdown condition, 326
  solution (of two-type screening problem), 212
Signal, 83
Signaling, 153, 181
  Bayes-Nash equilibrium, 184
  perfect Bayesian equilibrium of, 185
  pooling equilibrium, 183
  separating equilibrium, 183
Sim, Y. C., 255
Similarity transform, 51, 55
Sliding-mode solutions, 139
Smirnov, Georgi V., 75
Smith, Lones (1965–), 179
Smoothness condition (condition S), 106
Šmulian, Vitold L'vovich (1914–1944), 126
Sobolev, Sergei L'vovich (1908–1989), 243
Sobolev space, 243
Social welfare, 142, 145, 214
Socrates (469–399 B.C.), 6
Soltan, Petru S., 141
Solution to ODE, 20
Sonnenschein, Hugo Freund, 168
Sontag, Eduardo Daniel (1951–), 87, 88, 140
Sorger, Gerhard, 203

Index

Sorting condition, 218
Spence, Andrew Michael (1943–), 153, 181, 203, 218
Spence-Mirrlees condition, 218
s.t. (subject to), 231
Stability analysis, 45
 linearization criterion, 54
Stackelberg, Heinrich (Freiherr von) (1905–1946), 197
Stackelberg Leader-Follower Game, 197
State constraints, 105
State control constraint
 regularity, 104
State equation, 89
State space, 39, 84
State space form, 83
State transition matrix, 69
Stationary point. See Equilibrium (point)
Steady state. See Equilibrium (point)
Steiner, Jakob (1796–1863), 6
Stephens, Philip A., 79
Stern, Ron J., 15, 140
Stiglitz, Joseph Eugene (1943–), 181
Stochastic discounting, 176
Stokey, Nancy Laura (1950–), 168
Stole, Lars A., 227
Strang, William Gilbert (1934–), 251
Strange attractor, 63
Strategy
 behavior, 184
 Markovian, 190
 mixed, 157, 159
 open-loop vs. closed-loop, 154, 190, 191
 trigger, 154
Strategy profile, 155
 augmented, 175
 grim-trigger, 176
 mixed, 174
Structural stability, 19, 75
Strutt, John William (1842–1919), 63
Struwe, Michael (1955–), 10
Subgame, 169, 190
 perfection, 4, 169, 172, 190–192
Subgame-perfect folk theorem, 178
Subramaniam, V., 255
Subspace, 235
Successive approximation, 34, 240
Successor function, 169
Supergame, 173. See also Game, repeated
 history, 174
 Nash equilibrium, 174

Superposition, 25
Sutherland, William J., 79
Sutton, Richard S., 258
Sydsaeter, Knut, 140
System, 83
 autonomous (see System, time-invariant)
 controllable, 84
 discrete-time, 83
 observable, 84
 reachable, 84
 state space representation, 83
 time-invariant, 18, 84
System function, 17, 18, 48, 83
System matrix, 51

Tautochrone, 8
Taxation principle, 220
Taylor, Angus Ellis (1911–1999), 131
Taylor, Brook (1685–1731), 61
Terminal/salvage value, 92
Theorem Π, 44
Thom, René Frédéric (1923–2002), 75
Thomas, Ivor (1905–1993), 6
Thompson, Gerald Luther (1923–2009), 140
Tikhomirov, Vladimir Mikhailovich (1934–), 101, 140, 246, 252
Tikhonov, Andrey Nikolayevich (1906–1993), 28
Time consistency, 190, 191
Time derivative, 17
Time inconsistency, 168
Time-optimal control problem, 141
Time t history, 175
 augmented, 175
Tirole, Jean Marcel (1953–), 202
Total derivative, 242
Transversal (of autonomous system), 63
Transversality condition, 96, 107, 110
Trigger strategy, 154, 199
Tromba, Anthony J., 15
Tucker, Albert William (1905–1995), 249
Turnpike, 115
Type, 208
Type space, 5, 162, 166, 182, 209, 220
Tzu, Sun (ca. 500 B.C.), 149

$\mathcal{U}$-controllable states, 87
Ultimatum bargaining, 150
Uniform boundedness, 242

Uniform convergence ($\Rightarrow$), 242
Upper contour set, 149
Upper semicontinuity, 251
Utility function, 149
  quasilinear, 218

Value function, 4, 13–15, 82, 97, 119, 140
Vandenberghe, Lieven, 252
Van Loan, Charles F., 73
Van Long, Ngo, 203
Variation-of-constants method, 24
Vectogram, 136
Vector field, 39
Vector space, 234
  normed, 235
    complete (Banach space), 237
Verifiability. *See* Contractability
Verri, Pietro (1728–1797), 7
Vickson, Raymond G., 106
Vidale, Marcello L., 25, 118
Vidale-Wolfe advertising model, 47, 118
Vinter, Richard B., 15, 140
Virgil (Publius Virgilius Maro) (70–19 B.C.), 6
Virtual surplus, 223
Volterra, Vito (1860–1940), 13, 44

Walras, Léon (1834–1910), 10
Walter, Wolfgang (1927–), 75
Warder, Clyde Allee (1885–1955), 79
Warga, Jack (1922–), 140
Watt, James (1736–1819), 12
Weber, Robert J. (1947–), 203
Weber, Thomas Alois (1969–), 118, 140, 141, 227
Weierstrass, Karl Theodor Wilhelm (1815–1897), 6, 15, 141, 236, 237, 246
Weierstrass theorem, 246
Weitzman, Martin L. (1942–), 11
Welfare. *See* Social welfare
Well-posed problems, 28–39
Wets, Roger J.-B., 141
Wiener, Norbert (1894–1964), 12
Wiggins, Stephen Ray, 75
Willems, Jan C. (1939–), 8
Williams, Steven R. (1954–), 227
Wilson, Robert Butler (1937–), 161, 167, 168, 227
Wilson's oddness result, 161
Wolenski, Peter R., 15, 140

Wolfe, Harry B., 25, 118
Wronskian (determinant), 69

Young, William Henry (1863–1942), 101
Young-Fenchel transform (dual), 101

Zenodorus (ca. 200–140 B.C.), 6
Zermelo, Ernst Friedrich Ferdinand (1871–1953), 172
Zorich, Vladimir Antonovich (1937–), 42, 242, 244, 252